Plasma Engineering

Michael Keidar dedicates this book to his wife Victoria for her tremendous support and sacrifice allowing him to focus on research over many years. He also dedicates this book to the blessed memory of his mother, Dina. Isak I. Beilis would like to thank his wife Galina for support. Special dedication of this book is to the blessed memory of his parents and his brother.

Plasma Engineering
Applications from Aerospace to Bio- and Nanotechnology

Michael Keidar and Isak I. Beilis

AMSTERDAM • BOSTON • HEIDELBERG • LONDON
NEW YORK • OXFORD • PARIS • SAN DIEGO
SAN FRANCISCO • SINGAPORE • SYDNEY • TOKYO

Academic Press is an imprint of Elsevier

Academic Press is an imprint of Elsevier
32 Jamestown Road, London NW1 7BY, UK
225 Wyman Street, Waltham, MA 02451, USA
525 B Street, Suite 1800, San Diego, CA 92101-4495, USA

Notice

Library of Congress Cataloging-in-Publication Data
A catalog record for this book is available from the Library of Congress

British Library Cataloguing-in-Publication Data
A catalogue record for this book is available from the British Library

ISBN: 978-0-12-385977-8

For information on all Academic Press publications
visit our website at www.store.elsevier.com

Typeset by MPS Limited, Chennai, India
www.adi-mps.com

Printed and bound in the United States of America

13 14 15   10 9 8 7 6 5 4 3 2 1

Contents

Preface

Plasma science and applications have been seeing great progress over the last few decades. This progress is the consequence of development of modern plasma sources based on plasma generation in electrical discharges in vacuum, in low- and high-pressure gases, RF-discharges, magnetrons, etc. On the other hand, it is the result of novel plasma applications in plasma processing of materials, in space propulsion and, especially, in plasma-based nano- and biomedical technology. An important characteristic of the plasma research realm is the strong overlap between plasma science, technology, and application. This interplay of the plasma science, technology, and application is referred as the *plasma engineering*.

This book is an attempt to present aspects of plasma engineering by describing the physics of plasmas, plasma generation in different conditions, and technique of plasma applications from a unified point of view through the theoretical and experimental prism. The book consists of seven chapters considering plasma fundamentals, plasma diagnostics and methodology of the plasma engineering and various plasma applications. While describing the state of the art of plasma applications, authors often included their own research results. The material in the book is self-contained and it was our intention to make the presentation as simple and easily understandable as possible.

In the introductory part of the book (Chapter 1), basic plasma concepts and foundation of plasma engineering are introduced. Fundamental plasma phenomena including the different types of plasma particle collisions, waves, and instabilities are described. The boundary effects such as plasma-wall transition, electron emission mechanisms, and ablation of the walls contacted with relatively hot plasmas are detailed and explained. This is followed by the introduction of various diagnostic tools used to characterize plasmas in engineering systems. Fundamental principles and experimental methodology of plasma diagnostics are reviewed. The probe diagnostic description includes the planar, spherical, and emissive probes as well as probes operating in a collision-dominated plasma and in a magnetic field. Furthermore, electrostatic analyzer, interferometric technique, plasma spectroscopy, optical measurements, fast imaging, and others were explained giving appreciation of their advantages as well as limitations.

Physics of different types of electrical discharges is considered. The description begins with the classical Townsend mechanism of gas electrical breakdown and the Townsend discharge followed by the streamer mechanism and glow discharges. The nature of gas breakdown according to Paschen law is detailed. A broad range of high-current discharges including atmospheric and vacuum arcs are described taking into account recent developments. New results of simulation of very complicated cathode phenomena in a vacuum arc are presented.

Basic approaches and theoretical methodologies for plasma modeling and, in particular, approaches that are based on numerical simulations are described. The analysis begins with analysis of the behavior of a single particle in electric and magnetic fields. It is followed by the description of two basic approaches. The first one is based on the fluid description of plasma solving numerically magnetohydrodynamic (MHD) equations. The second one is the kinetic model particle techniques that take into account kinetic interactions among particles and electromagnetic fields. This simulation is computationally extensive as it is able to resolve local parameters of the rarified plasmas.

A significant part of this book is devoted to plasma engineering application in space propulsion. Space propulsion is required for satellite motion in outer space. Plasma physics and engineering of thrusters based mainly on the electromagnetic plasma acceleration are described. Hall thruster, pulsed plasma thrusters, and microthruster are some examples of plasma thrusters considered. In the case of a Hall thruster, the basic mechanisms of electron transport are considered. In the framework of the pulsed plasma thruster, this book covers the ablation phenomena in the presence of dense plasma.

The important part of the book covers the plasma effects in nanoscience and nanotechnology. *Nanoscience* and *Nanotechnology* study nanoscopic objects used across many scientific fields, such as chemistry, biology, physics, materials science, and engineering. Application of low-temperature plasmas in nanoscience and nanotechnology is a relatively new and quickly emerging area at the frontier of plasma physics, gas discharge, nanoscience, and bioengineering. The description involves recent original experimental and theoretical results in the field of plasma-based techniques of nanomaterial synthesis. Particular emphasis is given on the carbon-based nanoparticle synthesis such as single-walled carbon nanotubes and graphene which are fundamental building blocks. A magnetically based novel approaches to control length and electric properties of nanoparticles in plasma-based synthesis are described.

Plasma medicine is an emerging field combining plasma physics and engineering, medicine, and bioengineering to use plasmas for therapeutic applications. This field is emerging due to advance in the cold atmospheric plasma (CAP) technology. The latest original results on CAP applications in medicine are presented in the last part of the book. Physics of cold plasmas and diagnostics employed such as fast imaging, microwave scattering, and so on are covered. The effects associated with CAP interaction with cells such as cell migration, apoptosis, and integrins activation are explained. The therapeutic potential of CAP with a focus on selective tumor cell eradication capabilities and signaling pathway deregulation is shown.

We should note that the aforementioned topics cover an extensive research field and we certainly understand that the present book does not exhaustively answer all questions. While we tried to address plasma engineering issues in both width and depth, we could not avoid just "scratching the surface" by considering some aspects. However, we hope that the wide range of research areas described will be very useful for understanding the physics and the plasma engineering applications and ultimately will stimulate future research.

This book can be used as a text for courses on Plasma Engineering or Plasma Physics in Departments of Aerospace Engineering, Electrical Engineering, and Physics. It can also be useful as a reference for early career researchers and practicing engineers.

The authors would like to acknowledge colleagues and friends, their encouragement, fruitful discussions, and support.

Firstly, we would like to thank our colleagues Raymond Boxman and Samuel Goldsmith with whom the original models of the plasma jet expansion in vacuum arc were developed. We are particularly thankful to Ian Brown who guided one of us (MK) during his tenure as a postdoctoral scientist at Berkeley and with whom we collaborated on some important aspects of multiple-charged ion transport. We thank Andre Anders with whom we worked on problems related to plasma transport in curved magnetic field and ion implantation. Very special thanks goes to Iain Boyd with whom we worked for a number of years and who introduced one of us (MK) to the world of particle simulation of rarefied gases and plasmas. Results of our collaborative work served as the foundation of a significant part of Chapter 5 dealing with modeling of plasma propulsion devices. The work on Hall thruster modeling and simulation would not be possible without the experimental insight and experience of Yevgeny Raitses with whom we have long-term collaboration of various aspects of plasma propulsion physics and most recently plasma-based nanotechnology. Our long-term collaboration with Nathaniel Fisch produced many important results used in this book. One of us (MK) would like to thank Michael Schulman with whom we collaborated on fundamental aspects of high-current vacuum arc interrupters. Michael Schulman, Paul Slade, and Eric Taylor contributed greatly in developing models of the high-current vacuum arcs and the interruption process. The work on carbon arcs for synthesis of carbon nanotubes performed with Anthony Waas triggered our consequent work on the plasma-based nanotechnology. We are particularly very thankful to Igor Levchenko and Kostya (Ken) Ostrikov with whom we have long-term research collaboration on various topics related to plasma nanoscience and nanotechnology. Our joint work with Mary Ann Stepp on cold plasma−controlled cell migration planted seeds for development of the cold plasma application in medicine. Most recent work on cold plasma cancer therapy would be impossible without contributions from Barry Trink, Anthony Sandler, Jonathan Sherman, Alan Siu, and Jerome Canady. We are very grateful to Mikhail Shneider who contributed a lot to understanding on the cold plasma physics. Our great appreciation and thanks to our colleague Alexey Shashurin for his contribution to a number of original works described in this book. We are also grateful to our coauthors (in alphabetic order) Robert Aharonov, Erik Antonsen, Rodney Burton, Jean Luc Cambier, Yongjun Choi, Uros Cvelbar, Jeffrey Fagan, Alec Gallimore, Brian Gilchrist, Rafael Guerrero-Preston, Terese Hawley, Michael Kong, Mahadevan Krishnan, Richard Miles, Othon Monteiro, Anthony Murphy, Leonid Pekker, Frederick Phelan, Claude Phipps, Jochen Schein, Vladimir Sotnikov, Gregory Spanjers, and John Yim who made very significant contributions to the original publications used in this book. We are particularly grateful to our former and current graduate students Andrew Porwitzky, Minkwan Kim, Madhu Kundrapu, Jarrod Fenstermacher, Therese Suaris, Lubos Brieda, Jian Li, Taisen Zhuang, and Olga Volotskova who contributed to some original work used in the book.

Last but not least, we owe very special gratitude to our families for their support and encouragement.

Michael Keidar,
Washington DC

Isak I. Beilis
Tel Aviv, Israel
November 2012

Plasma Concepts

1.1 Introduction

When a neutral gas is ionized, it behaves as a conductive media. Ionization process is the phenomenon associated with striping electrons from the atoms thus creating the pair of negatively and positively charged particles. Electrical properties of such ionized gas depend on the charged particle density. One of the most important distinctions between the ionized gas and the neutral media is that Coulomb interaction between charged particles determines the dynamic of the gas. Ionized gas is able to conduct the current. This property is of particular interest in the presence of the magnetic field when the interaction of the current and magnetic field leads to electromagnetic body force thus altering its flow dynamics. There are weakly ionized gases and strongly ionized gases. Weekly ionized gas is characterized by a relatively small fraction of charged particles and its behavior can be largely described by neutral gas laws while one needs to invoke electrodynamics to describe appropriately the strongly ionized media. We shall call a physical state of an ionized gas in which the densities of positively and negatively charged particles are approximately equal as a quasi-neutrality state. Plasma is defined as an ionized gas, which satisfies the quasi-neutrality condition.

Development of the plasma physics was always associated with particular applications. Starting from lighting sources, current interrupters, thermonuclear fusion, and plasma accelerators, nowadays plasma applications range from plasma processing, space propulsion, nanotechnology, and plasma medicine.

Prominent physicists contributed to developing the field of the plasma physics and engineering. Irving Langmuir (1881−1957) initiated an active study of the plasma as a new direction of science. The term "plasma" was introduced in 1928 in his article describing the positive column of low-pressure gas discharge. While Langmuir introduced the term plasma, the matter in the plasma state was known to human since much early times. Lighting, northern light, solar wind, and Earth ionosphere are some examples of plasmas. Irving Langmuir received the Nobel Prize in Chemistry, 1932. Mott-Smith indicated in his letter [1] that Langmuir takes the term by analogy between "the blood plasma carries around red and white corpuscles and germs" and the multicomponent ionized gas. The great success in developing the foundation of the plasma science was achieved by Langmuir due to effective collaboration with his famous coworkers Compton, Tonks, Mott-Smith, Jones, Child, and Taylor.

Hannes Alfven (1908−1995) is widely known, as a father of the plasma magnetohydrodynamics. He developed theories regarding the nature of the galactic magnetic field and space plasmas. Prof. Alfven received a Nobel Prize in Physics in 1970 for "fundamental work and discoveries in magnetohydrodynamics."

Plasma Engineering.
© 2013 Elsevier Inc. All rights reserved.

Plasma physics, as it is known today, was developed over last 50 years and encompasses many areas ranging from the high-temperature plasmas of thermonuclear fusion to the low-temperature plasma in material processing. The plasma fundamentals and configurations for thermonuclear fusion applications were formulated and developed by Igor Tamm, Andrei Sakharov, Lev Artzimovich, Marshall Rosenbluth, Lyman Spitzer, and many others. Science of the interstellar ionized medium and astrophysical plasmas was by Yakov Zeldovich and Vitaly Ginsburg. Gas discharge plasma physics was introduced by A. von Engel, M. Steenbeck, and then developed by Loeb, Townsend, Thomson, Kaptzov, Granovsky, and Raizer.

In the following as a way of introduction to plasma physics, we will discuss some basic plasma properties. It can be indicative of what kind of plasma is considered by analyzing two main characteristics of plasma behavior in time and space-plasma, i.e. plasma oscillation and Debye length. These two parameters quantitatively describe the plasma and depend on plasma density and temperature.

Let us first focus on understanding of the plasma quasi-neutrality phenomena and its importance. Any perturbation in plasma such as shift of electrons with respect to ions leads to charge separation. The charge separation produces the electric field that works to restore the unperturbed plasma. Let us consider the quasi-neutrality phenomena in the plasma of the high-current vacuum interrupter. As an example, fully ionized plasma is formed with the electron density of about 10^{22} m^{-3} and in the volume with radius of about 1 mm. If plasma quasi-neutrality is violated due to charge separation at the characteristic distance of about 1 mm, the electric field of about 10^{11} V/m will appear. This means that the voltage drop of about 10^{8} V will be set at the distance of about 1 mm. It is clear that such large electric field will work to restore the charge neutrality. However, if relatively small plasma volume is considered, such electric field and potential drop might not be strong enough to affect the particle motion and restore the quasi-neutrality. Thus, quasi-neutrality condition can be violated at the small scale. The characteristics scale where charge separation can exist is called the Debye length.

1.1.1 Debye length

The electrical neutrality or quasi-neutrality is preserved over some characteristic length scale. Let us determine this length scale.

The characteristic length can be obtained from the following argument. The potential energy of the charged particle in the case of full charge separation by distance of about L_D is of the order of the particle thermal energy kT_e. Maximum possible potential energy due to full separation of charges can be estimated in the planar capacitor case as $(e^2 L_D^2 N_0^2/\varepsilon_0)$. Thus one can obtain that

$$e\varphi \sim \frac{e^2 L_D^2 N_0^2}{\varepsilon_0} \sim kT_e$$

where N_0 is the charge particle density. From this equality, one can obtain that the characteristic distance of the charge separation is $L_D = \sqrt{(\varepsilon_0 kT_e/e^2 N_0)}$. This distance we shall call the Debye length.

The same expression for the Debye length can be obtained by considering potential shielding in plasmas. We shall examine initially electrically neutral plasma with density N_0. To this end it can be supposed that an electric field disturbs the plasma equilibrium state by, for example, immersing the transparent sheet having the negative potential Φ_0 with respect to the plasma as shown in

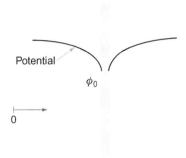

Potential

ϕ_0

0

FIGURE 1.1

Debye length definition from the electric field shielding argument.

Figure 1.1. We will consider one-dimensional plane geometry. As response to the charge perturbation ion and electron distribution will be rearranged to a new state that will correspond to the disturbed electric field. Since ions are heavy particle, their response time is much larger than that of electrons. By taking this into account, we can assume that ions will not be moving on the timescale of interest. This allows us to assume that ion density N_i in the entire plasma region will remain the same as before, i.e., equal N_0.

On the other hand, electrons with temperature T_e will respond to the repelling electric field and their density will decrease. Electron density can be calculated from the Boltzmann relation as

$$N_e = N_0 \exp(e\varphi/kT_e) \tag{1.1}$$

here φ is the potential. The potential φ distribution in the perturbed region can be calculated using the Poisson equation:

$$\frac{d^2\varphi}{dx^2} = \frac{e}{\varepsilon_0}(N_e - N_i) \tag{1.2}$$

Substituting Eq. (1.1) into Eq. (1.2) will lead to the following:

$$\frac{d^2\varphi}{dx^2} = \frac{e}{\varepsilon_0}N_0(\exp(e\varphi/kT_e) - 1) \tag{1.3}$$

Assuming that the disturbed energy is small in comparison with the temperature yielding expansion in the Taylor series:

$$\exp(e\varphi/kT_e) \approx 1 + \frac{e\varphi}{kT_e} + \cdots \tag{1.4}$$

Taking into account only first-order terms in the Taylor series (Eq. (1.4)), the Poisson equation will have the following form:

$$\frac{d^2\varphi}{dx^2} = \frac{e^2\varphi}{\varepsilon_0 kT_e}N_0 \tag{1.5}$$

Using the following denote $L_D = \sqrt{(\varepsilon_0 kT_e/e^2 N_0)}$, the solution of the above equation can be expressed as

$$\varphi = \Phi_0 \exp(-x/L_D) \tag{1.6}$$

Parameter L_D is named the Debye length and it is the fundamental characteristics of plasmas. The solution for potential distribution Eq. (1.6) shows that disturbed φ significantly decreased at distance of length L_D and demonstrates the screening length over which the plasma neutrality will be preserved. Thus electric field will be shielded at the Debye length scale and plasma will remain quasi-neutral far from the perturbed region.

1.1.2 Plasma oscillation

In this section we will discuss temporal behavior of the plasma. Since plasma is electrically neutral, any violation from quasi-neutrality leads to the formation of a high electric field that restores it. Such electric field appears due to large difference between ion and electron masses. Any electron motion affects the microscale electric field and, as a result, electrons oscillate around ions. Let us describe a simplified model of plasma oscillations assuming uniform cold plasma with density N_e and neglecting the electron thermal motion. When such uniform system will be perturbed, the electrons will be shifted with respect to ions as shown in Figure 1.2.

As electrons move, a net positive charge of ions will be left. It is plausible to assume that at the electron timescale ions, will not be moving since their mass is much larger. Due to charge separation, the induced electric field E will act to reduce this charge separation and to accelerate electrons back. Electron will gain kinetic energy and their inertia will cause them passing their original position. As a result, the plasma became again charge separated and an electric field will be formed

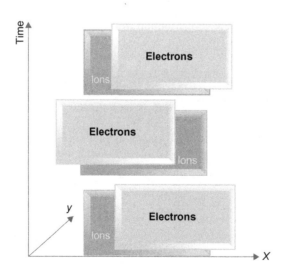

FIGURE 1.2

Schematics of a simple plasma slab.

again in the opposite direction. If there is no damping mechanism (for instance, collisions), these oscillations will continue forever. To calculate the frequency of these oscillations, an electric field can be evaluated using the Gauss theorem that is applied to rectangular region:

$$\oint E \cdot ds = \frac{\oint q \, dV}{\varepsilon_0} \tag{1.7}$$

where q is the charge in the volume V, s is the surface area, and ε_0 is the permittivity:

$$\oint q \, dV = eN_e x \, \Delta y \, \Delta z \quad \text{and} \quad \oint E \, ds = -E_x \, \Delta y \tag{1.8}$$

Thus one can see that the electric field can be calculated as:

$$E_x = -\frac{eN_e x}{\varepsilon_0} \tag{1.9}$$

The equation of motion for electrons:

$$m_e \frac{d^2 x}{dt^2} = eE_x \tag{1.10}$$

After substituting Eq. (1.9) into Eq. (1.10), the following will be obtained:

$$m_e \frac{d^2 x}{dt^2} + \frac{e^2 N_e}{\varepsilon_0 m_e} x = 0 \tag{1.11}$$

where m_e is the electron mass and t is the time. Designate new parameter as

$$\omega_p = \sqrt{\frac{e^2 N_e}{\varepsilon_0 m_e}}$$

The equation of oscillation can be written as

$$m_e \frac{d^2 x}{dt^2} + \omega_p x = 0 \tag{1.12}$$

where ω_p is the plasma frequency. This equation has the following solution:

$$x = A_1 \cos(\omega_p t) + A_2 \sin(\omega_p t) \tag{1.13}$$

where A_1 and A_2 are the amplitudes of oscillations which can be determined as constants of integration using the initial condition. It is interesting to note that these oscillations are described by the frequency ω_p, which is one of the major characteristics of the plasma. In the absence of collisions and thermal effect, this frequency uniquely determines the oscillation in the plasma. These oscillations are localized and do not propagate. Thermal effect leads to propagation of the oscillation, i.e., generation of the plasma wave or Langmuir wave that will be described in Section 1.2. Collisions can lead to the damping of the oscillation and Langmuir wave propagation that is also subject to a separate analysis described in Section 1.2.

1.1.3 Plasma types

Our daily life involves plasmas in various ways. The spark that jumps between the contacts represents the plasma associated with electrical spark discharge in air, plasma displays have captured the attention of the industry due to the quality of the picture they can produce. Nevertheless, under the ordinary conditions on the Earth, plasma is rather a very rare phenomenon. But in Universe, cold solid bodies are an exception. Most of the matter in the Universe is ionized and thus in the plasma state (about 99%). Plasma in the Universe is produced by various mechanisms. In the stars, the neutral atoms are ionized due to high temperature. Interstellar gases are ionized due to the ultraviolet radiation from the stars.

Plasmas produced in Nature and laboratory plasmas are characterized by a wide range of temperatures and pressures. Using the Debye length and plasma frequency which depend on the electron number density and electron temperature, it is possible to classify plasmas as rarified, dense, classical, and quantum. Figure 1.3 demonstrates the diagram of various objects with their plasma parameters. It spans from high-temperature plasmas in thermonuclear fusion reactors to room-temperature cold plasmas in some biomedical application. Let us consider several examples of the plasmas in controlled thermonuclear fusion, arc discharges, and cold plasmas.

1.1.3.1 Thermonuclear fusion

High-temperature thermonuclear plasma research was started in 1950 in both the United States and the Soviet Union. The main directions of research were concerned with heating and control of the plasma stability. Since then, for more than half a century, scientists from all over the globe are working on turning the promise of controlled fusion into a practical reality. In fusion reactions that are of interest for energy production, nuclei of light elements combine together to form heavier elements releasing enormous amount of energy. The reactions of greatest interest for fusion energy involve

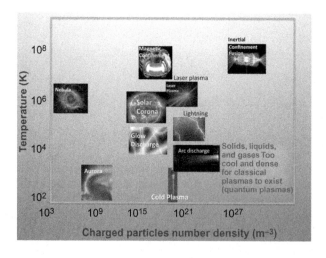

FIGURE 1.3

Diagram of electron temperature—electron density showing various natural and man-made plasmas.

deuterium, a stable isotope of hydrogen whose nucleus contains one proton and one neutron. It should be pointed out that deuterium occurs naturally, making up about 0.015% of all hydrogen.

Steady-state reaction can be supported if power input is balanced by the losses. Power input is supplied externally or from the reaction itself, while power losses are typically due to particle convection to the walls and radiation. Lawson [2] demonstrated that a simple power balance leads to two separate requirements in terms of ion temperature that has to be about 5 keV and a product of the density and confinement time which has to be about 10^{20} s/m^3.

Such temperatures are necessary in order to overcome Coulomb repulsive forces and fuse the deuterium (D = 2H) and tritium (T = 3H) into helium:

$$D + T \rightarrow_2 He^4 + n + 17\,MeV$$

Recall that the fusion reaction produces much more energy than the nuclear fission. For instance, U235 split releases about 0.8 MeV per nucleon, while D–T fusion into He produces about 3.5 MeV per nucleon.

Lawson's criterion demonstrates the importance of the plasma confinement. Among various possible confinement approaches, magnetic confinement has been the mainstream approach from the beginning. Several practical magnetic confinement schemes were proposed such as the stellarator (Lyman Spitzer) and tokamak (that was proposed in the former Soviet Union). Tokamak configuration demonstrated the best confinement results. However, it became obvious that further progress will require resources of the international community. An example of this is the Joint European Torus (JET) in Culham, UK, which has been in operation since 1983. In 1991, the JET tokamak achieved the world's first controlled release of fusion power. Scientists have now designed the next-step device—ITER—that will produce more power than it consumes: for 50 MW of input power, 500 MW of output power will be produced.

1.1.3.2 Vacuum arcs

The vacuum arc is an electrical discharge, which occurs in the vapor ejected from the electrodes. The vapor that forms the conductive path is supplied from small luminous areas on the cathode surface called cathode spots. The concept of the cathode spot unites two physically different regions: the metal surface, which can be heated by the plasma spot to extremely high temperatures, and the near-cathode plasma, which is generated as a result of cathode vaporization and atom ionization. The cathode surface phenomena (electron emission, heating) in the spot area are mutually depended on the phenomena occurring in the spot plasma. Consequently, the combination of these two interacting regions shall be determined by the cathode spot parameters. The near-cathode plasma in the spot is characterized by high particle density, current density, and temperature, which depend on the type of spot and the form of the discharge.

Various theoretical estimates and experiments have determined the following values of near-cathode plasma parameters: heavy particle density $N \sim 10^{24}-10^{26}$ m^{-3}, degree of ionization is about 0.1–100%, current density $J \sim 10^9-10^{12}$ A m^{-2}, electron temperature $T_e \sim 1-7$ eV, ion temperature $T_i \sim 0.5-3$ eV.

Research on vacuum arc was always closely related to applications. Arc discharges were considered primarily as a lighting source. Over time it was realized that vacuum arc in mercury behaves differently dependent on the electrode polarity. This led to idea that vacuum arc can be used as AC current rectifier. Significant development can be attributed to research at General Electric (Lafferty,

Greenwood). Substantial contribution to the physics and application of the mercury vacuum arc was done by Weintraub, Child, Kesaev. Cathode spot is the most complicated phenomenon in the vacuum arc. Ecker, Kesaev, and most recently Beilis developed modern understanding of the complicated cathode spot mechanism.

One of first applications of the vacuum arc was a switching medium in power circuit breakers. The arc is an essential element in the current interruption process. When a pair of current-carrying contacts is separated, current is not interrupted immediately but continues to flow through an arc, which is at once established. One of the earliest definitive works on the subject was done by Sorenson and Mendenhall in 1926, but vacuum switches became available commercially for power system applications only in the beginning of 1960s.

The vacuum arc plasma jet is an excellent source of energetic ions for the deposition of high-quality films. The advantages of this technology are:

good quality films over a wide range of deposition conditions;
high deposition rates;
low substrate temperature;
retention of alloy composition from the source to the coating.

The applications include the deposition of hard coatings such as TiN and TiAlN, on cutting tools to improve their lifetime and performance, transparent conducting thin films, amorphous semiconducting silicon thin films, and diamond-like carbon film.

In addition, vacuum arcs are used in metal ion accelerators and recently for spacecraft micropropulsion. Microcathode thruster based on the vacuum arc was recently developed. Figure 1.4 illustrates the operation of this thruster in the laboratory conditions.

The discharge at the anode is distributed diffusely over the anode surface and this electrode does not suffer erosion. The material used to form the plasma stream is evaporated at microscopic spots on the cathode surface, i.e., on average, a cold, solid surface.

1.1.3.3 Cold plasma

Recent progress in the atmospheric plasmas led to production of the cold plasmas with ion temperatures close to room temperature. This makes these plasmas different from typical low-temperature plasma of a typical electrical discharge. Such plasmas represent very useful tool enabling interaction with biological tissue without thermal damage. Cold plasma technology development resulted in rapid formation of a new field of plasma medicine [3,4]. Typical cold plasma source is shown in Figure 1.5. The plasma source is equipped with pair of high-voltage (HV) electrodes—central electrode (which is isolated from the direct contact with plasma by ceramics) and outer ring electrode as shown schematically in Figure 1.5. Electrodes are connected to a secondary of HV resonant transformer (voltage U up to 10 kV, frequency ~ 30 kHz). A typical photograph of the plasma jet is shown in Figure 1.5 for $U \sim 5$ kV and helium feeding corresponding to the flow rate of about $v_{fl} = 5-10$ l/min. The visible plasma jet had a length of approximately 5 cm and was well collimated along the entire length.

While effects of the electric field on cell was known for a long time [5], cold plasma offers much larger array of possible pathways due to the presence of various chemically active species and charged particles. Cold nonthermal atmospheric plasmas can have tremendous applications in biomedical technology. In particular, plasma treatment can potentially offer a minimum-invasive

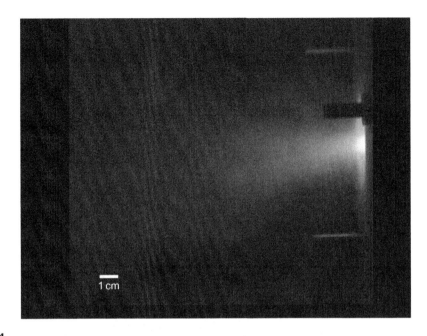

FIGURE 1.4

Microcathode thruster for spacecraft micropropulsion operating in vacuum chamber.

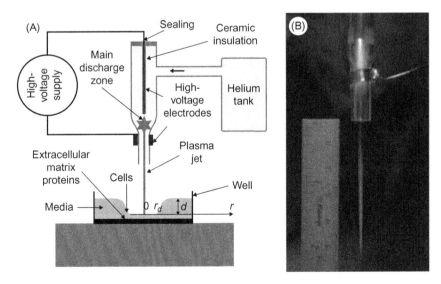

FIGURE 1.5

(A) Schematic view of the plasma gun and (B) typical photograph of plasma jet at $U = 5$ kV and $v_{fl} = 10$ l/min.

Source: *Reprinted with permission from Ref. [68]. Copyright 2008, American Institute of Physics.*

FIGURE 1.6

Atmospheric lightning.

surgery that allows specific cell removal without influencing the whole tissue. Conventional laser surgery is based on thermal interaction and leads to accidental cell death, i.e., necrosis and may cause permanent tissue damage. In contrast, nonthermal plasma interaction with tissue may allow specific cell removal without necrosis. In particular, these interactions include cell detachment without affecting cell viability, controllable cell death, etc. It can be also used for cosmetic methods of regenerating the reticular architecture of the dermis. The aim of plasma interaction with tissue is not to denaturate the tissue but rather to operate under the threshold of thermal damage and to induce chemically specific response or modification. In particular, presence of the plasma can promote chemical reaction that would have desired effect. Chemical reaction can be promoted by tuning the pressure, gas composition, and energy. Thus the important issues are to find conditions that produce effect on tissue without thermal treatment. Overall plasma treatment offers the advantage that can never be thought of in most advanced laser surgery [6]. For instance, cold plasma demonstrated great promise in cancer therapy [7]. In recent years, cold plasma interaction with tissues becomes very active research topic due to aforementioned potential.

1.1.3.4 Plasma in nature

Atmospheric lightning is probably the most known phenomenon to humans since the early days of humankind. Lightning (shown in Figure 1.6) is an electric discharge, which typically occurs during the thunderstorms. Luminous flashes above thunderstorms have been reported by eyewitnesses for over a century although these flushes were registered only two decades ago [8]. Among them are

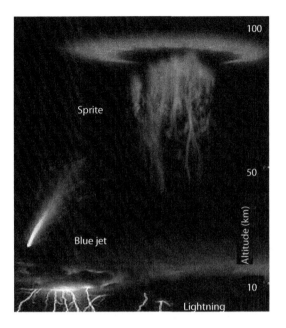

FIGURE 1.7

Discharges in upper atmosphere.

"red sprites"—optical flashes predominantly in the red located at 50–90 km above ground and associated with giant thunderstorms. Optical observation made with photometers of high spatial resolution revealed that red sprites start as a luminous cloud which then propagates mostly downward developing a highly branched structure as shown schematically in Figure 1.7. Recently upward-propagating stratospheric flashes were identified, named blue jets (BJ) primarily due to blue color. At present, a number of BJ as well as gigantic blue jets (GBJ), propagating into the mesosphere/lower ionosphere, were captured. It appears that the GBJ short-circuit the thundercloud to the ionosphere and thus may have implications for affecting the global electric circuit [9].

1.2 Plasma particle phenomena
1.2.1 Particle collisions
1.2.1.1 Definitions
Plasma particles change their momentum, energy, and states of excitation or charge through particle collisions known as elementary processes. Two types of collisions in plasmas can be distinguished, namely, *elastic* and *nonelastic*. *Elastic collisions* can be broadly defined as events in which the total kinetic energy of particles is conserved and particles retain their original charges. Elastic collisions in which momentum is redistributed between particles involved are described by the following formula:

$$A_{fast} + B_{slow} \rightarrow A_{slow} + B_{fast}$$

Nonelastic collisions are those in which kinetic energy is redistributed between particles and is transferred into some internal mode of one or more colliding particles or into creation of a new particle(s). Nonelastic collisions lead to direct and reverse processes, which include particle excitation and deexcitation, ionization and recombination, charge exchange, dissociation, and charge neutralization. These elementary processes are described by the following formulas:

Excitation and deexcitation:

$$A_{fast} + B \leftrightarrow A_{slow} + B^*$$

Several major ionization pathways can be distinguished in the typical plasma system:

- *Direct ionization* by electron impact is the ionization of neutrals from the ground state by electrons having high enough energy to cause ionization in a single act. The reverse process is the recombination by which ions capture the free electrons to neutralize the ion into atom. When the ion captured two electrons, the process named three-body recombination and with one electron is the radiative recombination.

$$A + e \leftrightarrow A^+ + 2e \quad \text{or} \quad A + e \leftrightarrow A^+ + e + e \,(hv)$$

- *Stepwise ionization* by electron impact is the ionization of previously excited neutral species.

$$A + e \rightarrow A^* + e \quad \text{and then} \quad A^* + e \rightarrow A^+ + 2e$$

- *Ionization by collisions of heavy atoms* in which kinetic energy can lead to ionization process. If chemical energy is important, such process is called associative ionization process.

$$A_{fast} + B \rightarrow A_{slow} + B^+ + e \,(\text{ionization})$$

- *Photoionization* process in which ionization is caused by photons interaction with neutral atoms. The reverse process is the photo recombination.

$$A + hv \leftrightarrow A^+ + e$$

When a molecular gas is ionized, the reaction can go into excitation of the molecule as well as the molecule can be dissociated. Dissociation is the process when the molecule decayed into separate atoms.

Charge-exchange collisions. Positive ions can collide with atoms and capture a valence electron, resulting in transfer charge from atom to the ion. It can be described as follows:

$$A^+ + B \rightarrow A + B^+$$

Associative ionization is a gas phase chemical reaction in which two atoms/molecules collide and via interaction form an ion. Energy that is released as a result of the chemical reaction is transferred into the electron energy. Ionization occurs if this energy exceeds the ionization potential. One or both of the colliding species may be in the excited state. Such reaction can be written as

$$A + B \rightarrow AB^+ + e$$

where species A interacts with B to form the ion AB^+ and electron.

Penning ionization involves reaction between neutral atoms/molecules. This process is named after Dutch physicist Frans Michel Penning (effect reported in 1927) [10]. The term penning ionization refers to the interaction between an excited-state atom/molecule A and a target atom/molecule B and results in formation of an ion B, electron, and neutral gas atom/molecule A, i.e.,

$$A^* + B \rightarrow B^+ + e + A$$

Penning ionization occurs when the target atom/molecule B has an ionization potential lower than the internal energy of the excited-state atom/molecule A*. As an example, helium has excited states at very higher energy just below ionization potential, such as energy levels of 19.82 eV for the 2^3S state and 20.6 eV for 2^1S state. Such energetic levels are higher than the ionization potential for some gaseous and metal atoms. For instance, nitrogen has ionization potential of 14.5 eV, hydrogen of about 13.5 eV, and titanium of about 6.8 eV.

Coulomb collisions are elastic collisions between charge particles, representing electron−electron, electron−ion, and ion−ion collisions.

A superelastics collision is a process when the internal energy is transferred into the kinetic energy of an emitted particle. An example of this is the Auger effect in which *energetic* electrons are emitted from an excited atom after a series of internal relaxation events. Deexcitation of metastable atoms and molecules with simultaneous release of an electron is the surface scattering process. The probability for electron release by such process is characterized by the overall secondary electron emission (SEE) coefficient γ_e, which is the ratio of emitted electrons to number of metastable atom/molecules deexcitation at the surface. γ_e is about $10^{-4} - 10^{-2}$ dependent on the atom/surface pair [11−13]. The Auger effect is the basis of the surface-sensitive electron spectroscopy.

1.2.1.2 Cross section: mean free path

The collision probability of each elementary process mentioned above is characterized by particle density and a parameter named cross section. Let us define this parameter in case of elastic collisions considering a single test particle moving through a cloud with a density n_a of particles at rest (as shown in Figure 1.8) and having only elastic encounters with target particles.

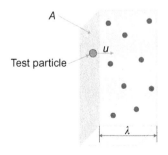

FIGURE 1.8

Schematics of particle collisions and cross-section definition.

The number of target particles in the volume V is $n_a V$ and the cross section σ of the target particle is defined as $\sigma = (\pi/4)d_p^2$, where d_p is the diameter of particle. Let us introduce the mean free path for collisions which is the average distance between consecutive collision events as

$$\lambda = \frac{1}{n_a \sigma} \tag{1.14}$$

The frequency of collisions with target particles can be defined as

$$\nu = \frac{u}{\lambda} \tag{1.15}$$

where u is the test particle velocity.

Let us introduce the collisional mean free path in detail. To that end we consider a simple gas consisting hard-sphere particle. The distance along the line joining the center of any two particles is equal to d (Figure 1.9).

Consider a single test particle z moving in randomly distributed particles at rest as shown in Figure 1.10A. It is assumed an oversimplified situation in which the test particle z moves at a uniform speed equal to the mean molecular speed $\langle C \rangle$ and all the other molecules are standing still.

Along the molecule path (Figure 1.10B), the volume trace by the sphere of influence per unit time is $\pi d^2 \langle C \rangle$. For a density of fixed particles, n_a, the number of particle centers lying within cylinder volume is $\pi d^2 \langle C \rangle n_a$. Since each of these centers corresponds to a collision, this product must also represent the number Σ of collisions per unit time for test molecule z, i.e.,

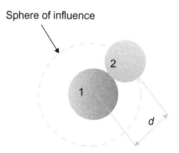

FIGURE 1.9

Sphere of influence of molecule among like molecules.

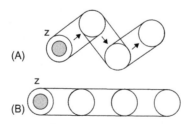

FIGURE 1.10

(A) Path of molecule z among stationary molecules: actual path and (B) path of molecule z among stationary molecules: imaginary straightened path.

$$\Sigma = n\pi d^2 \langle C \rangle \tag{1.16}$$

In essence, parameter Σ characterizes the collision frequency. The distance travel in time between the collisions characterizes the mean free path λ of the molecule with average velocity $\langle C \rangle$ as

$$\lambda = \frac{\langle C \rangle}{\Sigma} = \frac{1}{\pi d^2 n} = \frac{m}{\pi d^2 \rho} \tag{1.17}$$

More complicated analysis based on the particle kinetics produces very similar results with just a factor $\sqrt{2}$ in denominator (1.17):

$$\lambda = \frac{1}{\sqrt{2}n\pi d^2} \tag{1.18}$$

The factor $\sqrt{2}$ in the denominator arises from the fact that the correct speed to use for the evaluation of Σ in Eq. (1.17) is really *the mean relative speed* of the molecules, and this can be shown to be $\sqrt{2}\langle C \rangle$. Equation (1.18) can also be written in terms of the mass density $\rho = mn$, where m is the mass of the assumedly identical molecules. This gives

$$\lambda = \frac{m}{\sqrt{2}\pi d^2 \rho} \tag{1.19}$$

Thus for a given value of ρ, the mean free path for the rigid-sphere model is *independent of temperature* and depends only on the *gas density*, *mass*, and *diameter of the molecules*.

In case of assembly of test particles, the flux density of the particles can be defined as $\Gamma = nu$, where n is the particle density (see Figure 1.8). Let us assume that collisions remove particles from the beam. The change of the particle density can be calculated as follows:

$$dn = -\sigma n n_a\, dx \tag{1.20}$$

and the reduction of the particle flux can be calculated as

$$\frac{d\Gamma}{\Gamma} = -n_a \sigma\, dx \tag{1.21}$$

Integration of Eq. (1.21) will give the flux dependence on distance x as the following:

$$\Gamma(x) = \Gamma(0)\exp(-x/\lambda) \tag{1.22}$$

After a collision, the test particle scattered by an angle θ of scattered particle trajectory with respect to its incident direction as shown in Figure 1.11. This kind of collisions is characterized by another parameter defined as differential cross section $\sigma(\theta)$. Impact parameter (b, see Figs. 1.11, 1.12) is the perpendicular distance from the original center of a scattering particle to the original line of motion of a particle being scattered.

The probability of the particle emerging into the solid angle is $d\Omega = \sin\theta\, d\theta\, d\varphi$. Integrating over the full solid angle yields the so-named total cross section for a particular collision in the form

$$\sigma_{sc} = \int_0^{2\pi} \int_0^{\pi} \sigma(\theta)\sin\theta\, d\theta\, d\varphi \tag{1.23}$$

For many practical applications, it is important to consider interactions that lead to scatter by $\theta = 90°$ or more as shown in Figure 1.12. The elastic collision with $\theta = 90°$ occurs on $\chi = 45°$. Thus the cross section in this case is equal to cross section for scattering with angle larger than $\sigma(90) = (\pi a_{12}^2/2)$. The collision cross in the hard-sphere model is $\sigma = \pi a_{12}^2$. One can see that $90°$ is smaller than that of the hard-sphere model by factor of 2.

1.2.1.3 Charge-exchange cross section

Both resonant and nonresonant charge-exchange collisions can take place. Resonant charge exchange occurs between the same atoms, while nonresonant occurs between different atoms. Consider reaction

$$A^+ + B \rightarrow A + B^+$$

Electron transfer from B to A^+ occurs in two steps: first, release from B and then capture by A^+. By calculating the potential energy of electron in the electrostatic field of A^+ and B^+ and

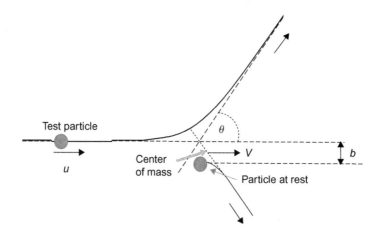

FIGURE 1.11

Schematics of scattering collisions in the laboratory frame; b is the impact parameter.

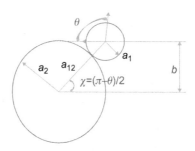

FIGURE 1.12

Schematics of the hard-sphere collision. a_1, a_2 are radii of the spheres and a_{12} is the distance between centers of the colliding spheres.

considering kinetic energy of electron, one can calculate the distance between A and B leading to electron transfer. For the ground state, resonant charge exchange cross section (i.e., A = B) can be estimated as:

$$\sigma_{cx} \approx 36\pi \left(\frac{e^2}{8\pi\varepsilon_0 E_i}\right)^2 \tag{1.24}$$

where e is the electron charge, ε_0 is the permittivity of vacuum, and E_i is the ionization potential. Detailed theoretical analysis and experiment show that cross section for charge exchange has weak dependence on the kinetic energy. While most of data on charge-exchange cross section are obtained experimentally, there are several simplified theories developed.

Sakabe and Izawa [14] calculated cross section for charge-exchange collisions by solving the time-dependent Shrodinger equation. The cross sections of resonant charge transfer between atoms and positive ions for all nontransition elements have been calculated by the impact parameter and close-coupling methods. The calculated results are in good agreement with the experimental results for the elements for which data are available. Figure 1.13 shows the calculation result and the experimental data on the cross section of helium atoms as a function of impact velocity. The calculated results are in good agreement with the experimental results, in particular in the low-velocity range less than 10^8 cm/s.

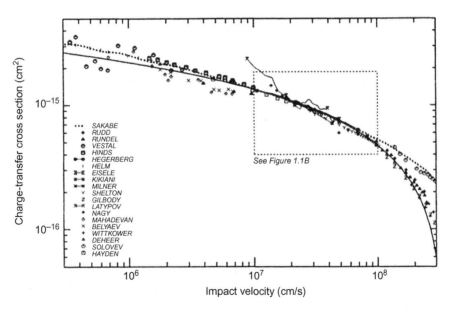

FIGURE 1.13

Cross section—impact—velocity curve for resonant charge transfer of a helium atom. Experimental data are given by symbols and thin lines [14]. A thick dashed line corresponds to calculation result.

Source: *Reprinted with permission from Ref. [14]. Copyright 1992 by the American Physical Society.*

1.2.1.4 Coulomb collision cross section

Let us consider collisions of two particles with charges q_1 and q_2. This type of interactions occurs between electron−electron, ion−electron, and ion−ion. Particles interact with one another via electrostatic potential:

$$\Psi(r) = \frac{q_1 q_2}{4\pi\varepsilon r} \qquad (1.25)$$

Coulomb forces are long-range interaction as it depends inversely on distance r^2. However, potentials are screened in plasma at the scale of about Debye length L_D. Thus the maximum impact parameter can be taken equal to the Debye length since the Coulomb electric field force decays exponentially at the distances larger than L_D. In other words, the maximum distance for interactions is $b_{max} \sim L_D$. Yet, a collision cross section based on the Debye length, i.e., $\sigma = \pi L_D^2$, is still very large. Taking into account that the electric field is partially screened in the Debye sphere, a collision cross section based on the Debye length includes scattering over very small angles that does not affect significantly transport coefficients. Thus we will limit our consideration of the scattering process by only scattering over large angles, e.g., $>90°$.

Scattering to large angles can be due to a single collision scattering over a large angle or due to cumulative effect of many small-angle collisions. In plasmas, because of the long-range nature of the electrostatic interaction, much frequent small-angle collisions led to larger effect than the fewer large angle events. In order to evaluate this effect, one needs to consider the cumulative effect of many scattering of electron by many ions with different values of impact parameter b (Figure 1.12).

Consider electron having initial velocity v that undergoes many small-angle scattering collisions. Each collision event will lead to small increment in electron velocity Δv in all directions. Recall that electron energy is nearly conserved in collision since electron is a light particle that is scattered off a heavy ion, while electron losses its momentum. Thus there is a reduction of the electron velocity in the original direction. The average velocity reduction can be estimated as

$$\frac{d\langle \Delta v \rangle}{dt} = -\frac{nZ^2 e^4 \ln \Lambda}{4\pi\varepsilon_0^2 m^2 v^2} \qquad (1.26)$$

where n is the ion density, Z is the ion charge, and $\ln \Lambda = \ln(b_{max}/b_{min})$ is called Coulomb logarithm. From this equation, one can define the electron−ion collisional frequency for loss of electron momentum as

$$\nu_{ei} = \frac{nZ^2 e^4 \ln \Lambda}{4\pi\varepsilon_0^2 m^2 v^3} \qquad (1.27)$$

To estimate the minimum impact parameter b_{min}, we need to point out that when the Coulomb potential energy becomes as large as the electron thermal energy, the scattering angle becomes the largest.

Thus b_{min} can be estimated from the condition that the Coulomb energy is equal to the thermal energy:

$$\frac{q_1 q_2}{4\pi\varepsilon b_{min}} = 2\frac{3}{2}kT \qquad (1.28)$$

The Coulomb logarithm represents the sum of all scattering angles by Coulomb collisions within the Debye sphere. Detailed analysis leads to the following expression for the Coulomb logarithm (if $T_i(m_e/m_i) < T_e < 10Z^2$ eV [15]):

$$\ln \Lambda = 23 - \ln (n_e^{0.5} Z T_e^{-3/2}) \tag{1.29}$$

where T_e is in eV and n_e is in cm^3. In plasma of interest is 5–10 eV. ln Λ is about 17 in the case of fusion plasma and about 10 in the case of arc discharge plasmas.

The total cross section for electron scattering on ions can thus be calculated using relationship between the collision frequency and cross section:

$$\sigma_c = \frac{\nu_{ei}}{nv} = \frac{Z^2 e^4 \ln \Lambda}{4\pi \varepsilon_0^2 m^2 v^4} \tag{1.30}$$

1.2.1.5 Ionization cross section

In general, the quantum mechanical analysis should be employed to describe the cross section of atom ionization. However, some qualitative analysis with reasonable quantitative results was performed using classical approach in physics. In this case, the ionization process was considered as interaction of impact electron with a valence electron of a neutral atom. It was assumed that the atom was ionized if the transferred energy to the valence electron exceeds the ionization potential E_i. This is the basis of the model that was developed by Thomson back in 1912 [16]. According to the Thomson's model, the ionization takes place if transferred energy is larger than the ionization potential, i.e., $E > E_i$. After integration over electron energies that is larger than E_i, the following expression for the ionization cross section was obtained [14]:

$$\sigma_i = \pi \left(\frac{1}{4\pi \varepsilon_0} \right)^2 \frac{1}{E} \left(\frac{1}{E_i} - \frac{1}{E} \right) \tag{1.31}$$

At very high energies, Thomson model predicts that ionization cross section fall as $\sim 1/E$. The cross section has its maximum at $E = 2E_i$. Thus the maximal cross section is

$$\sigma_{i\,max} = \pi \left(\frac{1}{4\pi \varepsilon_0} \right)^2 \frac{1}{4E_i^2} = \frac{1}{64\pi \varepsilon_0^2 E_i^2} \tag{1.32}$$

The dependence of the ionization cross section on the electron energy for different atoms is shown in Figure 1.14.

It should be pointed out that the electron temperature is typically in the order of few electron volts which is smaller than the ionization potential for most atoms of interest. However, the high-energy electrons from the tail of Maxwell distribution play a very important role in the ionization process. In order to obtain the collisional cross section in the case of electron energy distribution, one have to integrate over the velocity distribution:

$$\nu = n_a k_i(T) = n_a \int d^3 v_1 \, d^3 v_2 f_1 f_2 \sigma(v_1 - v_2)(v_1 - v_2) \tag{1.33}$$

where $k_i(T)$ is the ionization coefficient.

If Maxwellian distribution function can be assumed, we will arrive at the following expression:

$$k_i(T) = \left(\frac{m}{2\pi kT}\right)^{3/2} \int_0^\infty \sigma(v)v \exp\left(-\frac{mv^2}{2kT}\right) 4\pi v^2 dv \tag{1.34}$$

In order to calculate the cross section of atom ionization $\sigma(v)$, one can use the Thomson formula near $E = mv^2/2 \sim E_i$:

$$\sigma(E) = \sigma_0(E - E_i/E_i) \quad \text{if} \quad E > E_i$$
$$= 0 \quad\quad\quad\quad \text{if} \quad E \leq E_i \tag{1.35}$$

where $\sigma_0 = (\pi e^2/(4\pi\varepsilon_0 E_i)^2)$.

With this assumption, the above expression for ionization coefficient becomes the following simple form:

$$k_i(T) = \sigma_0\sqrt{\frac{8kT}{\pi m}}\left(1 + \frac{2T}{E_i}\right)\exp\left(-\frac{E_i}{T}\right) \tag{1.36}$$

Below the ionization cross sections for electron impact with different materials are demonstrated. An example of total cross section for electron scattering on nitrogen molecules is shown in Figure 1.15 and ionization collision cross section for several gases [18] is shown in Figure 1.16A and B. One can see that total scattering cross section can have several maxima. Peak in ionization cross section is clearly recognizable in both cases.

1.2.1.6 Plasma equilibrium

One particular thermodynamic state of the plasma is widely used to describe plasma chemical composition, so called plasma equilibrium. Let us first define the degree of ionization in the plasma. Consider plasma consisting of electrons, ions, and atoms. In the case of a low-ionized plasma, i.e., $n_e \ll n_a$, the ionization degree is defined as

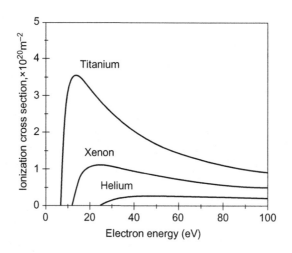

FIGURE 1.14

Calculated cross section according to the Thomson model (Eq. 1.31).

$$\alpha = n_e/(n_a + n_e) \approx n_e/n_a \qquad (1.37)$$

In the thermodynamic equilibrium, there are two reactions: direct reaction leading to ionization and a back reaction leading to recombination. In the case considered, the direct reaction is the electron impact ionization and back reaction is the recombination involving three particles. In other words, the stochiometric equation will have the form:

$$a + e \leftrightarrow i + 2e$$

with the rate of ionization being $k_i \cdot n_a \cdot n_e$ and the rate of recombination being $k_r \cdot n_i \cdot n_e^2$, where k_i and k_r are the ionization and recombination coefficient, respectively. We shall define the ionization equilibrium as a state in which ionization and recombination rates are equal.

The equilibrium constant $K(T)$ can be defined as follows:

$$K(T) = k_r/k_i = n_i \cdot n_e/n_a \approx n_e^2/n_a \qquad (1.38)$$

This equation can be solved for the electron density provided that the equilibrium constant dependent on temperature is known as a function of the pressure in the system. If equation of state for the plasma can be specified, this equation can be used to calculate the electron density as a function of the temperature.

Degree of ionization can also be determined from the thermodynamic equilibrium of plasma. We shall illustrate this below. This relationship that can be used to determine the plasma composition in equilibrium is called the Saha equation.

Let us derive the Saha equation from the statistical mechanics. We assume that plasma is in thermodynamic equilibrium, so that we can use a single temperature for all plasma components. The approach that will be used is quasi-classical taking into account statistical weights and quantum phase space. While being an approximation, this approach gives physically illustrative picture of the phenomena.

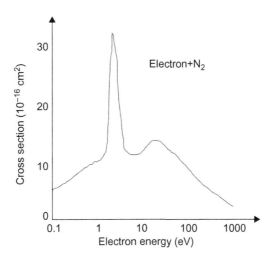

FIGURE 1.15

Total scattering cross section of molecular nitrogen (e + N_2).

Source: *Adapted from Ref. [17].*

According to the statistical mechanics, the probability of a particle to be at the energy level ε is determined by the Boltzmann relation:

$$p(\varepsilon) = C\,e^{-(\varepsilon/T)} \tag{1.39}$$

where T is the plasma temperature.

The number of particles can be calculated by multiplying the probability and number of states. The number of states assuming that particles do not have internal degrees of freedom is equivalent to the number of cells in the phase space h^3, where h is the Planck's constant. The volume in the momentum space for electrons having momentum between p and $p + dp$ equals:

$$d\Omega = 4\pi p^2\, dpV \tag{1.40}$$

Thus, the number of electrons in the volume V can be calculated as

$$dN_e = Vg_e \frac{4\pi p^2}{h^3} dp\, e^{-(\varepsilon/T)} \tag{1.41}$$

where g_e is the electron degeneracy.

We are considering the equilibrium state of electrons and neutrals and as such let us consider zero energy level corresponds to electrons in atom. It follows that free electron will have the energy

$$\varepsilon = E_i + \frac{p^2}{2m_e} \tag{1.42}$$

where E_i is the ionization potential. The total number of electrons can then be calculated by integrating over the entire spectrum:

$$N_e = Vg_e \frac{4\pi}{h^3} e^{-(E_i/T)} \int_0^\infty p^2 e^{-(p^2/2mT)} dp \tag{1.43}$$

After integration one can find that the electron density has the following expression:

$$N_e = Vg_e \frac{(2\pi mT)^{3/2}}{h^3} e^{-(E_i/T)} \tag{1.44}$$

Such quasi-classical approach cannot lead to the complete expression without considering additional assumptions. First of all, recall that in the integral (1.43) it was taken into account the number of electrons per a single atom. In order to calculate the full number of electrons, we have to multiply N_e by number of atoms in one quantum state, i.e., N_a/g_a. The resulting expression for the number of electrons is thus

$$N_e = N_a V \frac{g_e}{g_a} \frac{(2\pi mT)^{3/2}}{h^3} e^{-(E_i/T)} \tag{1.45}$$

Secondly, equilibrium state consists of electrons, neutrals, and ions. Thus, ionization balance must account for ions as well. This can be done by assuming that the volume corresponds to an ion in a given quantum state, i.e.,

$$V = \frac{1}{(N_i/g_i)} \tag{1.46}$$

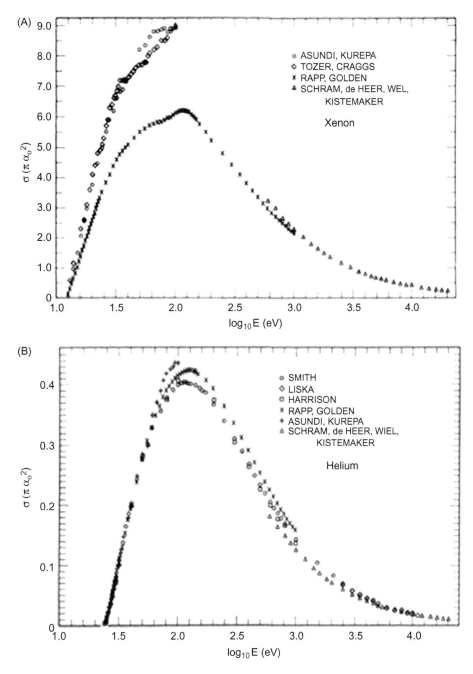

FIGURE 1.16

(A) Total cross sections for the ionization of atomic xenon. (B) Total cross sections for the ionization of atomic helium. *Note*: To convert cross sections in πa_0^2 units to 10^{16} cm^{-3} cross sections in πa_0^2 should be divided by 1.13673.

Source: *Reprinted with permission from Ref. [18]. Copyright 1966 by the American Physical Society.*

By supplementing this expression into Eq. (1.45), we arrive at the final expression for the Saha equation:

$$\frac{N_e N_i}{N_a} = \frac{g_e g_i}{g_a} \frac{(2\pi m T)^{3/2}}{h^3} e^{-(E_i/T)} \tag{1.47}$$

Equation (1.47) describes plasma that consists of electrons, atoms, and ions of a single type. However, it can also be used to calculate the composition of gas mixture. Let us demonstrate this on the following example.

EXAMPLE 1.1

Arc discharge operates in the chamber filled with helium at pressure of about 0.1 atm. Arc is supported by ablation of the anode, which is made of carbon. Calculate plasma composition of the mixture of helium and carbon as a function of plasma temperature assuming that plasma is in equilibrium state. Ratio of helium to carbon partial pressure is 3.

Let us start with equations for equilibrium for each individual species. In the case considered, these equations can be written as

$$\frac{n_e n_{Hei}}{n_{Hea}} = \frac{g_e g_{Hei}}{g_{Hea}} \frac{(2\pi m T)^{3/2}}{h^3} e^{-(E_{Hei}/T)} \tag{1.48}$$

$$\frac{n_e n_{Ci}}{n_{Ca}} = \frac{g_e g_{Ci}}{g_{Ca}} \frac{(2\pi m T)^{3/2}}{h^3} e^{-(E_{Ci}/T)} \tag{1.49}$$

where n_{Hei} is the density of helium ions and n_{Ci} is the density of carbon ions.

Arc plasma consists of the electrons, neutrals C and He, and ions C+ and He+. Thus, we have five unknown densities. In order to calculate an equilibrium composition, we can invoke two additional equations. One is the plasma quasi-neutrality:

$$n_e = n_{Ci} + n_{Hei} \tag{1.50}$$

Second equation is the equation for atom conservation, i.e., the ratio of total density of atoms and ions of helium to the total density of atoms and ions of carbon:

$$\frac{n_{Hei} + n_{Hea}}{n_{Ci} + n_{Ca}} = 3 \tag{1.51}$$

Finally, the total pressure is equal to the sum of partial pressures of all species.
Plasma composition calculation using Saha equilibrium is shown in Figure 1.17.

One can see that significant ionization starts at temperatures exceeding 12,000 K and it is dominated by ionization of carbon. Ionization of helium atoms starts at about 20,000.

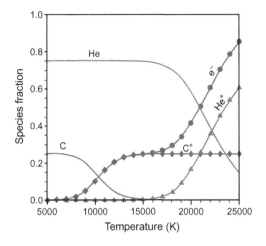

FIGURE 1.17

Plasma composition for carbon—helium mixture. Pressure is 0.1 atm.

1.3 **Waves and instabilities in plasmas**

1.3.1 **Electromagnetic phenomena in plasma**

1.3.1.1 *Conservation law for electric charge and current: electromagnetic waves*

Each charged particle in plasma has an electric charge. Electron has charge of about 1.6×10^{-19} C and ion charge being either single or multiple electron charges. While typically ions have a positive charge, negatively charged ions can be created due to electron attachment to atom or molecule. Good example is the oxygen that has very large cross section for electron attachment.

The electrostatic force F_{ij} is determined by the interaction of charges q_i and q_j with distance r_{ij} between the charges in form:

$$F_{ij} = \frac{q_i q_j}{4\pi\varepsilon_0 r_{ij}^2}$$

(1.52)

Charge is the additive property, i.e., the total charge in a volume can be expressed as a sum of number particles of type N_i with charges q_i:

$$Q = \sum_i N_i q_i$$

(1.53)

where q_i is the charge of a single particles.

Density of charges per unit volume:

$$\rho = \frac{dQ}{dV}$$

(1.54)

Considering uniform charge distribution in the volume V, current density can be defined as

$$J = \rho \cdot v$$

(1.55)

where v is the average velocity of charges. When charges are distributed nonuniformly in the volume, the total charge can be calculated by integrating over the considered volume:

$$\frac{\partial Q}{\partial t} = \frac{\partial}{\partial t}\int_V \rho \, dV = \oint_S \rho v \, ds \tag{1.56}$$

By employing the Gauss theorem:

$$\oint_S \rho v \, ds = \int_V \nabla \cdot (\rho v) dV \tag{1.57}$$

one can arrive at the following equation which is the conservation of charges:

$$\int \left[\frac{\partial \rho}{\partial t} + \nabla \cdot (\rho v) \right] dV = 0 \quad \text{or} \quad \frac{\partial \rho}{\partial t} + \nabla \cdot J = 0 \tag{1.58}$$

Electric and magnetic fields in the plasma must satisfy the Maxwell equations. These equations were proposed by James Maxwell to describe electromagnetic wave propagation in a media:

$$\nabla \times E = -\frac{\partial B}{\partial t} \tag{1.59}$$

$$\nabla \times H = J + \varepsilon \frac{\partial E}{\partial t} \tag{1.60}$$

$$\nabla \cdot B = 0 \tag{1.61}$$

$$\nabla \cdot E = \frac{\rho}{\varepsilon_0} \tag{1.62}$$

Introducing the electric potential, one can define the electric field as $E = -\nabla\varphi$, as a result the last equation will have the following form:

$$\Delta\varphi = -\frac{\rho}{\varepsilon_0} \tag{1.63}$$

Equation (1.63) is called the Poisson equation. If $n_e = n_i$, i.e., quasi-neutrality is fulfilled, the Poisson equation will reduce to $\Delta\varphi = 0$, which is equivalent to assuming the vacuum electric field. Note that the Poisson equation in reduced form can be used to calculate potential distribution only if plasma is uniform and in the absence of a magnetic field. In a general case, potential distribution in plasma can be calculated invoking the generalized Ohm's law as will be described in Chapter 4.

1.3.1.2 Electromagnetic wave propagation

Waves can transport energy to and from a media. In order to describe the wave propagation, we will use Eqs (1.59) and (1.60). Two limited cases can be considered, namely, a media with high conductivity and a media with low conductivity.

1.3.1.2.1 Propagation in a media with high conductivity (i.e., metal)
In this case it can be safely assumed that

$$J \gg \varepsilon_0 \frac{\partial E}{\partial t} \tag{1.64}$$

For a high conductive media, the set of Maxwell equations can be written as follows:

$$\nabla \times E = -\frac{\partial B}{\partial t} \tag{1.65}$$

$$\nabla \times H = J \tag{1.66}$$

$$\nabla \cdot B = 0 \tag{1.67}$$

$$\nabla \cdot E = 0 \tag{1.68}$$

The Ohm's law reads

$$J = \sigma E \tag{1.69}$$

where σ is the conductivity of metal. Employing the wave in the form of $E, H \sim \exp(j[kr - \omega t])$, where j is the imaginary unit, k is called the propagation constant, and ω is the frequency of the wave. Substituting the waveform into Eqs (1.65) and (1.66) leads to

$$\nabla \times E = \mu j \omega \tilde{H} \exp(-j\omega t)$$

where $\tilde{H}$ is the amplitude of the magnetic field. Apply $\nabla \times$ operator to both sides of this equation:

$$\nabla \times \nabla \times E = \mu j \omega \nabla \times H$$

Using operator properties and Eq. (1.66), this equation can be rewritten as

$$\nabla \times \nabla \cdot E - \Delta E = \mu j \omega \sigma E$$

Taking into account Eq. (1.68) i.e., $\nabla \cdot E = 0$, this equation can be reduced to the following equations for electric fields:

$$\Delta E + \gamma^2 E = 0 \tag{1.70}$$

The same approach can be applied to the magnetic field:

$$\Delta B + \gamma^2 B = 0 \tag{1.71}$$

where $\gamma = \sqrt{j \omega \mu \sigma}$.

Considering the problem of wave propagation in one-dimensional approximation, the amplitude of the electric and magnetic fields can be written as

$$E = E_0 \exp(-\gamma x) \tag{1.72}$$

$$B = B_0 \exp(-\gamma x)$$

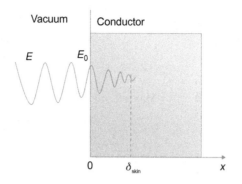

FIGURE 1.18

Electric field in the conductor.

where E_0 and B_0 are the amplitudes of the electric and magnetic fields at the boundary as shown in Figure 1.18. Let us analyze the parameter γ:

$$\gamma = \sqrt{j\omega\mu\sigma} = \frac{j+1}{\sqrt{2}}\sqrt{\omega\mu\sigma} \qquad (1.73)$$

The real part of the parameter γ is equal to

$$\alpha = \sqrt{\frac{\omega\mu\sigma}{2}}$$

Thus the expression for the electric field amplitude has the following form:

$$E = E_0 \exp(-\alpha x)\exp(-j\alpha x) \qquad (1.74)$$

One can see that the amplitude $E = E_0 \exp(-\alpha x)$ of the electric field decreases with distance. The parameter $1/\alpha$ is the characteristic length, which is called a skin layer. The skin layer thickness can thus be calculated as

$$\delta_{skin} = \sqrt{\frac{2}{\omega\mu\sigma}} \qquad (1.75)$$

1.3.1.2.2 Propagation in a media with low conductivity (i.e. dielectric)

In this case one can assume that $J \ll \varepsilon_0(\partial E/\partial t)$, i.e., in this case only a displacement current can be considered. Using the same approach as in the case of a wave propagation in the conductor, we arrive at the following equations:

$$\Delta E - \gamma^2 E = 0 \qquad (1.76)$$

$$\Delta B - \gamma^2 B = 0 \qquad (1.77)$$

where $\gamma = \sqrt{-\omega^2\mu\varepsilon_0} = j\sqrt{\omega^2\mu\varepsilon_0}$. Thus in this case the parameter γ has only an imaginary part. This means that there is no damping of the wave. This is shown schematically in Figure 1.19.

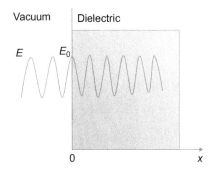

FIGURE 1.19

Electric field in the dielectric.

1.3.2 **Waves in plasma**

Any periodic motion of plasma can be decomposed into a superposition of sinusoidal oscillations with frequency ω and wavelength λ. In the following, we will consider that any quantities (density n, velocity v, electric field E) have sinusoidal waveform. Thus, the plasma density is represented as

$$n = \bar{n} \exp(j[kr - \omega t]) \tag{1.78}$$

where $\bar{n}$ is the density amplitude, $k \cdot r = k_x x + k_y y + k_z z$ is the vector of wave propagation. In a simple one-dimensional case is $n = \bar{n} \exp(j[k_x x - \omega t])$. Recall that the measurable quantity is in the real part:

$$\text{Re}(n) = \bar{n} \cos(k_x x - \omega t) \tag{1.79}$$

A point of constant phase of the wave moves so that $d/dt(k_x x - \omega t) = 0$ or in other words

$$\frac{dx}{dt} = \frac{\omega}{k} = v_\varphi \tag{1.80}$$

where v_φ is the phase velocity. Similarly electric field can be expressed:

$$E = \bar{E} \cos(k_x x - \omega t + \delta) \tag{1.81}$$

where δ is the phase shift with respect to the density oscillations as given in Eq. (1.79).

Note: Phase velocity can exceed the speed of light! However, this does not violate the theory of relativity because wave of constant amplitude does not carry any information. Information is carried only when wave is modulated. The modulation is traveled with velocity v_g, which is the group velocity of the wave.

Let us define the group velocity. For this reason we will consider the two added "beating" waves with the following conditions:

$$E_1 = \bar{E_0} \cos([k + \Delta k]x - [\omega + \Delta \omega]t) \tag{1.82}$$

$$E_2 = \bar{E_0} \cos([k - \Delta k]x - [\omega - \Delta \omega]t) \tag{1.83}$$

Using the following new denotation:

$$a = kx - \omega t$$

$$b = \Delta kx - \Delta\omega t$$

Summation of two electric fields will give the following:

$$E_1 + E_2 = \overline{E_0} \cos(a + b) + \overline{E_0} \cos(a - b) = \overline{2E_0} \cos(a) \cos(b) \tag{1.84}$$

$$E_1 + E_2 = \overline{2E_0} \cos(\Delta kx - \Delta\omega t)\cos(kx - \omega t) \tag{1.85}$$

It follows that the envelope of such wave is given by $\cos(\Delta kx - \Delta\omega t)$. Equation (1.85) can be interpreted as a simple wave with modulated amplitude $2\cos(\Delta kx - \Delta\omega t)$. In other words, the amplitude of the wave is itself a wave, and the phase velocity of this modulation wave is $\Delta\omega/\Delta k$. The propagation of information or energy in a wave occurs as a change in the wave. The simplest example is transition from the wave being absent to being present, which propagates at the speed of the leading edge of a wave train. In a more general sense, some modulation of the frequency and/or amplitude of a wave is required in order to convey information, and it is this modulation that represents the signal content. Thus the speed of information propagation can be described as $\Delta\omega/\Delta k$. This is the phase velocity of the wave amplitude, but since each amplitude wave may contain a group of internal waves, this speed is called the group velocity:

$$v_g = \Delta\omega/\Delta k$$

It should be pointed out that in accordance with theory of relativity, the group velocity v_g is always smaller than the speed of light, i.e., $v_g \leq c$.

1.3.3 Plasma oscillations

Any disturbance of an electric field in the plasma can displace the electrons from the uniform background of ions. As a result, an internal electric field will appear to restore the neutrality of plasma by pulling electrons back. Because of negligible inertia, electrons are overshooting and as a result will oscillate around their equilibrium position. On the timescale considered, ions are not responded to the oscillations. Such oscillations will occur at the so-called plasma frequency ω_p.

Consider wave propagation in such plasma under the following conditions: no magnetic field; all particles are cold (i.e., $k_B T = 0$); ions are not moving; plasma is infinite.

Under these conditions, a 1D equation for electron motion and continuity can be written as follows:

$$mn_e\left[\frac{\partial V_e}{\partial t} + (V_e\nabla_x)V_e\right] = -en_e E \tag{1.86}$$

$$\frac{\partial n_e}{\partial t} + \nabla_x \cdot (n_e V_e) = 0 \tag{1.87}$$

Poisson's equation reads

$$\varepsilon_0 \frac{\partial E}{\partial x} = e(n_i - n_e) \tag{1.88}$$

This system describes plasma state including its temporal evolution. Equations (1.86)–(1.88) will be employed in order to analyze the electron oscillations in plasma. To this end, the system of equations (1.86)–(1.88) will be linearized assuming that the amplitude of oscillation is small and using the standard approach of small perturbation, the plasma parameters from their respected equilibrium state:

$$n_e = n_0 + n_1$$
$$V_e = V_0 + V_1 \tag{1.89}$$
$$E_e = E_0 + E_1$$

where subscript "0" corresponds to the equilibrium part and subscript "1" corresponds to the perturbation part. In the framework of oscillations with small amplitude, the following condition take place:

$$n_0 \gg n_1$$
$$V_0 \gg V_1 \tag{1.90}$$
$$E_0 \gg E_1$$

In addition, we consider uniform quasi-neutral plasma is at rest:

$$\nabla n_0 = 0; \quad V_0 = 0; \quad E_0 = 0 \tag{1.91}$$

$$\frac{\partial n_0}{\partial t} = \frac{\partial V_0}{\partial t} = \frac{\partial E_0}{\partial t} = 0 \tag{1.92}$$

Taking into account the above conditions and assumption, and neglecting the second-order terms we arrive at the following simplified system of equations:

$$mn_e \left[\frac{\partial V_1}{\partial t}\right] = -en_e E_1 \tag{1.93}$$

$$\frac{\partial n_1}{\partial t} + \nabla \cdot (n_0 V_1) = 0 \tag{1.94}$$

or after implementing the assumption that plasma is uniform we arrive at the following:

$$\frac{\partial n_1}{\partial t} + n_0 \nabla \cdot (V_1) = 0 \tag{1.95}$$

The condition that all oscillating properties behave sinusoidal is considered, i.e.,

$$V = V_1 \exp(j[kx - \omega t])$$
$$n = n_1 \exp(j[kx - \omega t]) \tag{1.96}$$
$$E = E_1 \exp(j[kx - \omega t)$$

Let us substitute these expressions into Eqs (1.93) and (1.95). After differentiating these equations, the system of equations of interest (Eqs (1.93) and (1.95)) is reduced to the following algebraic equations:

$$-jm\omega V_1 = -eE_1$$
$$-j\omega n_1 = -n_0 jk V_1 \qquad (1.97)$$
$$-jk\varepsilon_0 E_1 = -en_1$$

The system of equation (1.97) can be solved for frequency ω and result in

$$\omega^2 = \frac{n_0 e^2}{m\varepsilon_0} \qquad (1.98)$$

For instance, in the case of a density of about 10^{18} m^{-3}, plasma frequency $(\omega/2\pi)$ is about 9 GHz. It should be pointed out that group velocity for plasma oscillations, i.e., $V_g = \partial\omega/\partial k = 0$ and therefore disturbance associated with plasma oscillation does not propagate, i.e., localized. While plasma oscillations are standing, the thermal motion of electrons can cause plasma wave to propagate. In the next section, we will consider electron plasma wave, which is developed due to electron thermal motion.

1.3.4 Electron plasma wave

Consider the case that electrons are streaming from the oscillating region due to presence of density gradient and their thermal motion. Such electron flux carries information about oscillation from the oscillating localized area thus creating the wave. This wave is called a plasma wave.

In order to account for the effect of wave formation, one needs to consider an additional term in the electron momentum (1.93), i.e., the electron pressure gradient ∇P_e. For our analysis, we will use the adiabatic index $\gamma = 3$ in the one-dimensional case:

$$\nabla P_e = 3kT_e \nabla n_e = 3kT_e \nabla(n_0 + n_1) = 3kT_e \frac{\partial n_1}{\partial x} \qquad (1.99)$$

Now the equation of motion for electrons will have the following form:

$$mn_0 \left[\frac{\partial V_1}{\partial t}\right] = -en_e E_1 - 3kT_e \frac{\partial n_1}{\partial x} \qquad (1.100)$$

After substitution of the sinusoid waveform, we arrive at the following:

$$-jm\omega n_0 V_1 = -eE_1 n_0 - 3kT_e jk n_1 \qquad (1.101)$$

If we express both perturbed density n_1 and electric field E_1 as a function of perturbed velocity V_1, we arrive at the following relation:

$$\omega^2 V_1 = \left(\frac{n_0 e^2}{m\varepsilon_0} + \frac{3kT_e}{m} k^2\right) V_1 \qquad (1.102)$$

This equation can be modified and solved for the frequency leading to dispersion relation:

$$\omega^2 = \omega_p^2 + \frac{3}{2}k^2 V_{th}^2 \tag{1.103}$$

where $V_{th}^2 = (2kT_e/m)$.

It follows from this dispersion relation that frequency depends on the wave number. Differentiating the last expression leads to

$$2\omega \, d\omega = 3k \, dk V_{th}^2$$

From here one can find that:

$$V_g = \frac{d\omega}{dk} = \frac{3}{2}\frac{k}{\omega} V_{th}^2 = \frac{3}{2}\frac{V_{th}^2}{V_\varphi} \tag{1.104}$$

Diagram of the plasma wave (correspondent to Eq. (1.103)) is shown in Figure 1.20. At any point on the dispersion curve, it is possible to define the group velocity and the phase velocity. From the diagram, one can note that in the case of large k (small wavelength λ), plasma wave travels at V_{th}.

It should be pointed out that existence of plasma waves was known since the time of Langmuir. Sometimes these waves are called Langmuir waves. Complete theory of these waves was developed by Bohm and Gross [19].

1.3.5 Sound waves in plasma

Sound waves propagate in gas due to collisions. Sound waves in plasmas can propagate even in absence of collisions by electron−ion coupling via the electric field. Since heavy particles are involved, such plasma oscillations will be in a low-frequency range. Let us illustrate these oscillations considering plasma quasi-neutrality condition, i.e., $n_e = n_i$.

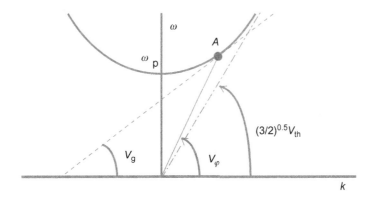

FIGURE 1.20

Dispersion relation for the plasma wave. Group and phase velocity are shown.

Starting with ion momentum equation:

$$Mn\left[\frac{\partial V_i}{\partial t} + (V_i \nabla)V_i\right] = enE - \nabla P = -en\,\nabla\varphi - \gamma kT_i\,\nabla n \qquad (1.105)$$

Linearization of this equation leads to

$$j\omega Mn_0 V_{i1} = -en_0 jk\varphi_1 - \gamma kT_i jkn_1 \qquad (1.106)$$

Boltzmann relation for electrons is employed:

$$n_e = n_0\,\exp\left[\frac{e\varphi_1}{k_B T_e}\right] \qquad (1.107)$$

Assuming that perturbations are small, one can use Taylor expansion:

$$n_e = n_0\,\exp\left[\frac{e\varphi_1}{k_B T_e}\right] \approx n_0\left(1 + \frac{e\varphi_1}{k_B T_e}\right) \qquad (1.108)$$

From the linearized ion continuity (Eq. (1.97)), it can be found:

$$j\omega n_1 = n_0 jk V_{i1} \qquad (1.109)$$

Substituting expression for the fluctuating density n_1 into the ion momentum (1.106), the following relationship is obtained:

$$j\omega Mn_0 V_{i1} = \left(en_0 jk \frac{k_B T_i}{en_0} + \gamma kT_i jk\right)\frac{n_0 jk V_{i1}}{j\omega} \qquad (1.110)$$

Solving this equation for the frequency, one can arrive at the following dispersion relation:

$$\omega^2 = k^2\left(\frac{k_B T_e}{M} + \frac{\gamma k_B T_i}{M}\right) \qquad (1.111)$$

As it is seen from Eq. (1.111), the considered wave has the linear dispersion relation and the same phase and group velocity, i.e.,

$$V_s = V_g = V_\varphi = \sqrt{\frac{k_B T_e + \gamma k_B T_i}{M}} \qquad (1.112)$$

where V_s is the acoustic velocity. It is interesting to note that ion wave still exist even if ion temperature $T_i = 0$. This can be understood as follows. In ion oscillations, electrons are not fixed; electrons are pulled along with ions shielding the electric field that can arise from ion motion. However, shielding of the electric field is not perfect and an electric field about kT_e/e can leak due to electron thermal motion as it was discussed above. As an ordinary sound wave, ions form regions of rarefaction and compression as shown schematically in Figure 1.21. The compression region expands into the rarefaction region due to ion thermal motion (this is the second term in Eq. (1.112)) and electric field due to electrons (first term in Eq. (1.112)). The sound speed depends on electron temperature since electric field is proportional to it and ion mass since ion inertia depends on it.

In other words, ion wave propagates due to two reasons: ion thermal motion that makes it similar to the sound wave in neutral gas and presence of the electric field in a quasi-neutral plasma that couples ions and electrons motion.

Note that we used a quasi-neutrality condition to derive the ion wave. If quasi-neutrality condition is relaxed and thus Poisson equation is used to calculate the electric field, one can obtain the modified dispersion relation:

$$\frac{\omega}{k} = \sqrt{\frac{k_B T_e}{M} \frac{1}{1 + k^2 \lambda_D^2} + \frac{\gamma k_B T_i}{M}} \tag{1.113}$$

One can see that this equation is the same as Eq. (1.112) except for the factor $(1 + k^2 \lambda_D^2)$. Since Debye length is small, the quasi-neutral approximation is usually valid except for shortest wavelength (large k). The generalized dispersion relation is shown in Figure 1.22.

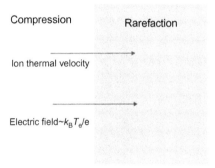

FIGURE 1.21

Schematics of ion sound propagation.

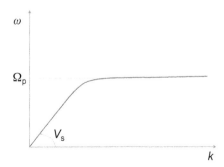

FIGURE 1.22

Dispersion relation for the ion sound wave.

If oscillations are short wavelength, i.e., $(k\lambda_D)^2 \gg 1$, one can find (from Eq. (1.113)):

$$\omega^2 = k^2 \frac{n_0 e^2}{\varepsilon M k^2} = \frac{n_0 e^2}{\varepsilon M} = \Omega_p^2 \tag{1.114}$$

where Ω_p is the ion plasma frequency. Thus in the case of a short wavelength, the ion acoustic wave becomes a constant frequency wave. Ion sound was first measured by Wong et al. in 1964 [20].

1.3.6 Waves in plasma with magnetic field

When magnetic field is applied, many types of waves are possible (Alfven waves, ion cyclotron wave, electron cyclotron wave, etc.) and plasma becomes anisotropic with waves propagating differently along and across the magnetic field. In this section, the electron oscillations perpendicular to magnetic field as shown in Figure 1.23 will be considered.

This case can be described by the following set of equations.

Momentum equation:

$$mn_e \left[\frac{\partial V_{e1}}{\partial t} \right] = -en_e E_1 - eV_{e1} \times B_0 \tag{1.115}$$

Continuity equation:

$$\frac{\partial n_1}{\partial t} + n_0 \nabla_x \cdot V_1 = 0 \tag{1.116}$$

Poisson equation for electric field:

$$\varepsilon_0 \frac{\partial E}{\partial x} = -en_{e1} \tag{1.117}$$

It was assumed that there is no thermal electron motion i.e. electron temperature is zero and ions are considered as a background.

Furthermore a longitudinal wave with wave vector parallel to the direction of electric field, i.e., $k \parallel E_1$ will be considered. In this case Eq. (1.115) after linearization reduces to the following system of equations for velocity components:

$$-j\omega m V_x = -eE_1 - eV_y B_0 \tag{1.118}$$

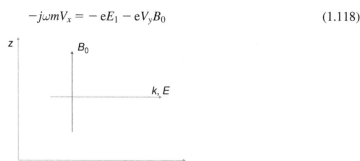

FIGURE 1.23

Electron wave in a magnetic field.

$$-j\omega m V_y = e V_x B_0 \qquad (1.119)$$

$$-j\omega m V_y = 0 \qquad (1.120)$$

Some algebraic manipulation allows obtaining the following expression for the electron velocity in x-direction:

$$V_x = \frac{eE/jm\omega}{1 - (\omega_c^2/\omega^2)} \qquad (1.121)$$

where ω_c is the electron cyclotron frequency. According to continuity equation (1.116), the perturbation density is

$$n_1 = \frac{k}{\omega} n_0 v_x \qquad (1.122)$$

Substitution this expression into the Poisson equation (1.117) and after linearization, the following can be arrived:

$$\left(1 - \frac{\omega_c^2}{\omega^2}\right) E = \frac{\omega_p^2}{\omega^2} E \qquad (1.123)$$

The dispersion relation for the wave in a magnetic field is thus

$$\omega^2 = \omega_p^2 + \omega_c^2 = \omega_h^2 \qquad (1.124)$$

where ω_h is the upper hybrid frequency. Physical explanation of these oscillations is as follows. Similar to plasma oscillations, electrons form region of compression and rarefaction. There are two restoring forces acting on electrons, the electric field force and the Lorentz force due to presence of the magnetic field. As a result, frequency of these oscillations is larger than that of the plasma oscillations as can be seen from Eq. (1.124).

Taking into account ion motion in this analysis leads to following dispersion relation:

$$\omega = (\Omega_c \omega_c)^{1/2} = \omega_l^2 \qquad (1.125)$$

where ω_l is the lower hybrid frequency, which is the combination of the ion and electron cyclotron wave. In this case, electron and ion oscillations are coupled so that the frequency is lower than in the case of the upper hybrid frequency.

The complete dispersion equation for the magnetized plasma is typically solved numerically to obtain the propagation constant for each of the waves at an arbitrary angle with respect to the magnetic field. Typically it is presented in Clemmow–Mullaly–Allis (CMA) diagram in which dispersion ω versus k for the principal waves in magnetized plasma is plotted [22].

1.3.7 Plasma instabilities

In the previous section, we considered waves in plasma that is equilibrium unperturbed state. Particles in plasma had equilibrium (Maxwellian) velocity distribution and density, and magnetic and electric fields were uniform. In this case, there is no free energy available and thus waves were excited by external means. In this section we will consider situation in which plasma is not in a perfect equilibrium and thus free energy is available to excite waves and drive system to an unstable state. Instabilities can be classified by the free energy that drives their development. Here we will follow classification that was proposed by Chen [21].

1. *Streaming instability*: The beam of energetic particles travels through the plasma. The drift energy is utilized to excite waves. An example of this instability is the two-stream instability.
2. *Rayleigh–Taylor instabilities*: In this case, instability is driven by presence of the density gradient and external nonelectromagnetic forces.
3. *Universal instabilities*: Even if plasma is in the equilibrium, plasma pressure gradient leads to the expansion. All confined plasmas defuse slowly due to particle collisions. This expansion energy can drive the instability. This kind of instability always present in plasma and this is why they called universal. A known example is the drift instability.
4. *Kinetic instability*: If the velocity distribution of plasma particle is not Maxwellian, there is always deviation from equilibrium and the instability can be driven by energy distribution anisotropy.

In the following section, we consider a particular example of streaming instability, namely, the two-stream instability. For more details analysis of the waves in plasmas and plasma instabilities, we refer to several textbooks [21,22].

1.3.7.1 Two-stream instability

Consider cold plasma with $T_e = T_i = 0$ and without a magnetic field, i.e., $B = 0$.

The linearized equations of motion for ions (Eq. (1.93)) and electrons (Eq. (1.100)) will have the following form:

$$Mn_0 \frac{\partial v_{i1}}{\partial t} = en_0 E_1 \tag{1.126}$$

$$mn_0 \left[\frac{\partial v_{e1}}{\partial t} + (v_0 \nabla) v_{e1} \right] = -en_0 E_1$$

where v_0 is the velocity of electron stream. Electrostatic wave is considered:

$$E_1 = E \exp(j(kx - \omega t)) \tag{1.127}$$

Employing the form (1.127) for the electric field in Eq. (1.126), we arrive at

$$-j\omega Mn_0 v_{i1} = en_0 E_1 \tag{1.128}$$

Solving this equation for v_{i1} yields

$$v_{i1} = -\frac{je}{M\omega} E \tag{1.129}$$

After employing the electric field in the form (1.127), the equation for electron motion will have the following form:

$$mn_0(-j\omega + jkv_0)v_{e1} = -en_0 E_1 \tag{1.130}$$

Solving this equation for the electron velocity

$$v_{i1} = -\frac{jeE}{m(\omega - kv_0)} \tag{1.131}$$

Under considered conditions, the ion continuity equation is

$$\frac{\partial n_{i1}}{\partial t} + n_0 \nabla \cdot v_{i1} = 0 \tag{1.132}$$

One can solve this equation for the ion density:

$$n_{i1} = \frac{jen_0k}{M\omega^2}E \tag{1.133}$$

Electron continuity equation is

$$\frac{\partial n_{e1}}{\partial t} + n_0 \nabla \cdot v_{e1} + (v_0 \cdot \nabla)n_{e1} = 0 \tag{1.134}$$

After linearization, this equation becomes

$$(-j\omega + jkv_0)n_{e1} + jkn_0v_{e1} = 0 \tag{1.135}$$

Solving this equation for the electron density yields

$$n_{e1} = \frac{n_0k}{\omega - kv_0}v_{e1} = -\frac{jekn_0}{m(\omega - kv_0)^2}E \tag{1.136}$$

We can now plug Eqs (1.133) and (1.36) into the Poisson's equation:

$$\varepsilon_0 \nabla \cdot E_1 = e(n_{i1} - n_{e1}) \tag{1.137}$$

which looks like:

$$jk\varepsilon_0 E = e(jen_0kE)\left[\frac{1}{M\omega^2} + \frac{1}{(\omega - kv_0)^2}\right] \tag{1.138}$$

The dispersion relation is thus

$$1 = \omega_p^2\left[\frac{m/M}{\omega^2} + \frac{1}{(\omega - kv_0)^2}\right] \tag{1.139}$$

Let us now examine whether these oscillations are stable or they can become unstable. To this end we will find a root of ω assuming that it is a complex variable, i.e.,

$$\omega = \alpha + j\gamma$$
$$\alpha = \mathrm{Re}(\omega); \quad \gamma = \mathrm{Im}(\omega) \tag{1.140}$$

Taking into account Eq. (1.140), the electric field becomes

$$E_1 = E\exp(j(kx - \omega t)) = E\exp(\gamma t)\exp(j(kx - \alpha t)) \tag{1.141}$$

Let us analyze the dispersion equation (1.139) using the following normalized variables:

$$x_1 = \omega/\omega_p \text{ and } x_2 = kv_0/\omega_p$$

The dispersion equation becomes

$$F(x_1, x_2) = \frac{m/M}{x_1^2} + \frac{1}{(x_1 - x_2)^2} = 1 \qquad (1.142)$$

The solution of Eq. (1.142) can be shown graphically in Figure 1.24.

The function $F(x_1, x_2)$ is shown with two possible cases considered. It can be seen that in Figure 1.24A, there are four real roots, while in Figure 1.24B, there are two real and two imaginary roots with one of the roots being positive. Thus for sufficiently small kv_o plasma becomes unstable. In the case of a given v_0, plasma is unstable for long-wave oscillations.

The maximum growth rate can be estimated from Eq. (1.139) for $m/M \ll 1$:

$$\text{Im}\left(\frac{\omega}{\omega_p}\right) \approx \left(\frac{m}{M}\right)^{1/3} \qquad (1.143)$$

This instability is sometimes called the "Buneman" instability. Physics of this instability can be explained as follows. The natural electron frequency is ω_p, while the ion frequency is $\Omega_p = (m/M)^{0.5}\omega_p$. Because of the Doppler shift of oscillations moving with electrons, these two frequencies could be equal for given kv_o. As a result, the energy can be transferred from electrons to ions providing conditions for wave growth.

1.3.7.2 Kinetic instabilities

So far we considered instabilities and waves using the fluid approximation and as such assuming that particles in plasma obey equilibrium (Maxwellian) velocity distribution. However, in many

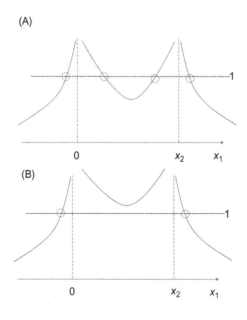

FIGURE 1.24

Dispersion relation in the case of two-stream instability.

cases, the energy distribution function of the particles in plasmas can deviate from Maxwellian distribution due to imbalanced conditions. The deviation from the equilibrium distribution can be maintained due to rare collisional events in rarefied plasmas.

Let us consider small deviation from equilibrium as follows:

$$f(r, v, t) = f_0(r, v, t) + f_1(r, v, t) \tag{1.144}$$

where $f_0(r,v,t)$ is the Maxwellian distribution function. Linearized, the collisionless Boltzmann equation leads to

$$\frac{\partial f_1}{\partial t} + v \cdot \nabla f_1 - \frac{e}{m} E_1 \frac{\partial f_0}{\partial v} = 0 \tag{1.145}$$

In accordance with previous analysis, we assume

$$f_1 \sim \exp(j(kx - \omega t)) \tag{1.146}$$

By substituting Eq. (1.146) into Eq. (1.145), we arrive at

$$-j\omega f_1 + jk v_x f_1 - \frac{e}{m} E_x \frac{\partial f_0}{\partial v} = 0 \tag{1.147}$$

Solving Eq. (1.147) for f_1 yields the following:

$$f_1 = j \frac{e}{m} E_1 \frac{\partial f_0/\partial v}{\omega - k x_x} \tag{1.148}$$

Using the Poisson equation, the electric field E_1 can be expressed as

$$jk\varepsilon_0 E_1 = -en_1 = -e \int f_1 \, dv \tag{1.149}$$

After substituting f_1 and some algebra, dispersion relation can be obtained:

$$1 = -\frac{\omega_p^2}{k^2} \int \frac{\partial f_0/\partial v}{\omega/k - v} \, dv \tag{1.150}$$

One can see that integral (1.150) has a singularity. Landau treated this integral using the contour integration. By doing that, one can arrive at

$$\omega = \omega_p \left(1 + j \frac{\pi}{2} \cdot \frac{\omega_p^2}{k^2} \left[\frac{\partial f_0}{\partial v}\right]_{v=v_\phi}\right) \tag{1.151}$$

where $v_\phi = (\omega/k)$ is the phase velocity.

Taking into account that the f_0 is the Maxwellian distribution function, one can obtain expression for the decay rate of this oscillations of distribution function.

$$\mathrm{Im}(\omega) = -\omega_p 0.22\sqrt{\pi} \left(\frac{\omega_p}{k v_{\text{th}}}\right)^3 \exp\left(-\frac{1}{2k^2 L_D^2}\right) \tag{1.152}$$

Since imaginary part is negative, wave is damping. This effect is called Landau damping.

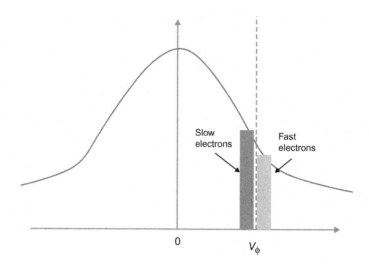

FIGURE 1.25

Electron distribution function and distortion of a Maxwellian distribution function near phase velocity due to Landau damping.

Note: This is very surprising result because the wave is damped without collisions! It was first discovered by Landau [23] theoretically and then was confirmed experimentally by Malmberg and Wharton [24].

In order to understand the underlying physical mechanism, let us consider the electron velocity distribution function as shown schematically in Figure 1.25. In particular we are interested in particles with velocities near the phase velocity as shown in Figure 1.25. The particles with velocity close to the phase velocity of the wave travel with wave and thus can interact efficiently with the wave. If decreasing branch of the velocity distribution function is considered as shown in Figure 1.25, it can be seen that more particles have velocity smaller than the phase velocity (slow electrons). In other words energy, more particles have energy smaller than the wave energy than particles with energy larger than the wave energy. In this case, the wave energy is transferred into particles and thus wave is damping.

Using analogy with waves in the sea can provide intuitive explanation of the Landau damping. In this case, we consider the particles as surfers trying to catch the wave as shown in Figure 1.26. If the surfer is moving on the water surface at a velocity slightly less than the waves, he will eventually be caught and pushed along the wave (and will gain the energy), while a surfer moving slightly faster than a wave will be pushing on the wave as he moves uphill (and will lose the energy to the wave).

In summary, in this section, we considered wave propagation in the plasma and plasma–wave interactions. In particular, we considered Langmuir wave, ion sound wave, as well as waves in a magnetic field. Plasma instability associated with either drift or gradients was considered for conditions relevant for some aerospace applications. In addition, we consider kinetic effect such as Landau damping which is relevant for rarefied plasmas considered in subsequent chapters.

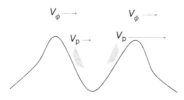

FIGURE 1.26

Physical analogy of Landau damping.

1.4 **Plasma—wall interactions**

This section describes phenomena appeared in plasma—wall regions including electrostatic sheath formation and associated with particle—wall interaction, heat flux formation from the plasma, electron emission, and vaporization.

1.4.1 **Plasma—wall transition: electrostatic phenomena**

Typically laboratory plasmas contact the wall and such interaction leads to formation of the transitional region. The plasma-wall transition is characterized by subregion having different space charge, electric field particle densities, and velocities. Sheath is the strongly space charged subregion that appears as a result of charge particle separation in the immediate vicinity of the wall of a strong electric field. The subdivision of a discharge into a plasma bulk region and a sheath was introduced by Langmuir and Tonks [25]. Sheath region solution requires boundary conditions at the plasma—sheath interface assuring the monotonic potential distribution. Such boundary condition at the interface was considered by Bohm [26] who formulated a criterion for a stable potential distribution inside of the sheath, known as a Bohm criterion for the ion speed at the mentioned interface. It should be pointed out that while condition for the ion velocity is established, the sheath thickness is not well defined.

The transition between quasi-neutral plasma and collisionless sheath is continuing to be a subject of great interest having both academic and practical importance [27—30]. A controversial discussion on this subject was intensified in the last decade [31—36]. One of the major discussed issues is how solution of quasi-neutral plasma and sheath can be joined together using either matching or patching mathematical technique. Matching means that all variable and their derivatives are continuous; patching require continuity of variables only. A comprehensive review which addresses the problem of joining plasma and sheath solution over a wide range of physical conditions was reported by Franklin [37].

In general, the modern computer technique allows obtaining the continuity solutions without subdivisions of the plasma—wall transition layer. However, there are physical situations in which plasma region required multicomponent and multidimensional analysis, while sheath region can still be treated as a one-dimensional system. Moreover, sheath analysis has significant methodical value of illustrating the physics of the plasma—wall transition. For example, the low-pressure discharges in microdevices and plasma thrusters were studied with essentially two-dimensional plasma region and one-dimensional sheath region [38]. Thus it is often convenient to obtain the plasma and the sheath solutions separately. The essence of the Bohm criterion and an example of a solution in plasma—sheath regions will be analyzed below.

1.4.1.1 Condition for stable sheath: Bohm criterion

Let us discuss the necessity of certain ion velocity at the sheath edge to have a monotonic potential distribution in the sheath. To this end, a collisionless sheath with quasi-neutral presheath can be considered schematically presented in Figure 1.27. The ion velocity at the sheath edge can be calculated using the equations of energy particle flux conservation[69]:

$$e\varphi(x) + \frac{1}{2}MV^2 = \frac{1}{2}C_s^2 M \tag{1.153}$$

where $\varphi(x)$ is the potential distribution function, M, V, and C_s are the ion mass, velocity, and sound velocity, respectively. The left-hand side corresponds to the ion kinetic energy evaluated in the sheath at boundary $x = 0$.

Ion continuity equation can be written as

$$n_s C_s = V(x)n(x) \tag{1.154}$$

where $n(x)$ is the density distribution function. From Eqs (1.153) and (1.154), one can derive:

$$V(x) = \sqrt{C_s^2 - \frac{2e\,\varphi(x)}{M}} \tag{1.155}$$

According to Eqs (1.153) and (1.154), the ion density distribution is therefore:

$$n(x) = n_s C_s \Big/ \sqrt{C_s^2 - \frac{2e\,\varphi(x)}{M}} \tag{1.156}$$

Electrons obey the Boltzmann distribution:

$$n_e(x) = n_{es} \exp\left(\frac{e\varphi(x)}{T_e}\right) \tag{1.157}$$

Plasma is quasi-neutral at the presheath—sheath, i.e., $n_{es} = n_{is} = n_s$, where n_{es} and n_{is} are, respectively, the electron and ion densities at the sheath edge.

The Poisson's equation:

$$-\nabla^2\varphi = \frac{e(n_i - n_e)}{\varepsilon_0} \tag{1.158}$$

Substituting expressions (1.156) and (1.157) for the ion and electron densities in the Poisson's equation:

$$-\nabla^2\varphi = \frac{en_s}{\varepsilon_0}\left[\frac{C_s}{\sqrt{C_s^2 - (2e\varphi(x)/M)}} - \exp\left(\frac{e\varphi(x)}{T_e}\right)\right] \tag{1.159}$$

Considering the one-dimensional approach, Eq. (1.159) can be expressed as

$$-\frac{d^2\varphi}{dx^2} = \frac{en_s}{\varepsilon_0}\left[\frac{C_s}{\sqrt{C_s^2 - (2e\varphi(x)/M)}} - \exp\left(\frac{e\varphi(x)}{T_e}\right)\right] \tag{1.160}$$

Equation for the electric field can be obtained by first integration of Eq. (1.160):

$$\frac{1}{2}\left(\frac{d\varphi}{dx}\right)^2 = \frac{en_s}{\varepsilon_0}\left[2W_s\sqrt{1 - \frac{e\varphi(x)}{W_s}} + T_e\exp\left(\frac{e\varphi(x)}{T_e}\right)\right] + C$$

where $W_s = MC_s^2/2$ and C is the constant of integration.

Using the following boundary conditions at $x = 0$:

$$\varphi(0) = \frac{d\varphi}{dx}(0) = 0$$

it can be obtained:

$$C = -\frac{en_s}{\varepsilon_0}(T_e + 2W_s) \tag{1.161}$$

Using Eq. (1.161), the equation for the electric field becomes

$$\frac{1}{2}\left(\frac{d\varphi}{dx}\right)^2 = \frac{en_s}{\varepsilon_0}\left[2W_s\sqrt{1 - \frac{e\varphi(x)}{W_s}} - 2W_s + T_e\exp\left(\frac{e\varphi(x)}{T_e}\right) - T_e\right] \tag{1.162}$$

The Taylor expansion near the sheath edge reads

$$\exp\left(\frac{e\varphi}{T_e}\right) \approx 1 + \frac{e\varphi}{T_e} + \frac{1}{2}\left(\frac{e\varphi}{T_e}\right)^2 + \cdots$$

$$\sqrt{1 - \frac{e\varphi}{W_s}} \approx 1 - \frac{1}{2}\frac{e\varphi}{W_s} - \frac{1}{8}\left(\frac{e\varphi}{W_s}\right)^2 + \cdots \tag{1.163}$$

To reach the stable monotonic potential distribution in entire sheath region, the following condition at the external boundary of the sheath should be fulfilled:

$$\left(\frac{d\varphi}{dx}\right)^2 > 0 \tag{1.164}$$

Consider potential near the sheath edge, $e\varphi/T_e < 1$. Thus high-order terms (such as $(e\varphi/T_e)^2$ and higher) can be neglected. After some algebra, Eq. (1.162) becomes

$$\frac{(e\varphi)^2}{2T_e} - \frac{1}{4}\frac{(e\varphi)^2}{W_s} > 0 \tag{1.165}$$

This leads to the following condition, which is known as Bohm criterion:

$$\frac{1}{T_e} > \frac{1}{2}\frac{1}{W_s} \quad \text{or} \quad C_s^2 \geq \frac{T_e}{M} = V_B^2 \tag{1.166}$$

1.4.1.2 Monotonic solution for sheath—presheath region

Here we illustrate the sheath and in a more general sense plasma—wall interface by considering solutions of a quasi-neutral plasma and a sheath separately. By employing the patching technique, we will illustrate how to obtain the monotonic solution for entire plasma—wall transition region.

The fluid-based approach is considered below to illustrate the sheath problem and its solution. Schematically the transition regions are shown in Figure 1.27. According to Langmuir's analysis [25], the ions are accelerated toward the Bohm velocity in a plasma region with a relatively weak electric field and this region called as *presheath* extends from the sheath into the quasi-neutral plasma volume. The electric field, thus, becomes a continuous function increasing in the sheath from a boundary value up to a maximal value at the wall. This boundary field can be obtained considering a model for the presheath region. The position of the sheath—plasma interface is determined by patching the electric field, ion velocity, and potential. Thus a relationship between the ion velocity and the electric field at the presheath—sheath interface should be determined to choose these parameters as boundary conditions for a sheath region solution.

The procedure described involves the following conditions: (i) the degree of the quasi-neutrality violation near the presheath edge that may affect the solution will be estimated; (ii) potential distribution is spatially monotonic in all regions, and (iii) continuity of the electrical field and potential at the sheath—presheath interface.

1.4.1.3 Mathematical formulation

Let us consider the governing equations to describe the problem including the plasma, the sheath, and the sheath—presheath interface. The plasma flow in the quasi-neutral plasma region is

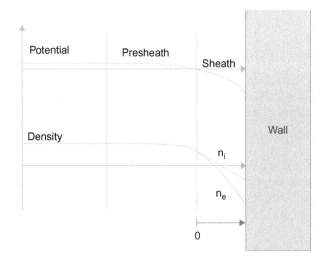

FIGURE 1.27

Plasma—wall transitional layer.

considered in the fluid approximation. Ion mass and momentum conservation taking into account and assuming that the electrons obey the Boltzmann distribution:

$$\frac{d(n_i V_i)}{dx} = \nu_i n_e \tag{1.167}$$

$$n_e = n_0 \, \exp^{(e\varphi/kT_e)} \tag{1.168}$$

$$n_i m V_i \frac{dV_i}{dx} = Z_i e n_i E - \nabla P_i - n_i m \nu_c V_i - \nu_i n_e m V_i \tag{1.169}$$

where n_e and n_i are electron and ion densities, respectively, ν_i is the ionization collision frequency, and ν_c is the charge-exchange collision frequency. In order to obtain main dependencies, we assume that collision frequency is constant, Z_i is the ion charge state (only single ionized ions will be considered, $Z_i = 1$), P_i is the ion pressure (cold ions are assumed, so $P_i = 0$) and $E = -d\varphi/dx$ is the electric field and φ is the potential. The potential is distributed according to Poisson's equation:

$$\frac{d^2\phi}{dx^2} = \frac{e}{\varepsilon_0}(n_e - n_i) \tag{1.170}$$

The system of equations (1.167)−(1.170) has following boundary conditions:

At the symmetry axis (as shown in Figure 1.28) ($x = 0$), $\varphi = 0$, $d\varphi/dx = 0$, and $n_e = n_i = n_0$.
At the wall ($x = X_w$), $\varphi = \varphi_w$ (wall potential).

The above system of equations is solved separately for the quasi-neutral collisional plasma region (presheath) and for collisionless space charge region (sheath). Finally it will be described how to patch these two solutions.

1.4.1.3.1 Presheath (quasi-neutral region, $n_e = n_i$)
Naturally we normalize these equations with $u = V_i/C_s$, $\chi = x/\lambda_c$, $\eta = e\varphi/kT_e$, where n_0 is the plasma density at the plasma center, $\lambda_c = C_s/(\nu_c + \nu_i)$, $C_s = (kT_e/m)^{1/2}$. The normalized system of equations (1.167)−(1.169) with quasi-neutrality assumption can be easily reduced to the following system of equations:

$$\frac{du}{d\chi} = \frac{(u^2 + \alpha)}{(1 - u^2)} \tag{1.171}$$

$$\frac{d\eta}{d\chi} = -\frac{(\alpha + 1)u}{1 - u^2} \tag{1.172}$$

where $\alpha = (\nu_i/(\nu_i + \nu_c))$. This system of equations and the boundary conditions determine the solution up to point $u \to 1$, where a large electric field and therefore a certain deviation, $\Delta n = n_e - n_i$, from the quasi-neutrality condition can be expected. Using the potential distribution (from Eq. (1.170)), this deviation can be described as follows:

$$\frac{\Delta n}{n_0} = \psi^2 \frac{d^2\eta}{d\chi^2} \tag{1.173}$$

where $\psi = (\lambda_D/\lambda_c)$ and λ_D is the Debye length at the plasma axis ($x = 0$) and λ_c is the collision mean free path. The relationship between V_o and E_o (V_o is the ion velocity normalized by C_s and E_o is the electric field normalized by $E^* = T_e/e\lambda_D$) can be obtained from Eqs (1.171) and (1.172). A monotonic sheath solution can be obtained by patching or matching the solution with normalized electric field and normalized velocity at the sheath−presheath interface.

1.4.1.3.2 Sheath

This is the plasma region where the volume space charge is relatively large and the plasma cannot be considered as a quasi-neutral substance. A collisionless sheath is assumed, i.e., $\nu_i = \nu_c = 0$. This allows description of the ion density distribution in the sheath using a simple form of the mass and momentum conservations (Eqs (1.167) and (1.170)) as

$$n_i/n_s = \frac{1}{\sqrt{1 + 2 \cdot ((\eta_o - \eta)/V_o^2)}} \tag{1.174}$$

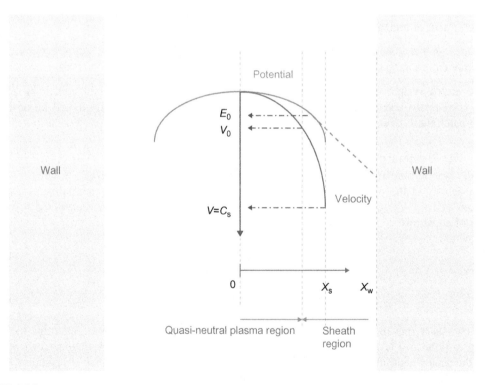

FIGURE 1.28

Schematics of the plasma−wall transition regions (not to scale). Ion velocity, V_o (normalized by C_s), and electric field, E_o (normalized by E^*), distribution in the neighborhood of the presheath−sheath interface. The broken line shows the potential distribution in the sheath after patching.

The electron density distribution in the sheath is still determined by the Boltzmann equilibrium:

$$n_e/n_s = e^{\eta - \eta_0} \tag{1.175}$$

where n_s is the density, η_0 is the potential, and V_o is the ion velocity at the sheath—presheath interface. The potential distribution in the sheath is calculated from the following normalized Poisson equation:

$$\psi^2 \frac{d^2\eta}{d\chi^2} = \frac{(n_e - n_i)}{n_0} \tag{1.176}$$

1.4.1.3.3 Direct numerical solution of the sheath—presheath regions
The above described approach can be compared with a direct numerical solution of the entire plasma—wall interface. This solution describes the entire transition region between the plasma and the wall without division between a quasi-neutral part and a sheath and taking into account collisions. Let us normalize Eqs (1.167)—(1.170) for the entire transition region from the plasma to the wall:

$$\frac{du}{d\chi} = -\frac{1}{u} \cdot \frac{d\eta}{d\chi} - 1 \tag{1.177}$$

$$\frac{dN_i}{d\chi} = -\frac{N_i}{u} \cdot \frac{du}{d\chi} + \alpha \frac{N_e}{u} \tag{1.178}$$

$$N_e = e^\eta \tag{1.179}$$

$$N_e - N_i = \psi^2 \frac{d^2\eta}{d\chi^2} \tag{1.180}$$

where $\psi = (\lambda_D/\lambda_c)$, $N_e = n_e/n_o$; $N_i = n_i/n_o$. Below, the exact solution will be compared with the approximate patching solution. Before the solution will be presented, the relationship between velocity and electric field at the sheath edge is analyzed in the next section.

1.4.1.4 Monotonic potential distribution in the sheath
Consider Poisson equation for the sheath potential distribution in dimensionless form:

$$\psi^2 \frac{d^2\eta}{d\chi^2} = \exp(\eta) - \left(1 - \frac{2\eta}{V_o^2}\right)^{-1/2} \tag{1.181}$$

The first integration of Eq. (1.181) in the vicinity of the sheath boundary with nonzero electric field at the sheath boundary, $E_o = -(d\eta/d\chi)_o$, gives

$$\frac{1}{2}\psi^2 \left(\frac{d\eta}{d\chi}\right)_s^2 = \frac{1}{2}\psi^2 E_o^2 + \exp(\eta_s) + V_o^2 \left(1 - \frac{2\eta_s}{V_o^2}\right)^{1/2} - 1 - V_o^2 \tag{1.182}$$

where η_s is the potential at some point in the sheath including region close to the sheath edge. For the purpose of simplifying analyses, we assumed here that potential at the sheath boundary equals zero, $\eta_0 = 0$. All parameters (density, velocity, and potential) are normalized by their values at the sheath edge. Near the sheath edge, one can make a series expansion in the small parameter $\eta_s \ll 1$:

$$\exp(\eta_s) \approx 1 + \eta_s + \frac{1}{2}(\eta_s)^2 + \cdots$$

$$\sqrt{1 - \frac{2\eta_s}{V_0^2}} \approx 1 - \frac{1}{2}\frac{2\eta_s}{V_0^2} - \frac{1}{8}\left(\frac{2\eta_s}{V_0^2}\right)^2 + \cdots \tag{1.183}$$

Neglecting small terms in the expansion with order higher than η_s^2 reduces Eq. (1.182) to the following:

$$\frac{1}{2}\psi^2\left(\frac{d\eta}{d\chi}\right)_s^2 = \frac{1}{2}\psi^2 E_o^2 + \eta_s^2\left(1 - \frac{1}{V_o^2}\right) \tag{1.184}$$

The condition for the monotonic potential distribution in the sheath would be that the right-hand side of Eq. (1.184) is positive. This condition has the following form:

$$\eta_s^2 - \frac{\eta_s^2}{V_0^2} + \frac{1}{2}E_o^2\psi^2 \geq 0 \tag{1.185}$$

Obviously, Eq. (1.185) has no real solution in the sheath if $V_0 < 1$ and $E_o = 0$. Therefore, for $E_o = 0$, the potential distribution in the sheath is monotonic only when $V_o \geq 1$, i.e., with Bohm condition.

Equation (1.185) has real solution (the monotonic potential distribution) in the sheath for $V_o < 1$ when $E_o > 0$. The requested E_o that supports the monotonic potential distribution was obtained from the numerical solution of the sheath equations and shown below.

To get more understanding of the above points, let us analyze the solutions of Eq. (1.182) as a function of two variables η and $d\eta/d\chi$, i.e., potential–electric field plane. In this case, the solution of Eq. (1.182) can be represented as dependences with parameter of $(1/2)\psi^2 E_o^2$. Mathematical analysis of Eq. (1.182) in the $(\eta - d\eta/d\chi)$ plane shows that the solutions could be centers or saddles. Centers solution corresponds to the periodic behavior of the mathematical system (oscillations), while saddles solution corresponds to spatially monotonic behavior.

The contours of $(1/2)\psi^2 E_o^2$ directly computed from Eq. (1.182) are shown in Figure 1.29A. It can be seen that both saddles and centers curves of the same level of initial electric field E_o are possible and those solutions depend on the initial velocity V_o. Figure 1.29B shows that smaller $(1/2)\psi^2 E_o^2$ corresponds to centers curves, while large $(1/2)\psi^2 E_o^2$ corresponds to saddles curves.

Recall that dependent on the boundary conditions at the sheath edge (i.e., electric field and ion velocity), spatially monotonic or oscillatory behavior of potential distribution is possible.

1.4.1.5 Solutions in plasma and sheath regions: procedure of patching

Firstly, the quasi-neutral (presheath) region (i.e., Eqs (1.171), (1.172)) is considered. The distribution of the velocity, electric field, and quasi-neutrality deviation is shown in Figure 1.30A and Figure 1.30B. The ion velocity variation in this scale length (fraction of a mean free path) is small, while the electric field and quasi-neutrality deviation noticeably increase in the presheath with distance toward the presheath edge and become singular (Figure 1.30A). As shown in Figure 1.30B, both the electric field and the quasi-neutrality deviation significantly increase when the ion velocity approaches to the Bohm velocity. This deviation becomes stronger with coefficient ψ decreasing.

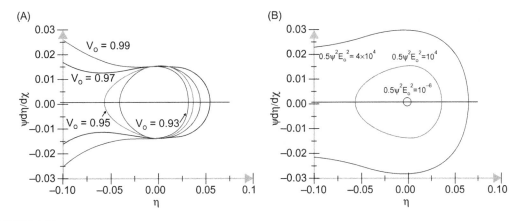

FIGURE 1.29

(A) Level curves of $(1/2)\psi^2 E_o^2$ for the case of $V_o = 0.95$. Centers curves correspond to oscillatory solution, while saddles curves correspond to the monotonic solution. (B) Level curves of $(1/2)\psi^2 E_o^2$ for the case of $V_o = 0.95$. Centers curves correspond to oscillatory solution, while saddles curves correspond to the monotonic solution.

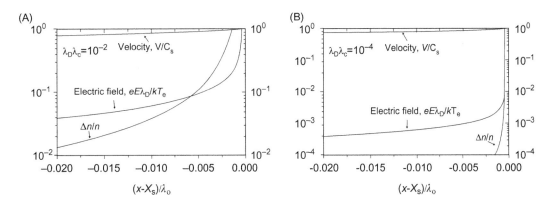

FIGURE 1.30

Velocity, electric field, and quasi-neutrality assessment spatial variation near the presheath edge with ψ as a parameter. Zero coordinate $((x - X_s)/\lambda_c = 0)$ corresponds to the presheath edge, i.e., condition: $V = C_s$. $\alpha = 1$. (A) $\psi = 10^{-2}$ and (B) $\psi = 10^{-4}$.

Now let us consider the procedure of patching the presheath and the sheath solutions. The existence of the solution with spatially monotonic potential distribution in the sheath region is examined in Section 1.4.1.4. The procedure of finding the monotonic solution is as follows. Having the boundary velocity V_o and corresponded electric field E_o, the potential distribution in the sheath (Eqs (1.174)–(1.176)) is calculated. Two types of potential distribution are possible as shown in Figure 1.31. When electric field E_o is small and $V_o < 1$, the only oscillating solution of the Poisson

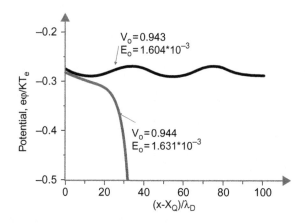

FIGURE 1.31

Potential distribution in the sheath with boundary velocity as a parameter. Boundary electric field E_o corresponds to the velocity V_o in the presheath solution.

equation is possible. A combination of the nonzero electric field E_o and velocity V_o (if $V_o < 1$) provides the spatially monotonic solution. These conditions are the boundary condition for the sheath.

In the above described approach, the velocity V_o is searched as a minimal velocity for which spatially monotonic potential distribution in the sheath is possible.

The calculations show that when ψ increases, the ion velocity at the sheath edge is smaller than the Bohm velocity and at the same time the obtained electric field increases with ψ (Figure 1.32). However, the electric field increase coexists with the quasi-neutrality violation, as it is also shown in Figure 1.32. In the opposite limit, i.e., when ψ is small, the required velocity approaches the Bohm velocity and at the same time the electric field approaches zero. One can see that up until $\psi = 10^{-2}$, the level of deviation from quasi-neutrality is small (about few percent) and can be tolerated in many cases. In the case when the parameter ψ increases and approaches unity, the collisions become important in the sheath and therefore Eq. (1.174) should be modified taking into account the change of momentum by the collisions. The tendency is shown in Figure 1.32; when parameter ψ increases, the velocity significantly decreases and the Bohm criteria requirement is relaxed. Also there is an intermediate range of ψ (where $10^{-4} <$ range $< 10^{-2}$) when the sheath can still be considered collisionless, but the sheath and presheath solutions can be patched with velocity smaller than the Bohm velocity and a nonzero electric field.

The accuracy of the proposed patching procedure for $\psi = 10^{-4}$ is compared with the continuous direct numerical solution (1.177)–(1.180) as shown in Figure 1.33, where the electric field as a function of the potential in the presheath and the sheath is shown. One can see a good agreement between continuous (direct) solution and solution using above developed patching approach.

The density distribution across the sheath region shown in Figure 1.34 illustrates that two approaches (collisionless sheath-solid curve (i.e., patching) and dotted curve—direct numerical solution) produce very close results. The patching solution is obtained using the appropriate boundary conditions for ion velocity and electric field necessary for continuous distribution of these

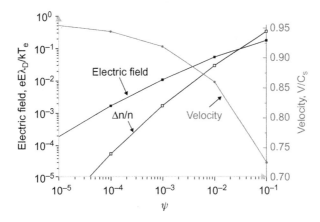

FIGURE 1.32

Velocity, electric field, and quasi-neutrality assessment at the patching point as a function of Debye length ratio to the collision mean free path ψ.

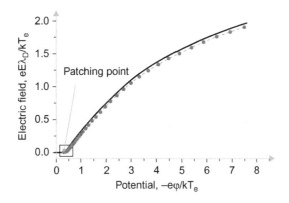

FIGURE 1.33

The variation of electric field as a function of potential. $\alpha = 1$; $\psi = 10^{-4}$. The direct numerical solution of Eqs (1.177)–(1.180) is shown with dotted line. $\varphi_w = -8kT_e/e$.

parameters at the sheath–presheath interface. The difference between those solutions is due to the fact that the collisions were neglected in the sheath solution. This implies that collisional effects in the sheath are not significant.

1.4.1.6 Typical electrostatic sheath
1.4.1.6.1 Child–Langmuir sheath
A sheath with large potential bias on the wall with respect to the plasma is considered schematically presented in Figure 1.35, where "0" corresponds to the sheath edge and "s" corresponds to the

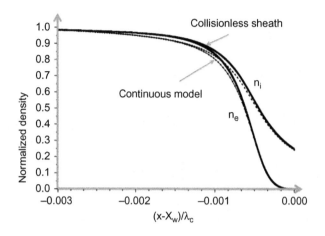

FIGURE 1.34

Ion and electron density distribution in the sheath. $\alpha = 1$; $\psi = 10^{-4}$. The direct numerical solution of Eqs (1.177)–(1.180) is shown with dotted lines. The calculated patching: velocity is $V_0 = 0.944$ and electric field, $E_0 = 1.631 \cdot 10^{-3}$.

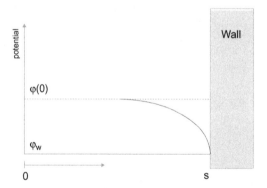

FIGURE 1.35

Schematics of the sheath region.

wall. As a negative voltage is applied, electrons will be repelled away from the wall leading to positive sheath formation. The ions are accelerated toward the wall by the electric field of the sheath.

We can obtain the thickness of the sheath by integrating the Poisson equation (Eq. (1.170)):

$$\frac{d^2 \varphi}{dx^2} = \frac{e n_s}{\varepsilon_0} \left[\exp\left(\frac{e\varphi}{T_e}\right) - \left(1 - \frac{2e\varphi}{MC_s^2}\right)^{-0.5} \right] \qquad (1.186)$$

Setting the following boundary conditions:

$$\varphi(0) = 0$$
$$\varphi(s) = \varphi_w$$

The potential at the wall is chosen to be much larger than the electron temperature. In this case in most of the sheath region:

$n_e = n_s \exp(e\varphi/T_e) \to 0$, and thus first term in Eq. (1.186) can be neglected. Also assuming that $n_i = \text{constant} = n_s$, the Poisson equation is then has the following form:

$$\frac{dE}{dx} = \frac{en_s}{\varepsilon_0} \tag{1.187}$$

After integration of Eq. (1.187), it can be obtained a simple solution which is called ion matrix sheath.

For the electric field:

$$E = \frac{e}{\varepsilon_0} n_s x, \tag{1.188}$$

The potential distribution can be obtained by integrating Eq. (1.188):

$$\varphi = -\frac{e}{2\varepsilon_0} n_s x^2 + C \tag{1.189}$$

Employing the boundary condition, i.e.,

$$\varphi = -U_0$$

at $x = s$,

$$\varphi = 0$$

at $x = 0$.

Thus one can find that sheath thickness as

$$s = \left(\frac{2\varepsilon_0 U_0}{en_s}\right)^{0.5} \tag{1.190}$$

The approach can be extended using the ion conservation equation, i.e., $n_i v_i = \text{constant}$ in the sheath. As it is done above, we also assume that electrons are depleted in the sheath due to high potential in comparison with electron temperature, i.e., $n_e \ll n_i$ and that the ion velocity in the sheath is much higher than that at the sheath edge.

General equation (1.186) for the sheath is reduced to

$$\frac{d^2\varphi}{dx^2} = \frac{en_s}{\varepsilon_0}\left[-\left(-\frac{2e\varphi}{MC_s^2}\right)^{-0.5}\right] \tag{1.191}$$

Following the definition of the ion current density, i.e., $J_0 = en_sC_s$, we can rewrite Eq. (1.191) as

$$\frac{d^2\varphi}{dx^2} = \frac{J_0}{\varepsilon_0}\left[-\left(-\frac{M}{2e\varphi(x)}\right)^{0.5}\right] = \frac{d}{dx}\left(\frac{d\varphi}{dx}\right) \tag{1.192}$$

After integration, we arrive at

$$\frac{1}{2}\left(\frac{d\varphi}{dx}\right)^2 = \frac{J_0}{\varepsilon_0}\left(\frac{M}{2e}\right)^{0.5}(-\varphi(x))^{1/2}2 + C \tag{1.193}$$

Using the boundary condition $\varphi(0) = 0$, it follows that constant $C = 0$.
After some arithmetical operations:

$$\left(\frac{d\varphi}{dx}\right) = \left[\frac{4J_0}{\varepsilon_0}\left(\frac{M}{2e}\right)^{1/2}\right]^{1/2}(-\varphi)^{1/4} \tag{1.194}$$

Second integration thus produces

$$\frac{4}{3}(-\varphi)^{3/4} = 2\left(\frac{J_0}{\varepsilon_0}\right)^{1/2}\left(\frac{M}{2e}\right)^{1/4}x \tag{1.195}$$

Using boundary condition at the wall, i.e., $-\varphi = V_0$ at $x = s$, it is

$$\frac{2}{3}(U_0)^{3/4} = \left(\frac{J_0}{\varepsilon_0}\right)^{1/2}\left(\frac{M}{2e}\right)^{1/4}s \tag{1.196}$$

Solving this equation for the current results in

$$J_0 = \frac{4}{9}\varepsilon_0\left(\frac{2e}{M}\right)^{1/2}\frac{U_0^{3/2}}{s^2} \tag{1.197}$$

This equation is well known as a Child–Langmuir law of space-limited current in a plane diode. When spacing s is fixed, the current between two electrodes can be calculated as a function of applied voltage. Child–Langmuir law is not valid when $V_0 \sim T_e$ and ion velocity at the positively charged electrode is not zero which is the case of a floating wall potential.

1.4.1.6.2 Sheath at floating wall
The potential drop across a sheath between the plasma and a floating wall can be calculated taking into account that total current to the wall must be equal to zero.
Ion current can be calculated as

$$I_i = en_sC_s \tag{1.198}$$

Electron current is calculated as

$$I_e = \frac{1}{4}n_s\left(\frac{8eT_e}{\pi m}\right)^{0.5}\exp(\varphi_w/T_e) \tag{1.199}$$

As $I_i = I_e$, the wall potential solving Eqs (1.198) and (1.199) is

$$\varphi_w = -T_e\ln\left(\frac{M}{2\pi m}\right)^{0.5} \tag{1.200}$$

1.4.1.6.3 Sheath with arbitrary ion distribution function: kinetic approach

By considering an arbitrary ion distribution function, Bohm criterion for a stable sheath can be generalized. Allen [39] demonstrated that the generalized Bohm criterion has the following form:

$$\int_0^\infty \frac{g(w)}{w^2} dw \le \frac{M}{kT_e} \tag{1.201}$$

where $g(w)$ is the normalized ion velocity distribution function, w is the ion velocity, M is the ion mass, k is the Boltzmann's constant, and T_e is the electron temperature.

We will follow the interpretation of the condition (1.201) proposed in [40].

Ion density inside the sheath region can be calculated using the distribution function as

$$n_i(\varphi) = n_s \int_0^\infty dw \frac{wg(w)}{(w^2 + (2e/M)[\varphi_s - \varphi])^{1/2}} \tag{1.202}$$

where φ_s is the potential at the sheath edge and n_s is the ion density at the sheath edge. After differentiating Eq. (1.202) with respect to φ_s, one arrives to

$$\frac{dn_i}{d\varphi_s} = -n_s \frac{e}{M} \int_0^\infty dw \frac{wg(w)}{(w^2 + (2e/M)[\varphi_s - \varphi])^{3/2}} \tag{1.203}$$

At the sheath edge, ion density gradient becomes

$$\frac{dn_i}{d\varphi_s} = -n_s \frac{e}{M} \int_0^\infty dw \frac{g(w)}{w^2} \tag{1.204}$$

If we assume that electrons obey Boltzmann distribution, the electron density derivative at the sheath edge is

$$\frac{dn_e}{d\varphi_s} = -n_s \frac{e}{kT_e} \tag{1.205}$$

Sheath stability is associated with requirement that electron density decreases faster than the ion density, i.e.,

$$\frac{dn_i}{d\varphi_s} \le \frac{dn_e}{d\varphi_s} \tag{1.206}$$

Let us now substitute expressions for ion and electron density gradient into Eq. (1.206):

$$n_s \frac{e}{M} \int_0^\infty dw \frac{g(w)}{w^2} \le n_s \frac{e}{kT_e} \tag{1.207}$$

This leads us to Eq. (1.201), which is the generalized Bohm criterion in the case of arbitrary ion density distribution.

1.4.1.6.4 Sheath with secondary electron emission (SEE)

Under typical steady-state conditions considered above, the potential drop between the plasma and the wall is negative in order to repel the excess of thermal electrons (see Section 1.4.1.6.2).

However, when the wall has substantial SEE, the floating potential drop may be different from that in the simple sheath. Hobbs and Wesson [41] were first to describe this situation in detail.

Considering a one-dimensional sheath model and taking into account the SEE from the wall, the electrostatic potential distribution in the sheath satisfies the Poisson equation:

$$\nabla^2 \varphi = e/\varepsilon_0(n_{e1} - n_i + n_{e2}) \tag{1.208}$$

where φ is the potential, n_{e1} is the density of plasma electrons, n_{e2} is the density of secondary electrons, and n_i is the ion density. The plasma electrons are assumed to obey the Boltzman distribution:

$$n_{e1} = (n_o - n_{e2}(0))\exp(-e\varphi/kT_e) \tag{1.209}$$

where $n_{e2}(0)$ is the density of the secondary electrons at the sheath edge and n_o is the electron density at the sheath edge that is equal to the ion density $n_i(0)$. The ions are assumed to be cold and have at the presheath–sheath interface the energy $U_{io} = (1/2)m_i V_o^2$, where V_o is the ion velocity at the sheath edge and m_i is the ion mass. Ions have free motion in the sheath and their density decreases according to

$$n_i = n_o \times V_o(V_o^2 + 2e\varphi/m)^{-0.5} \tag{1.210}$$

The current continuity equation can be written in the form

$$j_{e1} + j_i - j_{e2} = 0 \tag{1.211}$$

where j_{e1} is the flux of the primary electrons from the plasma ($j = nV$ is the flux definition for all species), j_i is the ion flux, and j_{e2} is the flux of the secondary electrons. From this equation, one can obtain that

$$n_{e2} V_{e2} = \delta/(1 - \delta)n_o V_o \tag{1.212}$$

where δ is the SEE coefficient determined as $\delta = j_{e2}/j_{e1}$. We will furthermore assume that the electrons emitted from the surface are monoenergetic and freely move in the sheath. The boundary conditions for the sheath problem are the following:

$$\varphi(0) = 0; \quad d\varphi/dr \, (r' = 0) = E_o \text{ and } V = V_o \tag{1.213}$$

From the current continuity equation (Eq. (1.211)), one can calculate the potential drop across the sheath $\Delta\varphi_w$ as

$$\Delta\varphi_w = kT_e \ln ((1 - \delta)/V_o(2\pi m_e/kT_e)^{0.5}) \tag{1.214}$$

One can see that the potential drop across the sheath decreases with the SEE coefficient. It should be noted that in this model, we have assumed that a monotonic potential distribution exists. However, when the SEE coefficient δ approaches unity, the solution in form (1.214) breaks down and when δ exceeds a critical value, a potential well forms such that a fraction of emitted electrons is returned to the wall. This happens when δ is less than 1 and it was obtained that the critical value of the SEE coefficient δ^* can be calculated as

$$\delta^* = 1 - 8.3(m_e/m_i)^{0.5} \tag{1.215}$$

From Eq. (1.215), it can be estimated that in the case of BN (commonly used dielectric material), the critical SEE coefficient is about 0.95.

1.4.1.7 Sheath in a magnetic field

Magnetic field affects the plasma—sheath transition. Schematically plasma—wall interface in a magnetic field is shown in Figure 1.36.

Daybelge and Bein [42] were the first to completely analyze a collisionless sheath, in which a magnetic field exists between fully ionized plasma and an infinite plane metal wall. The potential distribution in the sheath was obtained from the solution of Poisson equation, using the wall potential as a parameter and assuming a Maxwellian distribution for the charged particle velocities at the sheath—plasma edge. They obtained that sheath thickness is comparable to the ion Larmor radius in the case of negative wall potentials, whereas for positive wall potential it is reduced to the electron Larmor radius. However, Daybelge and Bein considered in their model the electrostatic sheath region only for the case $L_D \gg r_{Le}$, where L_D is the Debye length and r_{Le} is the electron Larmor radius. Therefore, the ionic presheath acceleration was not considered. Later, the effect of the magnetic field on the plasma presheath was studied by Chodura [43] in the collisionless limit. Plasma quasi-neutrality in the presheath and a Boltzmann distribution for the electron density were assumed in order to determine the electric field in the sheath. Chodura found that in order to satisfy the usual Bohm condition at the sheath—plasma edge, a new double-layer presheath was required. In the first double presheath region, the ions are accelerated to supersonic velocities parallel to the magnetic-field lines, which may be oblique with respect to the wall. In the second presheath region, which has the thickness of approximately the ion Larmor radius, the ion velocity vector is rotated so that the velocity normal to the wall is supersonic at the entrance to the sheath. Unlike the above-mentioned models, Riemann's analysis of the presheath region in a plasma placed in a magnetic field also accounted for collisions[70]. The assumptions of Riemann's fluid model are similar to those of Chodura, except that an ion collision term of constant collision frequency was added. A dependence of the potential distribution in the presheath on the magnetic field and collision frequency was found. The main effect of a strong magnetic field is to reduce the collisional presheath

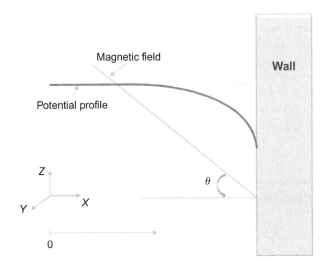

FIGURE 1.36

Schematics of the sheath region in a magnetic field.

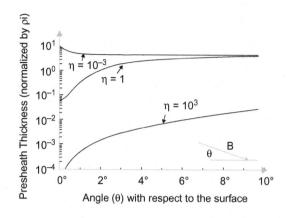

FIGURE 1.37

The magnetic presheath thickness (in units of the ion gyroradius) as a function of the intersection angle $(90 - \theta$, see Figure 1.36), with the electron to ion current ratio η as a parameter.

thickness, as it approximates now the ion Larmor radius and not the ion mean free path. Riemann also showed that in the collisional case, additional plasma presheath is not required, in contrast to the case studied by Chodura.

In all of the above-mentioned magnetic presheath models, a Boltzmann distribution for the electron density was assumed, and the magnetic field acted only on the ions. It should be noted that the Boltzmann distribution for the electron density is physically significant in the case of plasma flow parallel to the magnetic field. However, for plasma transport across a magnetic field, it is not obvious that this distribution will occur. It was shown by Beilis and coworkers [44,45] that the assumption about the electron distribution function influences strongly the presheath thickness. Effect of the angle of the magnetic field with respect to the wall on the presheath thickness is illustrated in Figure 1.37. One can see that when magnetic field is changed from being parallel to the wall (0°) to small incidence angle a significant change in the presheath thickness is occurred.

It was also shown that while density was calculated from the general system of equation (i.e., electron momentum equation that takes into account the magnetic field), the density profile does not deviate significantly from the one obtained from the Boltzmann equation as shown in Figure 1.38. However, neglecting the magnetic field term in the electron momentum equation affects the potential, density, velocity distribution in the presheath and the presheath thickness.

1.5 Surface phenomena: electron emission and vaporization

Plasmas in Laboratory settings are confined by physical boundaries. In this section, we consider interaction of plasma with solid materials by considering phenomena occurring at the material surface. In particular, we will consider emission of electron and material evaporation as shown schematically in Figure 1.39.

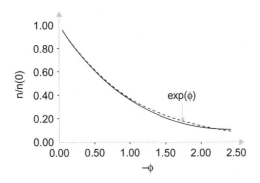

FIGURE 1.38

The plasma density as a function of potential in the presheath. The broken curve is the Boltzmann density distribution.

1.5.1 **Electron emission**

Electron emission is defined as a phenomenon of liberation of electron from the surface that is stimulated by temperature elevation, radiation, or by strong electric field. In a metallic crystal lattice, the outer electrons, valence electrons, and orbits overlap, and are shared by all the atoms in the solid. These electrons are not bound and are free to conduct current. Typically the free electron density in a metal is about 10^{23} cm^{-3}. In a semiconductor, valence electrons are bound in covalent bonds. Electrons are forming a sphere in the energy space and the surface of this sphere is at the Fermi energy level. Although valence electrons are free to move inside the metal, they are bound relatively to the vacuum. A potential barrier at the surface is formed due to polarization of the electron gas according to the Shottky theory. Therefore, an image force F_{im} appears at the surface:

$$F_{im} = \left(-\frac{e}{2x}\right)^2 \tag{1.216}$$

where x is the distance from the surface and e is the electron charge. Thus, it is necessary to spend a certain amount of energy to enable for electron to overcome the potential barrier at the surface and be emitted from the solids. The amount of this energy is called the work function (W_f). The work functions for typical metallic emitters are presented in Table 1.1. In order to calculate the electron emission current, the electron energy distribution in the solid body should be taken in account. For example, in DC discharges, the electron current is sustained by the emission of electrons from the surface of the cathode.

According to Fermi−Dirac statistics, the electron energy distribution in the metal is

$$f(w)dw = \frac{(\pi/2)((8me/h^2))^{3/2}w^{1/2}\,dw}{\exp(w - w_0/kT) + 1} \tag{1.217}$$

where w is the electron energy and w_0 is the Fermi level.

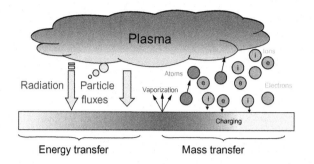

FIGURE 1.39

Schematics of plasma–wall interactions and surface phenomena.

Table 1.1 The Work Functions for Typical Metallic Emitters

Element	Work Function (eV)
Cu (copper)	4.4
Ti (titanium)	3.84
Ag (silver)	4.26
Au (gold)	4.8
W (tungsten)	4.5

$$w_0 = \frac{h^2}{2m_e}(3n_e/8\pi)^{2/3} \qquad (1.218)$$

where n_e is the electron density and m_e is the electron mass.

1.5.1.1 Thermionic emission

When the metal is heated by an energy source, the surface temperature increases allowing the electrons to overcome the potential barrier at the surface. Integrating the distribution function taking the minimal energy electron as W_f:

$$\int_{\sqrt{(2W_f/m_e)}}^{\infty} f(w)dw = AT^2 \exp\left(-\frac{W_f}{kT}\right) \qquad (1.219)$$

where $A = (4\pi e m_e k^2/h^3) \approx 1.2 \times 10^6$ (A/m² K²)

Presence of the electric field changes the potential distribution near the surface and lowers the potential barrier (shown in Figure 1.40) by the so-called Shottky effect reducing the work function by

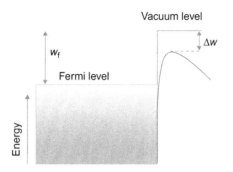

Schematics of the energy levels near the surface. With the external electric field, the potential barrier becomes finite.

$$\Delta W = \sqrt{\frac{e^3 E}{4\pi\varepsilon_0}} \tag{1.220}$$

Taking this into account the electron reflection due to quantum mechanical effect of tunneling, the expression for the current density by thermionic electron emission will have the following form:

$$j = DAT^2 \exp\left(-\frac{W_f - \Delta W}{kT}\right) \tag{1.221}$$

where D is the electron reflection coefficient. In general, electron reflection coefficient depends on material, surface conditions. As a result, effective thermoionic current varies significantly.

1.5.1.2 Field emission

The electron emission by electric field action is modeled assuming that surface temperature is zero. Presence of the electric field near the surface transforms the infinite thick potential distribution into a potential barrier of a finite width (Figure 1.41). As a result, electrons can escape the metal with probability determined by quantum mechanical tunneling mechanism. The current density due to the field emission, one needs to account for this quantum mechanical effect. Such treatment was done by Fowler–Nordheim, and the emission current density is given by the equation:

$$j = C_1 E^2 \exp\left(-\frac{C_2}{E}\right) \tag{1.222}$$

where E is the electric field,

$$C_1 = \frac{e^3 W_0^{0.5}}{2\pi h(W_f + W_0)W_f^{0.5}}$$

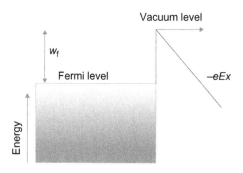

FIGURE 1.41

Schematics of the energy levels near the surface in the case of field emission. With the external electric field, the potential barrier becomes finite.

$$C_2 = \frac{8\pi(2m_e)^{0.5} W_f^{1.5}}{3eh}$$

1.5.1.3 T–F emission

When the metal temperature and electric field are sufficiently high, a combined effect of thermoionic and field action was takes place. This mechanism is known as T–F emission:

$$J_{TF} = \int_{-e\varepsilon_B}^{\infty} eN(\varepsilon, T)D(\varepsilon, E)d\varepsilon \tag{1.223}$$

where $e\varepsilon_B$ is the lowest potential inside the metal. According to Eq. (1.223), electron emission current is determined by two functions, namely, density of states N and emission probability D. Murphy and Good [46] performed the T–F emission analysis for some limited regions of T and E. The electron emission for wide region of T and E appropriate for conditions in a cathode spot was calculated in Ref. [47]. Results can be presented in the form of emission current density distribution as a function of energy (relative to Fermi level) of emitted electrons as shown in Figure 1.42. One can see that electrons can be emitted with energy that is lower or higher than the Fermi level dependent on electric field and surface temperature. Such general approach can recover both limits of thermoionic emission in the case of a small electric field and field emission in the case of smaller temperature.

1.5.1.4 Secondary electron emission

SEE is defined as a phenomenon in which primary electrons of sufficient energy induce the emission of secondary electrons by hitting the surface. The number of secondary electrons emitted per incident particle is called secondary emission yield. If the primary incident particles are ions, the effect is termed secondary ion emission. The coefficient of the SEE depends on the energy of incident electrons and the incident angle. Primary electron collisions with the surface of a target generate emissions of two main types of secondary electrons, namely, "true" secondary electrons and elastically reflected backscattered electrons. Measured electron emission yield is due to contribution

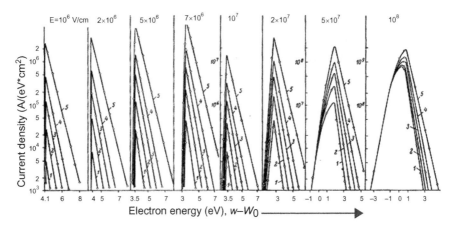

FIGURE 1.42

Dependence of the emission current density on the energy of emitted electrons with electric field and surface temperature as parameters (work function is 4.5 eV).

Source: *After Beilis [47].*

of all secondary electrons. In general, SEE is higher for dielectric materials in the low-energy range (up to 100 eV) and high in the case of metal target material in the case of a kV energy range. Dawson [48] measured yields for some dielectric matrials as shown in Figure 1.43. Lucalox alumina and sapphire are both alumina, but in the preparation of this Lucalox alumina, a small amount of magnesia was added. As shown in Figure 1.2, the yields from the highly polished sapphire were a little greater at all energies. However, those from the polished Lucalox had become similar to the sapphire in their variation with primary energy.

SEE yield $\delta(E_p)$ as a function of energy of the primary electrons was measured for several dielectric materials [49] as shown in Figure 1.44. One can see that SEE yield increases with energy and reaches unity at the energy of primary electrons of about $25-30$ eV in the case of boron nitride.

1.5.2 Vaporization

Vaporization is the phenomenon of emission of atoms or molecules from the surface exposed to heat. In general, vaporization describes any type of phase transition from solid/liquid into a gas. Below the direct transition from solid or liquid material into the gas phase will be considered.

An important issue for many applications is the vaporization of a hot surface due to interaction with discharge plasmas. This phenomenon occurs in different devices such as ablation-controlled arcs [50], pulsed plasma thrusters [51,52], high-pressure discharges [53], vacuum arcs [54], electro-guns [55], and metal evaporation by laser radiation action [56,71]. The mechanism of vaporization into vacuum was described by Langmuir relationship [57].

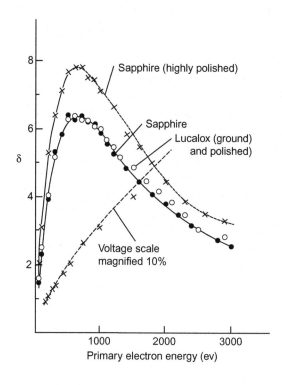

FIGURE 1.43

SEE yield for dielectric materials.

Source: *Reprinted with permission from Ref. [48]. Copyright 1966, American Institute of Physics.*

1.5.2.1 Langmuir model

This is the simplified model that is based on the equilibrium condition near the surface. Consider a rate at which molecules condense on a surface $\kappa\nu$, where κ is the sticking coefficient (ratio of the flux condensed/flux striking the surface) and ν is the ablated flux. Langmuir showed that for the case of vapor for same metal, the coefficient $\kappa = 1$.

On the other hand, the flux of evaporated molecules noted as μ equal to ν in the equilibrium. Taking into account that $\nu = (1/4)nV_a$, one arrives to the following rate of vaporization:

$$\Gamma = \mu = \frac{1}{4}n\sqrt{\frac{8kT}{\pi m}} = \frac{P_0}{\sqrt{2\pi mkT}} \tag{1.224}$$

The main Langmuir's assumptions:

1. Atomic emission rate depends on surface temperature and not on the gas properties.
2. Γ represents the gross atomic flux from the surface.
3. $P_\infty = 0$ and therefore Γ represents the net out flux.

Despite the fact that this model is not exact in case of $P_\infty \neq 0$, it is widely used [58]. In the next section, more general kinetic model will be described.

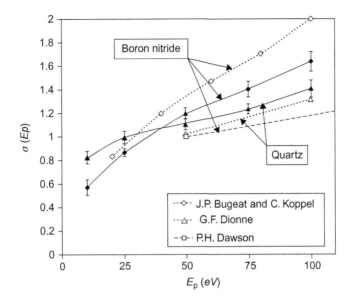

FIGURE 1.44

Total yield of SEE $\sigma(E_p)$ from boron nitride and quartz as a function of energy of primary electrons.

Source: *Reprinted with permission from Ref. [49]. Copyright 2003, American Institute of Physics.*

1.5.2.2 Kinetic models

Anisimov [59] considered a case of vaporization of a metal exposed to laser radiation using a bimodal velocity distribution function in the nonequilibrium (kinetic) or Knudsen layer. The main result of this work is the calculation of the maximal flux of returned atoms to the surface, which was found to be about 18% of the flux of vaporized atoms. This result was obtained under the assumption that the atom flow velocity is equal to the sound velocity at the external boundary of the kinetic layer. Schematically this layer is shown in Figure 1.45. In many physical situations, however, the expansion of the vapor is not by the sound speed since there is dense plasma in the volume discharge. The vaporization phenomena in case of plasma presence at the metal surface with velocity lower than the sound speed were developed by Beilis [60–62] who considered metal vaporization into discharge plasmas in the case of a vacuum arc cathode spot. He concluded that the parameters at the outer boundary of the kinetic layer are close to their equilibrium values and that the velocity at the outer boundary of the kinetic layer is much smaller than the sound velocity.

The nonequilibrium layer close to the evaporating surface can be also modeled using the particle method known as direct simulation Monte Carlo (DSMC) [63]. This method solves numerically the Boltzmann equation allowing calculation of the thickness of the nonequilibrium layer. Below the numerical results will be compared with the results of above-mentioned analytical approach for the case when the vapor velocity at the outer boundary of the kinetic layer is given as a parameter.

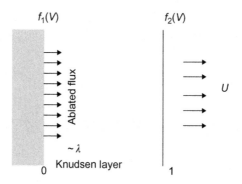

FIGURE 1.45

Schematics of the Knudsen layer.

1.5.2.3 Model of the nonequilibrium layer

1.5.2.3.1 DSMC particle approach

In the nonequilibrium layer near the surface, there are collisions between particles that eventually lead to change of the velocity distribution function. The DSMC method uses particle motion and collisions to perform a simulation of gas dynamics under nonequilibrium conditions. Each particle has spatial and velocity coordinates. Collision approach between particles is based on a probability model developed from the kinetic theory and commonly used in DSMC [64].

To perform the DSMC simulation, we have to specify conditions at two boundaries (see Figure 1.45). At the evaporating surface with density n_o and temperature T_0, the velocity distribution function for emitted particles is in the equilibrium form [59]:

$$f_0(\mathbf{v}) = n_o \left(\frac{m}{2\pi k T_o}\right)^{3/2} \exp\left(-\frac{mv^2}{2kT_o}\right) \quad V_x > 0 \tag{1.225}$$

At the outer boundary of the kinetic layer, the distribution function for particles is assumed to be

$$f_1(\mathbf{v}) = n_1 \left(\frac{m}{2\pi k T_1}\right)^{3/2} \exp(-m\frac{(v_x - U)^2 + v_{y^2} + v_{z^2}}{2kT_1}) \tag{1.226}$$

where U is the velocity at the outer edge of the Knudsen layer as shown in Figure 1.45, n_1 is the density, and T_1 is the temperature at the outer boundary. These boundary conditions are supplemented by using an empirical relation between T_o and n_o.

The DSMC model employed has the following strategy. Uniform cells with size of 0.5λ are employed and time step $\Delta t = 0.3\lambda/V_{m,s}$ [64], where λ and $V_{m,s} = (2kT_o/m)^{0.5}$ are the molecular mean free path and the most probable thermal speed at the ablated surface. Molecules enter the flow field successively from the surface due to evaporation and from the outer boundary due to Maxwellian velocity distribution that allows particle velocities in negative direction. Using the above assumed distribution functions at the surface and at the external boundary (Eq. (1.226)), we can calculate the flux of molecules entering from the surface, G_s, and from the outer boundary of the layer, G_b, as follows:

$$G_s = n_o (2kT_o/m)^{0.5}/2\pi^{0.5} \qquad (1.227)$$

$$G_b = n_1 (2kT_1/m)^{0.5}[\exp(-\alpha^2) - \pi^{0.5}\alpha\{1 - \mathrm{erf}(\alpha)\}]/2\pi^{0.5} \qquad (1.228)$$

where $\alpha = U/(2kT_1/m)^{0.5}$. The molecular interaction is described by the variable hard-sphere (VHS) model [62]. The VHS model is employed to select molecular collision pairs from cells and to distribute the postcollision velocities. This model assumes that the scattering from molecular collision is isotropic in the center of mass frame of reference. Both boundaries (wall and outer boundary of the kinetic layer) are assumed to be perfectly absorbing. The flow will arrive at a steady state when the sum of G_s and G_b is exactly balanced by the flux of molecules leaving from the outer boundary or sticking on the surface:

$$G_s + G_b = G_r + G_f \qquad (1.229)$$

where G_r is the flux of the particles returned to the surface during the time step and G_f is the flux of the particles crossing the outer boundary of the layer.

The location of the outer boundary was determined as a part of the solution. The parameters at the outer boundary (n_1 and T_1), the evolution of the velocity distribution function inside the kinetic layer, and the flux of returned particles are calculated as a function of the outer boundary location of the layer with given velocity U at this boundary as parameter.

1.5.2.3.2 Analytical approach

Let us consider the analytical approach, where the vapor parameters T_1 and n_1 at the outer boundary can be obtained using integral relations for the nonequilibrium layer. This means that the problem is reduced to integration of the conservation equations of mass, momentum, and energy across the layer [59,65]. We consider a nonequilibrium layer (thickness of about a mean free path λ) adjacent to the surface (as shown in Figure 1.45), where the velocity distribution function of the evaporated molecules reached equilibrium by the rarefied collisions with the background heavy particles and furthermore the vapor flow is described by a hydrodynamic approach. We have used Anisimov's assumption [59] that the velocity distribution function for the returned particles ($v_x < 0$) is $\beta f_1(\mathbf{v})$, where β is the proportionality coefficient. The relations of the heavy particle parameters at the outer boundary of the kinetic layer are

$$\frac{n_o}{2(\pi d_o)^{0.5}} = n_1 v_1 + \beta \frac{n_1}{2(\pi d_1)^{0.5}}\{\exp(-\alpha^2) - \alpha \pi^{0.5}\mathrm{erfc}(\alpha)\}$$

$$\frac{n_o}{4d_o} = \frac{n_1}{2d_1}\{(1 + 2\alpha^2) - \beta[(0.5 + \alpha^2)\mathrm{erfc}(\alpha) - \alpha \exp(-\alpha^2)/\pi^{0.5}]\} \qquad (1.230)$$

$$\frac{n_o}{(\pi d_o)^{1.5}} = \frac{n_1}{(d_1)^{1.5}}\pi^{-1}[\alpha(\alpha^2 + 2.5) - 0.5\beta\{(2.5 + \alpha^2)\alpha\,\mathrm{erfc}(\alpha) - (2 + \alpha^2)\exp(-\alpha^2)/\pi^{0.5}\}]$$

where $d_o = m/2kT_o$, $d_1 = m/2kT_1$, $\mathrm{erfc}(\alpha) = 1 - \mathrm{erf}(\alpha)$, $\mathrm{erf}(\alpha)$ is the error function. The system of equation (1.230) is obtained using the conservation laws of mass, momentum, and energy across the layer [59]. The flux of returned particles at the outer boundary was obtained calculating the parameters of the kinetic layer in the form

$$J_- = \int_{-\infty}^{0} \beta f_1(V)V \, dV = \beta \cdot N_1 \left(\frac{kT_1}{2\pi m}\right)^{0.5} \left\{\exp(-\alpha^2) - \alpha\pi^{0.5} \, \mathrm{erfc}(\alpha)\right\} \tag{1.231}$$

The system of equation (1.229) has four unknowns and therefore the solution can be found having one unknown as a parameter, which is the velocity U at the outer boundary of the kinetic layer in our case.

1.5.2.3.3 Examples of Knudsen layer calculation

The returned atom fluxes calculated according to DSMC and the analytical approaches with the thickness of the kinetic layer as a parameter are shown in Figure 1.46.

One can see that in the case of nomalized velocity ≤ 0.5 (Figure 1.46), all results agree well. This is the case when the thickness of the nonequilibrium layer is about one mean free path. In the case when evaporation occurs at about the sound velocity at the outer boundary, the DSMC calculations approach the analytical value at a layer thickness of $\sim 10-20$ mean free paths.

The calculated backflux (J_-/J_+) dependence on distance inside the layer is presented in Figure 1.47. It can be seen that the flux of the returned molecules depends upon the distance from the evaporating surface where the external boundary is placed. Thus, up to a layer thickness of about 20 mean free paths, the flux changed strongly and then it is saturated. Recall that varying of the position of the external boundary can be viewed as the method to find the thickness of the Knudsen layer. In this particular example, one can see that it is about 20 mean free paths. Thus one can determine the DSMC calculation predicts a 16% flux of returned particles, which is very close to the analytical result of 18%. The reason for this difference can be understood by analyzing the

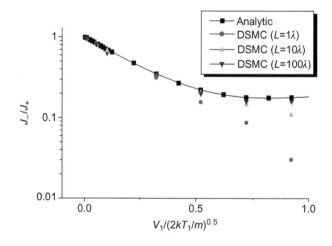

FIGURE 1.46

Comparison of the analytic and DSMC return flux as a function of velocity V_1 with the distance of location of the outer boundary of the kinetic layer L as a parameter.

Reprinted with permission from Ref. [65]. Copyright 2001, American Institute of Physics.

velocity distribution function of returned particles in the DSMC calculation. Let us now consider the velocity distribution.

The velocity distribution functions calculated according to DSMC and the analytical approximation of $\beta f_1(\mathbf{v})$ for the sound velocity at the outer boundary are shown in Figure 1.48. DSMC distribution function agrees well with the analytic approximation when the velocity at the outer boundary of the kinetic layer is small as shown in Figure 1.48B. Therefore, the calculated flux of returned particles is also found to be in good agreement with the analytical result.

The evolution of the particle distribution function within the Knudsen layer is shown in Figure 1.49. One can see that the velocity distribution function approaches a shifted Maxwellian at a distance of several mean free paths from the surface. The drift velocity slightly increases with further distance from the evaporating surface.

The results of calculation of the analytic system of equations 1.230 are presented in Figure 1.49 with the normalized velocity U as a parameter. The temperature T_1, density n_1, and the flux of returned particles J_- all decrease as the velocity at the outer boundary of the kinetic layer increases. In the limiting case of the sound velocity, the flux of returned particles is equal to 18% as was obtained by Anisimov [59]. In this case, the analytically predicted flux of returned particles is larger than that obtained by numerical simulations (16%, see Figure 1.47). It should be pointed out that the dependence of the flux of returned particles J_- on the velocity V_1 has a minimum near the sound speed (see Figure 1.50). The minimum corresponds to the sound speed with adiabatic index 1.3, i.e., the distribution function of the returned particles is proportional to the distribution function at the outer boundary of the kinetic layer [58] (Figure 1.50). Although Figure 1.49 shows results for velocities larger than the sound speed for methodic purposes, one should note that

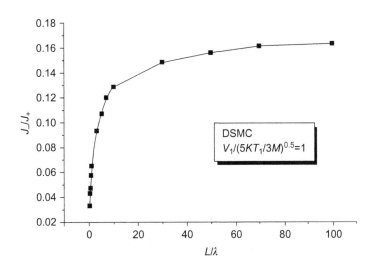

FIGURE 1.47

The DSMC calculated return flux as a function of the distance of location of the outer boundary of the kinetic layer L in the case of sound velocity at the boundary 1.

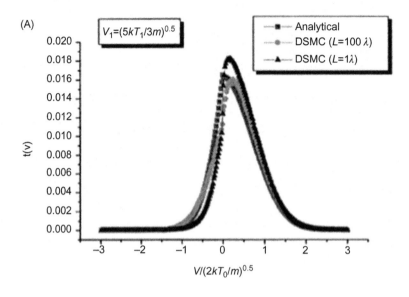

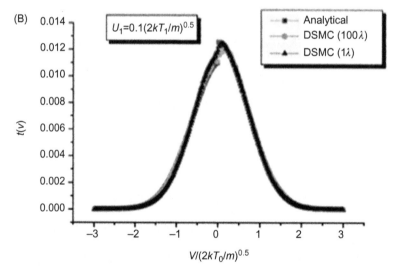

FIGURE 1.48

Variation of the velocity distribution function of the returned particles near the wall with the distance of location of the outer boundary of the kinetic layer L as a parameter.

Reprinted with permission from Ref. [65]. Copyright 2001, American Institute of Physics.

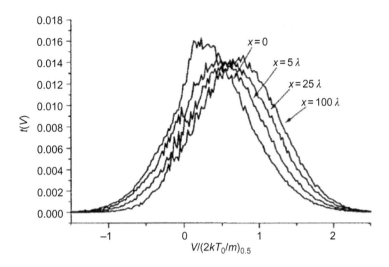

FIGURE 1.49

Variation of the velocity distribution function normal to the surface component with the distance from the evaporated surface as a parameter.

Reprinted with permission from Ref. [65]. Copyright 2001, American Institute of Physics.

solution with velocity higher than the sounds speed is not physical since there is no mechanism of supersonic acceleration in the model described.

1.5.2.3.4 Ablation of the Teflon into discharge plasma

It was shown in the previous section that the velocity at the outer boundary of the kinetic layer strongly affects the kinetic layer parameters. To find the velocity V_1, we apply the mass and momentum conservation equations for heavy particles in the hydrodynamic region (assuming a single fluid model) [66] between boundaries 1 and 2 as shown in Figure 1.51.

Assuming weakly ionized plasma in the hydrodynamic layer, the integration of the mass and momentum conservation equations yields the following relations between parameters at boundaries 1 and 2:

$$n_1 V_1 = n_2 V_2 \tag{1.232}$$

$$n_1 kT_1 + mn_1 V_1^2 = n_2 kT_2 + mn_2 V_2^2 \tag{1.233}$$

Combination of these two equations yields the following expression for the velocity at the outer boundary of the kinetic (Knudsen) layer:

$$\frac{V_1^2}{(2kT_1/m)} = \frac{(T_2 n_2/2T_1) - (n_1/2)}{n_1 - (n_1^2/n_2)} \tag{1.234}$$

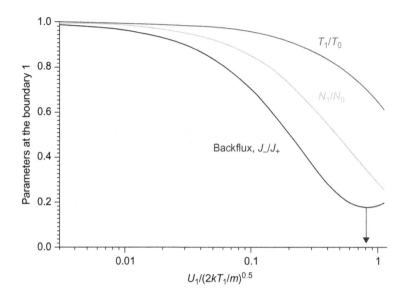

FIGURE 1.50

Temperature, density, and returned flux at the outer boundary of the kinetic layer as a function of velocity at the outer edge of the Knudsen layer.

This equation makes it possible to calculate the velocity at boundary 1 and therefore to calculate the ablation rate that is proportional to $V_1 n_1$. The system of equations is closed if the equilibrium vapor pressure can be specified. In the case of Teflon, the equilibrium pressure formula is used:

$$P = P_c \exp(- T_c/T_0) \tag{1.235}$$

where $P = n_0 k T_0$ is the equilibrium pressure, and $P_c = 1.84 \times 10^{15}$ N m^2 and $T_c = 20\ 815$ K are the characteristic pressure and temperature, respectively [67]. The solution of the problem depends upon plasma density n_2, plasma temperature T_2, and surface temperature T_0. The parameters n_2, T_2 are determined by the bulk plasma flow. It was estimated from various experiments that, under typical operation conditions of a pulsed plasma accelerator, the plasma density near the Teflon surface is about $10^{21} - 10^{24}$ m^{-3}, the plasma temperature is about $1-4$ eV, and the Teflon surface temperature T_0 is about $900-1000$ K [67].

Ablation rate contours in the plane with the plasma density and Teflon surface temperature as the coordinates are displayed in Figure 1.52. The same ablation rate that can be found in the high- and low-density range corresponds to the solution of the problem with small and large velocities at the outer boundary of the kinetic layer, respectively. There is no solution for the ablation rate in regions above the curve with ablation rate equal to zero. This region in the $n_2 - T_0$ plane corresponds to the case when the right-hand side of Eq. (1.234) is negative. The physical meaning of these results can be explained as follows. In the ablation-controlled discharge, the plasma density in the bulk is determined by the rate of ablation from the surface.

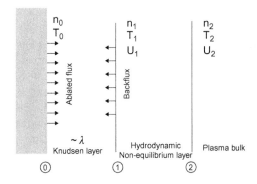

FIGURE 1.51

Schematic representation of the layer structure near the ablated surface.

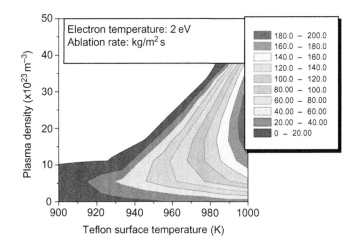

FIGURE 1.52

Contours of ablation rate in the plasma density (n_2)–Teflon surface temperature (T_0) plane.

In the case of small surface temperature, one can expect a smaller ablation rate and therefore high plasma densities in the discharge cannot be generated.

It is important to note that this model predicts the dependence of the ablation rate on the plasma bulk density, electron temperature, and the surface temperature. It is also found that the flow velocity at the outer boundary of the kinetic layer (that determines the ablation rate) is smaller than the sound velocity under typical conditions.

1.5.2.3.5 Outlook on evaporation analysis approached

Two kinetic approaches, numerical method DSMC and analytical approach, were employed to describe the particle parameters in the nonequilibrium kinetic layer near the evaporating surface.

DSMC approach makes it possible to determine the thickness of the kinetic layer and the evolution of the particle distribution function within the layer. The layer thickness, the vapor density, and temperature in the kinetic layer adjacent to the evaporating surface depend upon the velocity at the outer boundary of the layer. The thickness of the kinetic layer increases with the particle velocity at the outer boundary of the layer and vary from a few mean free paths λ for relatively small velocity up to about $10-20\lambda$ in the case of the evaporation with sound speed.

The flux of returned particle calculated using analytical model agrees well with that calculated by DSMC approach over a wide range of velocity at the outer boundary of the layer. The present model can be used for calculation of the evaporation rate of the heated surface by interacting with plasmas. The free parameter of this model, the velocity at the outer boundary of the layer, can be determined by coupling this model with a model of the hydrodynamic layer and the plasma bulk as shown in Section 1.5.2.3.4.

Homework problems
Section 1

1. **Debye length**

   A conductive sphere of radius a is immersed in an infinite uniform quasi-neutral plasma having density n_o, electrons in thermal equilibrium at temperature T_e, and ions in thermal equilibrium at temperature T_i. Only single charged ions are presented. A small negative DC potential $|V_0| \ll T_i, T_e$ is applied to the sphere.
   a. Start from Poisson's equation in spherical coordinates and using Boltzmann's relation for both the electrons and ions derive expression for potential distribution.
   b. Find expression for Debye length and show that Debye length is determined mainly by the temperature of the colder species.
   c. Calculate Debye length for arc discharge plasma having electron density of about 10^{14} m^{-3} and electron temperature of about 1 eV and ion temperature of 0.1 eV.
   Note 1: In spherical symmetry: $\nabla^2\varphi = (1/r)d^2(r\varphi)/dr^2$

2. **Plasma oscillation**

   Compute plasma frequency for the following cases:
   a. Vacuum arc with electron density of about 10^{21} m^{-3}.
   b. Hall thruster with electron density of about 10^{17} m^{-3}.
   c. Cold plasma with electron density of about 10^{20} m^{-3}.

3. Estimate whether following ionized gases are plasmas according to conventional definition (i.e., quasi-neutrality) if characteristic size of the system is:
   a. Interstellar space area (10 km)
   b. Magnetic fusion device (1 m)
   c. Internal confinement fusion (1 mm).

4. Compute the pressure (in atmospheres) exerted by arc plasma on its chamber. Assume that plasma is in equilibrium, i.e., $T_e = T_i = T_a = 3000$ K, $n = 10^{21}$ m^{-3}. Ideal gas assumption can be employed.

Section 2

1. **Ionization cross section**

    Plot ionization coefficient as a function of electron temperature (in eV) for nitrogen plasma. Ionization potential for nitrogen is 14.5 eV. Consider T_e range $1-10$ eV.

2. **Plasma equilibrium**

    Calculate electron density of helium plasma in arc discharge at 0.1 atm as a function of electron temperature. Assume that $g_i g_e / g_a = 2$. Consider range of electron temperature $1-5$ eV.

    Note: LTE can be assumed.

3. **Plasma composition**

    Calculate electron density of nitrogen−argon mixture. Pressure is 1 atm and ratio of partial pressures of nitrogen to argon is 0.5. Consider conditions for problem #2.

Section 3

1. Calculate the skin depth of electromagnetic wave in copper and iron material as a function of frequency. Consider frequencies in 10 kHz-1 MHz range.

2. Waves in plasmas. Consider plasma with density of about 10^{18} m^{-3} and electron temperature of about 5 eV. Plot the dispersion relation for electron plasma wave and ion plasma wave. Comment on results.

3. Calculate upper-hybrid and low-hybrid frequencies of arc discharge plasma having electron temperature 2 eV and electron density of about 10^{21} m^{-3}. Plot dependence of both frequencies as a function of a magnetic field which is less than 1 T. Comment on results. What is the lower limit of magnetic field for low-hybrid frequency?

    Note: Characteristic size of arc is 0.01 m.

4. A plasma consists of two uniform streams of ions with velocities V_x and ^-V_x and densities $1/3n_0$ and $2/3n_0$. The neutralizing unmobile electron fluid has density n_0. Derive a dispersion relation for instabilities in this system.

Section 4

1. Calculate the floating potential on the probe in contact with Argon plasma with electron temperature of about 10 eV. How floating potential will change in presence of secondary electron emission (yield is about 0.4).

2. Consider hydrogen ion diode with interelectrode distance of about 2 mm. The applied voltage is -1000 V. Calculate the ion current density in the diode.

3. Consider sheath between the quasi-neutral helium plasma having electron density of about 10^{18} m^{-3} and electron temperature of about 2 eV and the floating wall. Calculate the electron density at the wall. What is the ion velocity at the wall.

Section 5

1. Calculate the electron current density due to thermoionic emission from the copper electrode as a function of surface temperature in the range of $1500-4000$ K.

2. Equilibrium vapor pressure for copper (in mmHg) can be estimated as:
$P = \exp(A - (B/T))$, where $A = 9.232$; $B = 1.726 \times 10^4$.

Calculate ablation rate using the Langmuir formula in the case of surface temperature of about 2500 K.

3. Compare the ablation rate calculated by Langmuir formula with the one predicted by the kinetic model. Use assumption that the normalized velocity at the edge of the Knudsen layer is about 1. Consider copper surface and surface temperature in the rage of 2000−4000 K.

4. Calculate contribution of the Shottky effect on the thermionic current in the case of titanium electrode. Consider surface temperature in the rage of 2500−4000 K and electric field in the rage of 10^7−10^9 V/m.

References

[1] H.D. Mott-Smith, History of "Plasmas", Nature, 233 (1971) 219.

[2] J.D. Lawson, Some criteria for a power producing thermonuclear reactor, Proc. Phys. Soc. B 70 (1957) 6.

[3] G. Fridman, G. Friedman, A. Gutsol, A.B. Shekhter, V.N. Vasilets, A. Fridman, Applied plasma medicine, Plasma Proc. Polym. 5 (2008) 503−533.

[4] M.G. Kong, G. Kroesen, G. Morfill, T. Nosenko, T. Shimizu, J. van Dijk, et al., Plasma medicine: an introductory review, New J. Phys. 11 (2009) 115012.

[5] C.D. MsscCaig, A.M. Rajnicek, B. Song, M. Zhao, Controlling cell behavior electrically: current views and future potential, Physiol. Rev. 85 (2005) 943−978.

[6] E. Stoffels, I.E. Kieft, R.E.J. Sladek, L.J.M. van den Bedem, E.P. van der Laan, et al., Plasma needle for in vivo medical treatment: recent developments and perspectives, Plasma Sources Sci. Technol. 15 (2006) S169−S180.

[7] M. Keidar, R. Walk, A. Shashurin, P. Srinivasan, A. Sandler, S. Dapgupta, et al., Cold plasma selectivity and the possibility of a paradigm shift in cancer therapy, Br. J. Cancer 105 (2011) 1295−1301.

[8] E.M. Wescott, D. Sentman, D. Osborne, D. Hampton, M. Heavner, Preliminary results from the Sprites94 aircraft campaign, 2, blue jets, Geophys. Res. Lett. 22 (1995) 1209−1212.

[9] Yu.P. Raizer, G.M. Milikh, M.N. Shneider, Streamer and leader-like processes in the upper atmosphere: models of red sprites and blue jets, J. Geophys. Res. A (2010) 115.

[10] F.M. Penning, Die Naturwissenschaften, Über Ionisation durch metastabile Atome 15 (1927) 818.

[11] W. Sesselmann, B. Woratchek, J. Kuppers, G. Ertl, H. Haberland, Interaction of metastable noble-gas atoms with transition-metal surfaces: Resonance ionization and Auger neutralization, Phys. Rev. B 35 (1987) 1547.

[12] L.N. Kantorovich, A.L. Shluger, P.V. Sushko, A.M. Stoneham, The prediction of metastable impact electronic spectra (MIES): perfect and defective MgO(001) surfaces by state-of-the-art methods, Surf. Sci. 444 (2000) 31.

[13] P. Stracke, F. Wiegershaus, S. Krischok, V. Kempter, Formation of the nitrogen shape resonance in N + 2 and N*2(A$_3\Sigma$+u) collisions with metallic surfaces, Surf. Sci. 396 (1998) 212.

[14] S. Sakabe, Y. Izawa, Simple formula for the cross sections of resonant charge transfer between atoms and their positive ions at low impact velocity, Phys. Rev. A 45 (1992) 2086−2089.

[15] J.D. Huba, NRL Plasma Formulary, 2009.

[16] J.J. Thomson, Ionization by moving electrified particles, Phys. Mag. 23 (1912) 449.

[17] Y. Itikawa, Cross sections for electron collisions with nitrogen molecules, J. Phys. Chem. Ref. Data 35 (1) (2006).

[18] L. J. Kiefer, G. H. Dunn, Electron impact ionization cross-section data for atoms, atomic ions, and diatomic molecules: I. Experimental data, Rev. Mod. Phys. 38, 1 (1966).

[19] D. Bohm, E.P. Gross, Theory of plasma oscillations, Phys. Rev. 75 (12) (1949) 1864−1876.

[20] A.Y. Wong, R.W. Motley, N. D'Angelo, Landau damping of ion acoustic waves in highly ionized plasmas, Phys. Rev. 133 (1964) A436.

[21] F. Chen, Plasma physics and controlled fusion, Plasma Physics, vol. 1, Springer, New York, 2006.

[22] T.H. Stix, Waves in Plasmas, AIP, New York, NY, 1992.

[23] L. Landau, About oscillations of electron plasma, J. Exp. Theor. Phys. 16 (1946) 574.

[24] J.H. Malmberg, C.B. Wharton, Dispersion of electron plasma waves, Phys. Rev. Lett. 17 (1966) 175.

[25] L. Tonks, I. Langmuir, A general theory of the plasma of an arc, Phys. Rev. 34, 876, 1929.

[26] D. Bohm, in: A. Guthry, R.K. Wakerling (Eds.), The Characteristics of Electrical Discharges in Magnetic Field, McGraw-Hill, New York, NY, 1949.

[27] K.U. Reimann, The influence of collisions on the plasma−sheath transition, Phys. Plasmas 4 (11) (1997) 4158−4166.

[28] V. Godyak, N. Sternberg, On the consistency model of collisionless sheath model, Phys. Plasmas 9 (11) (2002) 4427−4430.

[29] N. Sternberg, V. Godyak, On asymptotic matching and the sheath edge, IEEE Trans. Plasma Sci. 31 (4) (2003) 665−677.

[30] M.S. Benilov, Method of matched asymptotic expansions versus intuitive approaches: calculation of space-charge sheaths, IEEE Trans. Plasma Sci. 31 (4) (2003) 678−690.

[31] R.N. Franklin, You cannot patch active plasma and collisionless sheath, IEEE Trans. Plasma Sci. 30 (1) (2002) 352−356.

[32] J.E. Allen, Comments on "On the consistency model of collisionless sheath model", Phys. Plasmas 9 (2002) 4427.V. Godyak, N. Sternberg, Response to "On the consistency model of collisionless sheath model", Phys. Plasmas 10 (2003) 1528.

[33] D.D. Tskhakaya, P.K. Shukla, Comments on "On the consistency model of collisionless sheath model", Phys. Plasmas 9 (2002) 4427.V. Godyak, N. Sternberg, Response to "On the consistency model of collisionless sheath model", Phys. Plasmas 10 (2003) 3437.

[34] K.U. Riemann, Comments on "On the consistency model of collisionless sheath model", Phys. Plasmas 9 (2002) 4427.V. Godyak, N. Sternberg, Response to "On the consistency model of collisionless sheath model", Phys. Plasmas 10 (2003) 3432.

[35] K.U. Riemann, Comments on "On asymptotic matching and the sheath edge", IEEE Trans. Plasma Sci. 32 (2004) 2265−2270.V. Godyak, N. Sternberg, Reply to comments on "On asymptotic matching and the sheath edge", IEEE Trans. Plasma Sci. 32 (2004) 2271−2276.

[36] R.N. Franklin, Where is the sheath edge? J. Phys. D Appl. Phys. 37 (2004) 1342−1345.

[37] R.N. Franklin, The plasma−sheath boundary region, J. Phys. D Appl. Phys. 36 (2003) R309−320.

[38] M. Keidar, I.D. Boyd, I.I. Beilis, Plasma flow and plasma−wall transition in hall thruster channel, Phys. Plasmas 8 (12) (2001) 5315−5322.

[39] J.E. Allen, A note on the generalized sheath criterion, J. Phys. D 9 (1976) 2331.

[40] R.C. Bissell, The application of the generalized Bohm criterion to Emmert's solution of the warm ion collisionless plasma equations, Phys. Fluids, 30 (1987), 2265.

[41] G.D. Hobbs, J.A. Wesson, Heat flow through a Langmuir sheath in the presence of electron emission, Plasma Phys. 9 (1967) 85.

[42] U. Daybelge, B. Bein, Electric sheath between a metal surface and a magnetized plasma, Phys. Fluids 24 (1981) 1190.

[43] R. Chodura, Plasma–wall transition in an oblique magnetic field, Phys. Fluids 25 (1982) 1628.

[44] I.I. Beilis, M. Keidar, S. Goldsmith, Plasma-wall transition: The influence of the electron to ion current ratio on the magnetic presheath structure, Phys. Plasmas 4 (1997) 3461.

[45] I.I. Beilis, M. Keidar, Sheath and presheath structure in the plasma–wall transition layer in an oblique magnetic field, Phys. Plasmas 5 (1998) 1545.

[46] E.L. Murphy, R.H. Good, Thermionic emission, field emission, and the transition region, Phys. Rev. 102 (1956) 1464−1473.

[47] I.I. Beilis, Emission processes at the cathode of electric arc, Sov. Phys. Tech. Phys. 19 (1974) 257−260.

[48] P.H. Dawson, Secondary electron emission yields of some ceramics, J. Appl. Phys. 37 (1966) 3644−3665.

[49] A. Dunavesky, Y. Raitses, N.J. Fisch, Yield of secondary electron emission from ceramic materials of hall thruster, Phys. Plasmas 10 (2003) 2574.

[50] C.B. Ruchti, L. Niemeyer, Ablation controlled arc, IEEE Trans. Plasma Sci. 14 (1986) 423.

[51] R. Burton, P. Turchi, Pulsed plasma thrusters, J. Prop. Power 14 (1998) 716.

[52] M. Keidar, I.D. Boyd, I.I. Beilis, Electrical discharge in the teflon cavity of a coaxial pulsed plasma thruster, IEEE Trans. Plasma Sci. 28 (2000) 376.

[53] M.I. Boulos, P. Fauchais, E. Pfender, Thermal Plasmas: Fundamentals and Applications, vol. 1, Plenum, New York, NY, 1995.

[54] R.L. Boxman, P. Martin, D. Sanders (Eds.), Vacuum Arc Science and Technology, Noyes Publications, Park Ridge, NJ, 1995.

[55] I.I. Beilis, V.E. Ostashow, Model for a high-current discharge moving between parallel electrodes, High Temperatures. Vol. 27, No. 6, 1989, pp. 817−821.

[56] I.I. Beilis, Laser plasma generation and plasma interaction with ablative target, Laser Part. Beams 25 (01) (2007) 53−63.

[57] I. Langmuir, The Vapor pressure of metallic tungsten, Phys. Rev. 2 (1913) 329.

[58] R. Li, X. Li, S. Jia, A.B. Murphy, Z. Shi, Study of different models of the wall ablation process in capillary discharge, IEEE Trans. Plasma Sci., 38 (2010) 1033−1041.

[59] S.I. Anisimov, Evaporation of a light-absorbing metal, Sov. Phys. JETP 27 (1968) 182.

[60] I.I. Beilis, On the theory of the erosion processes in the cathode region of an arc discharge, Sov. Phys. Dokl. 27 (1982) 150.

[61] I.I. Beilis, Parameters of kinetic layer of arc discharge cathode region, IEEE Trans. Plasma Sci. 13 (1985) 288.

[62] I.I. Beilis, Theoretical modeling of cathode spot phenomena, in: R.L. Boxman, P. Martin, D. Sanders (Eds.), Vacuum Arc Science and Technology, Noyes Publications, Park Ridge, NJ, 1995.

[63] G.A. Bird, Molecular Gas Dynamics and the Direct Simulation of Gas Flows, Clarendon Press, Oxford, 1994.

[64] E.S. Oran, C.K. Oh, B.Z. Cybyk, Direct Simulation Monte Carlo: Recent advances and applications, Annu. Rev. Fluid Mech. 30 (1998) 403.

[65] M. Keidar, J. Fan, I.D. Boyd, I.I. Beilis, Vaporization of heated materials into discharge plasmas, J. Appl. Phys. 89 (2001) 3095.

[66] M. Keidar, I.D. Boyd, I.I. Beilis, On the model of Teè on ablation in an ablation-controlled discharge, J. Phys. D 34 (2001) 1675.

[67] M. Keidar, I.D. Boyd, I.I. Beilis, Model of an electrothermal pulsed plasma thruster, J. Propul. Power 19 (2003) 424.

[68] A Shashurin, M. Keidar, S. Bronnikov, R.A. Jurius, M.A. Stepp, Living tissue under treatment of cold plasma atmospheric jet, Appl. Phys. Lett 93(2008) 181501.

[69] I.I. Beilis, V.A. Bityurin, U.A. Vasiljeva, et al, "MHD energy conversion. Physical and technical aspects", In: V.A. Kirillin and A.E. Sheyndlin. (Eds.), Moscow: Nauka, 1983, p. 368 (in Russian).

[70] K. U. Riemann, Theory of the collisional presheath in an oblique magnetic field, Phys. Plasmas 1, 552, 1994.

[71] I.I. Beilis, "Modeling of the plasma produced by moderate energy laser beam interaction with metallic target: Physics of the phenomena", Laser and Particle beams V30, N3, 2012, 341−356.

Plasma Diagnostics

2.1 Langmuir probes

Probe diagnostics is probably the simplest, most reliable, and widely used technique to measure plasma parameters. Probes or Langmuir probes are widely used for almost a century in various plasma conditions. Probe diagnostics is very well established although there is no general approach for all possible conditions and there are limitations for probe applications in some settings such as magnetic field, collision-dominated conditions, and transient plasmas.

We will start with considering probe in stationary plasma without a magnetic field. Presence of the probe in plasma leads to perturbation due to particle flux originated from the plasma and recombination on the surface. The particle flux to the probe consists of ion and electron fluxes. The probe current is the sum of electron and ion currents. Since electrons are much faster, it will be electron current that will determine the current to the probe in the case when the probe potential is the same as the plasma potential. Variation of the probe potential leads to variation in the current in the probe circuit. The potential drop between the probe and the plasma is concentrated in the vicinity of the probe in the electrical sheath which is about the Debye length if probe potential is relatively small. A small fraction of potential drop at the probe ($\sim 0.5 T_e$) penetrates into the quasi-neutral plasma forming a presheath. Ions are accelerated in this presheath to satisfy the sheath criterion (see Section 1.3).

The classical theory of the Langmuir probe is built on the assumption that the motion of the particle in the sheath is collisionless [1]. Note that the collision-dominated regime will be considered separately below. Typical probe current–voltage characteristic in the collisionless regime is shown in Figure 2.1 in the semilogarithmic scale. When the probe potential is negative with $|U| \gg k_B T_e$, the current becomes independent on the potential. It is due to the fact that almost the entire electron current is reflected and the probe current consists of ion current only. In general, ion current to the probe depends on the relation between the probe characteristic size R_p and the Debye length L_D. Theory for the planar probe corresponds to the case with $R_p \gg L_D$. This is the theory that will be described in this section. When plasma is rarified so that $R_p \ll L_D$, the ion collection area increases significantly and ion current is calculated in the framework of the so-called orbital motion limit (OML) theory (that will be described below in Section 2.2).

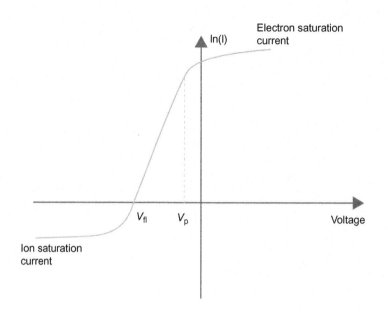

FIGURE 2.1

Current–voltage characteristics of the probe.

When the probe surface is large in comparison with the Debye length (i.e., thin sheath limit or planar probe), the ion current can be calculated according to Bohm sheath criterion (see Chapter 1):

$$I_i = 0.6enA\sqrt{\frac{k_B T_e}{M}} \tag{2.1}$$

where I_i is the ion current, e is the charge, n is the plasma density, and A is the collecting area, that is the same as the probe surface. As probe potential increases, more electrons start to contribute to the current. As a result, the probe current decreases and becomes zero. The potential that corresponds to zero probe current is called the floating potential (see Figure 2.1). This is the potential that insulated body in plasma will attain. The zero probe current means that the ion current I_i is equal to the electron current I_e, i.e.,

$$I_e = \frac{1}{4}en\left(\frac{8k_B T_e}{\pi m}\right)^{0.5} - \exp(-eV/k_B T_e) \tag{2.2}$$

Taking into account Eqs (2.1) and (2.2), the floating potential can be calculated as

$$V_{fl} = \frac{k_B T_e}{e}\ln\left[\left(\frac{M}{2\pi m}\right)^{0.5}\frac{1}{0.6}\right] \tag{2.3}$$

Further increase of the potential causes significant electron current to the probe which increases exponentially in accordance with Eq. (2.2). When probe potential reaches certain level, the

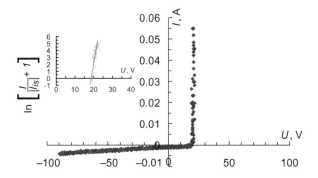

FIGURE 2.2

Voltage−current characteristics of single Langmuir probe at 0.1 Torr (helium, $I_{arc} = 60$ A, $t = 100$ ms after arc ignition r. The insert shows the magnified in vicinity of floating potential.

Source: *Reprinted with permission from Ref. [28]. Copyright (2011) by American Institute of Physics.*

current−voltage characteristics reduce its slope. This is the point when the potential difference between the plasma and the probe approaches zero, i.e., the probe potential approaches the plasma potential and the probe current consist of the saturated electron current. Thus, Langmuir probe can also be used to determine the plasma potential. Probe current continues to grow weakly with the probe potential increase due to increase in the effective collecting area of the probe as sheath thickness increases with the probe voltage. Let us now discuss the procedure of determining plasma parameters from the Langmuir probe characteristics.

Electron temperature can be obtained from the semilogarithmic plot (Figure 2.1) by taking logarithm of the expression for the electron current (Eq. (2.2)) in the form

$$\ln I = \ln\left(\frac{1}{4}en\left(\frac{8k_B T_e}{\pi m}\right)^{0.5} A\right) - eV/k_B T_e \tag{2.4}$$

Thus the slope of the current curve in the semilogarithmic plot (Figure 2.1) can be used to estimate the electron temperature. An example of current−voltage characteristic measured for helium arc discharge plasma is shown in Figure 2.2. The insert shows the semilogarithmic plot that allows estimating the electron temperature. The electron temperature determined from the slope of voltage−current characteristic in semilogarithmic scale was about 1−2 eV.

Electron density can be obtained from the ion saturation current, as the plasma is quasi-neutral. If probe potential is negative and much larger than the electron temperature, one can assure that the probe current is determined by the ion current only. In this case by using the measured probe current, one can calculate the electron density from Eq. (2.1). It is important to determine the probe collecting area. When the sheath thickness is much smaller than the characteristics probe size (i.e., flow is one dimension), the collecting area can be considered equal to the probe area and the planar probe approximation can be used.

2.2 Orbital motion limit

This is one of the important approaches that concerns with collection of the current by spherical and cylindrical probes. OML term was introduced by Langmuir and Mott-Smith [2]. They developed theory for particle collection in this limit for both monoenergetic and Maxwellian ion energy distributions. The theory is limited for a small probe with respect to the Debye length, i.e., $R_p \ll L_D$, where R_p is the probe characteristic dimension and L_D is the Debye length. The model considers the ion motion in the sheath around the probe by taking into account the ion and electron momentum equation coupled with Poisson equation. Those equations are solved to obtain the potential distribution and current−voltage characteristic of the probe. A method for solving the probe problem by analyzing all possible orbits of the charged particles and the effect of effective potential barriers in the sheath around spherical probes immersed in low-density plasmas was developed by Bernstein and Rabinowitz [3]. To make their calculation, they assumed of monoenergetic ions. Laframboise [4] solved this problem in the case of a Maxwellian velocity distribution and finite current collection for both ions and electrons.

The electron and ion densities and currents depend on the potential distribution in the sheath around the probe, which is governed by the Poisson equation:

$$\frac{d^2\varphi}{dr^2} = \frac{e}{\varepsilon_0}[n_e(r) - n_i(r)] \tag{2.5}$$

In this equation, n_e and n_i are electron and ion densities dependent on the radial position inside the sheath. The following boundary conditions are employed for this equation:

$$\varphi(r = R_p) = \varphi_p$$

$$\varphi(r \to \infty) = 0 \tag{2.6}$$

$$\frac{d\varphi}{dr}(r \to \infty) = 0$$

Electron and ion densities and fluxes depend on the energy distribution in the plasma bulk. Let us assume that both electrons and ions are described by a Maxwellian energy distribution function in the unperturbed plasma. Two constants of motion, i.e., total energy E and the angular momentum L determine the electron and ion dynamics in the sheath. Two-body central force problem can be reduced to an equivalent one-dimensional problem so that only radial component of the velocity has to be considered. This can be done when an effective potential energy is introduced as

$$\varphi_{eff} = \frac{L^2}{2mr^2} + q\varphi(r) \tag{2.7}$$

where L is the angular momentum of the ion, i.e., $L = mrV_\theta$, where V_θ is the angular velocity. Effective potential distribution is shown in Figure 2.3 with angular momentum L as a parameter.

Analysis of particle trajectories allows calculation of the local density. This is performed using the condition that particle can overcome the repelling potential if its kinetic energy is larger than

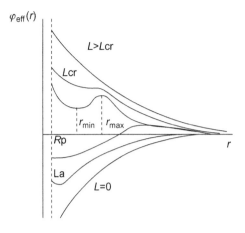

FIGURE 2.3

Effective potential distribution in the sheath around the probe. For detailed description, see Ref. [5].

the effective potential, i.e., $E \geq \varphi_{\mathrm{eff}}(r)$. This procedure allows calculation of the electron and ion densities as a function of the radial position. Both electron and ion density dependence on the potential and radial position can be calculated by integrating the distribution function:

$$n_\alpha = \int_{\varphi_{\mathrm{eff}}}^{\infty} f_\alpha(V) dV \tag{2.8}$$

$$j_\alpha = \int_{\varphi_{\mathrm{eff}}}^{\infty} f_\alpha(V) V \, dV \tag{2.9}$$

where $f_\alpha(V)$ is the velocity distribution function.

It should be pointed out that the integration depends on the potential distribution, i.e., whether the potential changes with distance from the probe r faster or slower than $1/r^2$. Detailed analyses of this situation were performed elsewhere [5]. Considering Maxwellian velocity distribution, ion current density in the case of the OML was calculated by integrating Eq. (2.9) using effective potential (Figure 2.3) as

$$j_\mathrm{i} = j_{\mathrm{i}0}\left(1 + \frac{eV_\mathrm{p}}{k_\mathrm{B}T_\mathrm{e}}\right) \tag{2.10}$$

where $j_{\mathrm{i}0}$ is the thermal ion current density from the plasma and V_p is the probe potential with respect to the plasma. One can see that in the case of the OML, the ion saturation current increases by factor that is approximately proportional to the ratio of the probe voltage to the electron temperature as shown in Figure 2.4.

As it was mentioned above, the orbital-motion-limited current corresponds to the current to the probe in the limit $R_\mathrm{p}/\lambda_\mathrm{D} \to 0$. Calculations performed demonstrated that this is a more precise approach for the case of a spherical probes than for a cylindrical probe.

Note: While the conditions in a plasma rarely correspond to the strong limit of the OML applicability, the OML results have still significance in that they provide an upper bound for the current that can be collected by the probe in the collisionless regime of probe operation.

2.3 Langmuir probes in collisional-dominated regime

There is certain controversy with respect to the application of Langmuir probes to the highly collisional atmospheric plasma when electron mean free path is much less than Debye length. The argument about applicability of the Langmuir probe technique for the plasma diagnostics in collision-dominated conditions is based on the fact that presence of collisions in the Debye sheath does not cancel Boltzmann distribution for plasma electrons in the sheath (for probe potentials in vicinity and lower than floating potential) [6−8]. As such the probe can be used to measure the electron temperature. In order to evaluate electron density from the probe current−voltage characteristic, one needs to invoke a model of the ion and electron transport across the sheath. In contrast with collisionless sheath in which particle transport across the sheath is determined by the flux from the plasma, transport across the collisional sheath is due to diffusion. In addition, the model should be coupled with the plasma model [9].

Typically local thermodynamical equilibrium (LTE) is reached at pressure of about 1 atm for most gases and any arc currents except for inert gases. For example, in helium (∼1 atm), arc electron temperature exceeds about twice the gas temperature for arc currents up to 100 A and LTE is reached only for arc currents larger than 200 A [9]. LTE allows simplifying significantly the analysis of Langmuir electrostatic probe operation due to availability of additional relation between plasma density and temperature (Saha's equation).

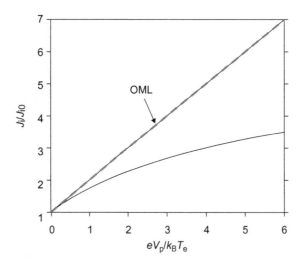

FIGURE 2.4

Dependence of the ion current on the voltage in OML limit and ion current calculated from Eq. (2.8).

2.4 **Emissive probe**

As it was mentioned above, the plasma potential can be determined by the change in the current curve slope as shown in Figure 2.1. However such definition sometimes has large uncertainty due to the expansion of the region with transient from linear dependence of $\ln I$ on V to saturated current dependence. The character of change the voltage−current curve slope depends on plasma parameters including plasma pressure. Such approach is not rigid as it is based on visualization.

The plasma potential with relatively high accuracy can be measured by using the emissive probe. The emissive probe consists of two cold and hot parts used to measure the current−voltage characteristics. When the probe potential is positive with respect to the plasma, there is no difference between the hot and cold probe currents since secondary electrons are trapped by the electric field in the sheath. However, when the probe potential is negative with respect to the plasma, secondary electrons escape the sheath and there is a difference between the hot and cold probes. The potential at which the current−voltage characteristics between the hot and cold probes depart from each other is the plasma potential. The details of plasma diagnostics using the emissive probe was described in a recent review [10]. In some instances, *double* or *triple* probes can be used to obtain plasma parameters. These techniques are described in Ref. [11].

2.5 **Probe in magnetic field**

Presence of the magnetic field can affect the probe characteristics. Magnetic field constrains the charged particles move at different directions—across and along the magnetic field. Thus, the study requires two-dimensional treatment. Chen [12] has given a general treatment of the probe in the presence of a magnetic field. Cohen [13] considered probes in a magnetic field and Lam [14] considered the in a collisionless strongly ionized plasma in a strong magnetic field.

Depending on the orientation of the magnetic field with respect to the probe collecting surface parallel or perpendicular, electron temperature can be estimated as shown in Figure 2.5. Typically

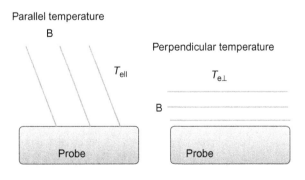

FIGURE 2.5

Probe in a magnetic field. If magnetic field is parallel to the wall, "perpendicular" temperature can be measured. If magnetic field intersects the wall, "parallel" electron temperature can be measured.

measured temperatures are different since electron distribution function in the magnetic field is typically anisotropic. Two-dimensional particle simulations of the probe in a strong magnetic field were performed [15] in order to describe magnetic field effect on the probe characteristics. The full particle orbits in the homogeneous magnetic field and the self-consistent electric potential are calculated with a particle-in-cell (PIC) code with two spatial coordinates and three velocity components ($2d$, $3v$). Simulations demonstrated that the ion trajectories are bent toward the normal to the probe surface so that the ion flow is focused on the edges of the probe. This leads to an enhancement of the ion current as compared to the current flowing in the flux tube subtended by the probe. As a consequence, the current does not saturate at large (negative) probe voltage and thus the effective probe size grows with increasing probe voltage.

It was illustrated that the magnetic field influences the electron density estimation from the probe measurements. A comparison was made of plasma density measurements obtained using cylindrical Langmuir probe and a highly sensitive microwave interferometer in a magnetic field as shown in Figure 2.6. The ratio of the plasma density measured by probe in a magnetic field to the plasma density obtained from interferometer is shown as a function of magnetic field and probe radius. One can see that this ratio decreases with the magnetic field increase. It is clear that when the plasma is magnetized, i.e., ion Larmor radius is comparable to the probe radius, the ion saturation current depends on the magnetic field [16]. If plotted against a product of magnetic field and probe radius, the ratio of the density obtained by probe to the interferometer-based measurement decreases and a systematic trend can be seen in Figure 2.6. One can conclude that probe measurements lead to underestimation of the electron density by a factor of 3 in a strong magnetic field.

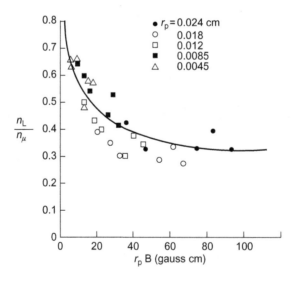

FIGURE 2.6

Ratio of the plasma density measured by probe to the plasma density obtained from interferometer.

Source: *Reprinted with permission from Ref. [16]. Copyright (1971) by American Institute of Physics.*

2.6 **Ion energy measurements: electrostatic analyzer**

To measure the ion energy distribution, one can use the electrostatic energy analyzer. The principle of the energy analyzer is very simple and it is based on the series of the retarding grids having different potential. Schematically energy analyzer is shown in Figure 2.7.

First grids are used to separate ion and electron components and to retard electrons. Grid 4 has a positive potential and is used to retard ions with energies lower than potential V on the grid. Last grid before collector is typically used to negate effect of the secondary electron emission.

A typical current−voltage characteristic obtained by retarding energy analyzer is shown in Figure 2.8.

In general, there are three ranges as depicted in Figure 2.8. The range A corresponds to the collector voltage much lower than the plasma potential. Thus all ions reaching the analyzer housing will be collected. In the transition range B, the ion current strongly decreases since the collector voltage approaches the plasma potential and ions are no longer accelerated toward the collector [17].

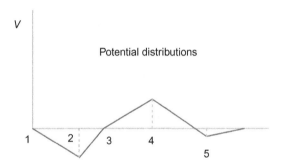

FIGURE 2.7

Potential distribution in retarding energy analyzer.

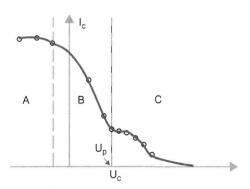

FIGURE 2.8

Typical current−voltage characteristics of the retarding energy analyzer.

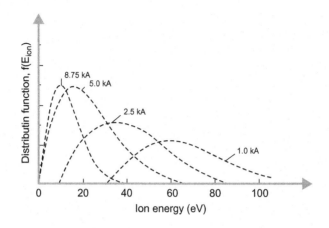

FIGURE 2.9

Ion energy distribution function measured by the REA with arc current as a parameter. After Ref. [17].

In the range C, ions are retarded since the collector potential is higher than the plasma potential. Only high-energy ions will reach the collector.

In general, ions are not monoenergetic and thus regarding curve appears in region C. The ion distribution function can be reconstructed by numerical differentiation of the current−voltage characteristics, i.e., [18]:

$$f(E_i) \sim \frac{\mathrm{d}I_c}{\mathrm{d}U_c} \tag{2.11}$$

Typical results are shown in Figure 2.9 illustrating ion energy distribution in a vacuum arc with arc current as a parameter.

2.7 HF cutoff plasma diagnostics

This diagnostics is based on the interaction of electromagnetic wave with plasma. Propagation of the electromagnetic wave in the plasma is determined by the dielectric constant of the plasma, ε which is in the absence of a magnetic field reads

$$\varepsilon = 1 - \frac{\omega_p^2}{\omega^2} \cdot \frac{1}{1 - j(\nu_c/\omega)} \tag{2.12}$$

where ν_c is the frequency of collisions, ω is the frequency of the electromagnetic wave.

Consider the range of parameters for which frequency of the electromagnetic wave is much larger than the collisional frequency in the plasma. In this case, dielectric constant of the plasma becomes

$$\varepsilon = 1 - \frac{\omega_p^2}{\omega^2} \tag{2.13}$$

where $\omega_p = \sqrt{(e^2 n_e/\varepsilon_0 m_e)}$ is the plasma frequency.

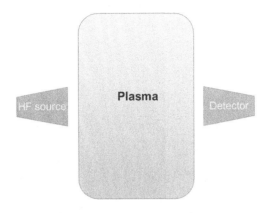

FIGURE 2.10

Schematics of the HF cutoff plasma density measurement.

One can see that dielectric constant of plasma ε become zero when $\omega = \omega_p$, and it is negative when plasma frequency is higher than the wave frequency. In other words, when electromagnetic wave reaches the plasma boundary, it will be reflected and does not penetrate the plasma. When high-density plasma is on the way of electromagnetic wave (i.e., if $\omega < \omega_p$), there is no signal that can be registered by detector (see schematics in Figure 2.10). Thus one can detect the plasma density when the signal stops reaching the detector. The electron density can be calculated as a critical density from the relationship $\omega = \omega_p$:

$$n_{cr} = \frac{\varepsilon_0 \omega^2 m_e}{e^2} \tag{2.14}$$

This technique is called the cutoff method for electron density detection.

Note: While this is the nonintrusive method (i.e., plasma does not be perturbed by the measurements), this is the technique that allows measurement of the average electron density without spatial resolution.

It should be noted that above described diagnostic technique can be used if the characteristic size of the plasma region is larger than the skin layer, i.e., $L \gg \lambda_{skin}$. The wave decay occurs within the skin layer (see Section 1.4).

2.8 Interferometric technique

Detailed information about the plasma density and plasma region size can be obtained using the interference of two electromagnetic energy beams produced by high-frequency (HF) source. The direct electromagnetic beam from the source is conducted through the plasma, while the second signal is received by detector via branching waveguide. Schematically this setup is shown in Figure 2.11. Attenuator is used to assure that the amplitude of both signals will be the same without the plasma. In addition, phase shift is set in the absence of plasma. If phase shift is π, the signal

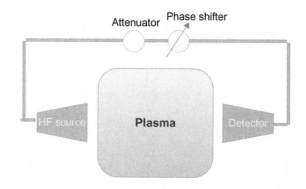

FIGURE 2.11

Schematics of the interferometer.

received by the detector will be zero. Phase shift setting can be controlled using the phase shifter as shown in Figure 2.11 or by changing the distance between the source and the detector.

Plasma leads to additional phase shift due to change of the reflection index of the media. In other words:

$$E_1 = E_0 \, \text{expi}(\omega t + \varphi(n)) \tag{2.15}$$

$$E_2 = E_0 \, \text{expi}(\omega t + \pi) \tag{2.16}$$

When plasma is generated between detector and source, the interferometric signal will appear. With plasma density increase, the amplitude of this signal decreases and, finally, when plasma density reaches the critical value defined by Eq. (2.14), there is only one signal from the direct waveguide.

The simplest model is to consider the uniform plasma with sharp boundaries. To this consideration or determination, the plasma density can be expressed as

$$n = n_{\text{cr}} \left[1 - \left(1 - \frac{\lambda k}{L} \right) \right] \tag{2.17}$$

where λ is the wavelength, k is the number of peaks in the interferometry signature, and L is the plasma dimension. It follows that near the critical plasma density corresponds to the cutoff, there will be maximum number of peaks, i.e., $k = L/\lambda$.

Note: It should be pointed out that using the interferometry, one needs to take into account that additional peaks appear due to interface with two different refraction indexes (i.e., class walls of the chamber). As such a simple interferometric approach introduces relatively large uncertainty in the plasma density measurements.

As an example, application of soft X-ray interferometer for measurements of the plasma density is shown in Figure 2.12. For a plasma with electron density, n_e, the index of refraction integrated along the column length in one arm of the interferometer will introduce a fringe shift in the interferogram as shown in Figure 2.12.

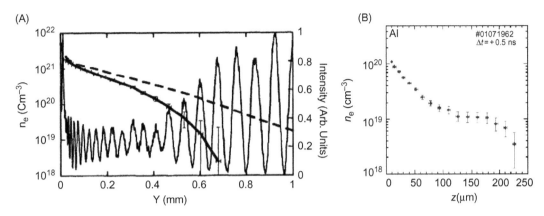

FIGURE 2.12

(A) X-ray laser interferometer probing a 0.5 cm Al plasma. (B) Electron density versus axial distance from the target obtained from interferogram shown in (A).

Source: Reprinted with permission from Ref. [29]. Copyright (1995) by the American Physical Society.

2.9 Optical measurements and fast imaging

Emission spectroscopy is a technique which examines the wavelengths of photons emitted by atoms or molecules during their transition. Each element emits a characteristic set of discrete wavelengths according to its electronic structure. Thus by observing these wavelengths, the elemental composition of the sample can be determined. Optical emission spectroscopy (OES) is widely used for characterization of the atmospheric pressure plasmas. Plasma composition and temperature at various modes (translational, rotational, vibrational) can be measured using this technique. Note that OES technique assumes that plasma obeying the LTE and thus possible deviations from the LTE will be detected by comparison of OES results with Langmuir probe measurements. An example of atmospheric plasma analysis using the OES is shown in Figure 2.13. Special resolution of various species can be achieved as well as shown in Figure 2.13B. Cold plasma jet that used helium as a carrier gas was studied. One can see that interaction of the helium jet with air leads to formation of reactive oxygen and reactive nitrogen species.

Fast imaging is another related diagnostic that is used to study atmospheric plasmas, in particular, fast phenomena. An example of imaging of the same atmospheric plasma jet is shown in Figure 2.14 using high-speed ICCD camera. One can see that propagation of the ionization front can be resolved. This phenomenon is dubbed in the literature as plasma "bullets."

2.10 Plasma spectroscopy

What distinguishes plasma is the glow which can be characterized by intensity, wavelength, etc. Radiation from the plasma provides important information about the plasma parameters and, thus, the goal of spectroscopic diagnostic is to interpret the emission from the plasma in terms of plasma properties. Two approaches, *passive* and *active* which are nonintrusive in nature, were used to study plasmas.

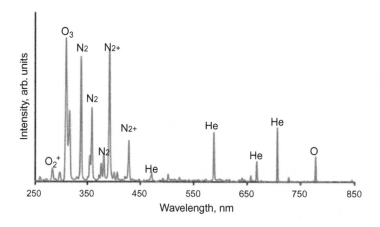

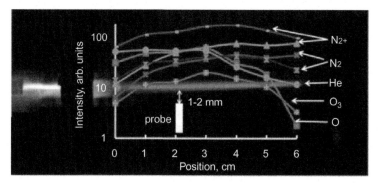

FIGURE 2.13

Optical spectrum of the atmospheric plasma jet and measurements set up.

Passive spectroscopy is based on the analysis of the radiation from plasma. The spectrum is composed of lines and continuum. The plasma that emitted this spectrum was composed of single- and multiple-charged ions, free electrons, and atoms. Atoms and partially stripped ions immersed in the plasma emit discrete spectra (lines) and continuous spectra. The structure of both is affected by interactions with the plasma particles and hence depends to some degree on the thermodynamic properties of those particles. Free electrons emit continuous spectra, which is also affected by interactions with plasma particles. An example of plasma emission signature is shown in Figure 2.15. It is shown that discrete spectrum is added to the free electron emission by recombination. Electron temperature can be estimated from the free-free electron emission spectra [19,20].

Laser-induced fluorescence (LIF) is an *active* spectroscopic technique that can be used to measure the structure of molecules, detection of selective species and flow visualization, and velocity measurements. Laser with a wavelength that has largest cross section is typically selected to be used to excite the species of interest. LIF technique works as follows. The beam from a tunable laser is passed through the plasma volume, exciting an electronic or vibrational transition of one of

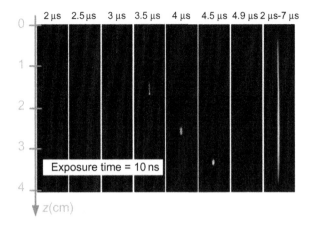

FIGURE 2.14

Series of instant photographs taken at different moments of time and indicating the propagation of ionization front. The photographs were taken with exposure time of 10 ns, except for the right one where 5 μs exposure time was used (2–7 μs exposure window). Moment of time $t = 0$ is associated with initiation of interelectrode discharge.

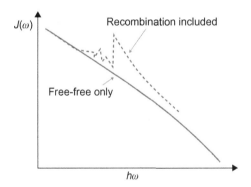

FIGURE 2.15

Emissivity of a plasma showing the addition of recombination to the free emission spectrum

Source: *After Ref. [19].*

the species. The fluorescence light is then collected. The velocity is determined by the Doppler shifted (see below) fluorescence peak emanating from the interrogated medium with respect to the commensurate lab-frame reference spectrum. The temperature is determined by the fluorescence spectra shape (width in particular) and the relative or absolute density is determined by the intensity of the fluorescence signal. The high spatial resolution of single-point LIF is essential in probing nonuniform plasma environments. Typically the signal-to-noise ratio of the fluorescence signal is very high, providing a good sensitivity to the process. It is also possible to distinguish between more species, since the lasing wavelength can be tuned to a particular excitation of a given species which is not shared by other species.

The motion of particles leads to the shift in the emission line. Let say that atom/ion has monochromatic line with wavelength λ_0. If the source of emission is moving toward the receiver, the receiver will register Doppler shortening of the wavelength by $\Delta\lambda = \lambda_0(v/c)$, where v is the velocity and c is the speed of light. Using LIF and Doppler broadening allows measuring velocity distribution. Lines splitting in an electric field (Stark broadening) can be used to measure the electric field distribution. Splitting of the lines in the presence of the magnetic field (Zeeman broadening) can be used to measure the magnetic field in plasmas.

2.11 Microwave scattering

Microwave diagnostic method for electron density measurements was recently predicted theoretically and then verified experimentally in Princeton University [21,22]. The method consists in the measurement of the scattering pattern from plasma objects being irradiated by the microwaves. The experimental setup uses 12.6 GHz Gunn-diode and homodyne detection system mixing the signals from local oscillator and from the scatterer. In the Rayleigh regime (when skin layer depth and microwave wavelength d_{skin}, $\lambda_{MW} >$ size of scatterer), the electric field amplitude of the scattered wave (E_s) is proportional to the conductivity and scatterer volume (V) or, if the scatterer is a collisional atmospheric plasma, to the total number of plasma electrons N_e: $E_s \propto \sigma V \propto N_e$. The method was successfully applied for temporally resolved measurements of small plasma objects having size of about 1 mm in length and about less than 10–100 mm in diameter, which correspond to the initial stages of laser-induced avalanche ionization in air and resonance-enhanced multiphoton ionization in argon. Microwave scattering method was also used to resolve temporal dynamics of cold atmospheric plasma jets [23]. Very recently scattering of microwave radiation on the nonequilibrium atmospheric plasma jet in order to measure the absolute plasma density was utilized [24].

The experimental microwave system is schematically presented in Figure 2.16. Two microwave horns with centers of exit sections located at $(x,y,z) = (6\ cm,0,0)$ and $(0,15\ cm,0)$ were used for radiation and detection of microwave signal. Microwave radiation was linearly polarized along the z-axis (12.6 GHz).

The output signal of the microwave system can be written as follows for the case of dielectric and conducting scatterers:

$$U = \begin{cases} A\sigma V, & \text{for conductor} \\ A\varepsilon_0(\varepsilon - 1)\omega V, & \text{for dielectric} \end{cases} \tag{2.18}$$

The coefficient A was determined to be about 11 V $\times$ W/cm^2 using the dielectric scatterers made of Teflon ($\varepsilon = 2.1$), alumina ($\varepsilon = 9.2$), polyethylene ($\varepsilon = 2.25$), and quartz ($\varepsilon = 3.8$) [25–27]. Then, plasma conductivity was determined based on Eq. (2.18), known coefficient A, and plasma volume. The temporal evolutions of average plasma density are presented in Figure 2.17 for two driven voltage amplitudes, 2.7 and 3.8 kV. It was observed that after discharge initiation, the plasma density reaches about $(5-10) \times 10^{13}$ cm^{-3} and then decays with characteristic times of few ms governed by electron attachment. The second peak of plasma density appears with a certain delay after decay of the first discharge (about 1 ms for $U_{HV} = 2.7$ kV) and indicates the presence of the second breakdown event.

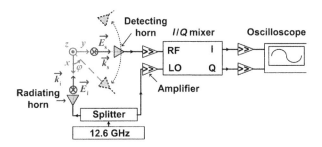

FIGURE 2.16

The schematics of Rayleigh microwave scattering (RMS) experimental setup.

Source: *Reprinted with permission from Ref. [24]. Copyright (2010) by American Institute of Physics.*

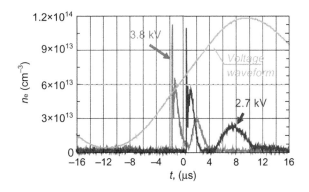

FIGURE 2.17

Temporal evolution of average plasma density in atmospheric plasma jet for U_{HV} = 2.7 and 3.8 kV.

Source: *Reprinted with permission from Ref. [24]. Copyright (2010) by American Institute of Physics.*

Homework problems

1. A probe with area 1 mm^2 is immersed in an argon plasma having electron temperature of about 3 eV. Calculate the plasma density if probe collects ion saturation current of about 10 mA.

2. Calculate the floating potential in the case considered in # 1.

3. Consider Xenon plasma with electron temperature of about 10 eV and plasma density of about 10^{16} m^{-3}. Spherical probe with radius of about 0.5 mm is used for measurements. Determine applicability of the OML approximation for plasma density measurements in such plasmas? Explain the answer.

4. Estimate the plasma density in the case of a spherical probe having radius of about 1 mm considering helium plasma with electron temperature of about 5 eV and probe saturation current of about 1 mA using the OML approach.

5. Using the cutoff plasma diagnostics plasma density was estimated to be 10^{17} m^{-3}. At what frequency of HF source this density was detected?

References

[1] P.M. Chung, L. Talbot, K.J. Touryan, Electric Probes in Stationary and Flowing Plasmas: Theory and Applications, Springer, New York, NY, 1975.

[2] H.M. Mott-Smith, I. Langmuir, The theory of collectors in the gaseous discharges, Phys. Rev. 28 (1926) 727−763.

[3] I.B. Bernstein, I.N. Rabinowitz, Theory of electrostatic probes in a low density plasma, Phys. Fluids 2 (1959) 112−120.

[4] J. Laframboise, Theory of cylindrical and spherical Langmuir probes in a collisionless plasma at rest, Rarified Gas Dynamics, vol. II, Academic Press, New York, NY, 1966 22−44.

[5] M. Keidar, I. Beilis, R.L. Boxman, S. Goldsmith, Non-stationary macroparticle charging in an arc plasma jet, IEEE Trans. Plasma Sci. 23 (6) (1995) 902−908.

[6] M.S. Benilov, Can the temperature of electrons in a high-pressure plasma be determined by means of an electrostatic probe? J. Phys. D Appl. Phys. 33 (2000) 1683−1696.

[7] M.T.C. Fang, J.L. Zhang, J.D. Yan, On the use of Langmuir probes for the diagnosis of atmospheric thermal plasmas, IEEE Trans. Plasma Sci. 33 (4) (2005) 1431−1442.

[8] E. Leveroni, E. Pfender, Electric probe diagnostics in thermal plasmas: double probe theory and experimental results, Rev. Sci. Instrum. 60 (12) (1989) 3744−3749.

[9] Y.P. Raizer, Gas Discharge Physics, Springer, Berlin, 1991.

[10] J.P. Sheehan, N. Hershkowitz, Emissive probes, Plasma Sour. Sci. Technol. 20 (2011) 063001.

[11] S. Chen, T. Sekiguchi, Instantaneous direct-display system of plasma parameters by means of triple probe, J. Appl. Phys. 36 (8) (1965) 2363−2375.

[12] F.F. Chen, Plasma Diagnostics Techniques, Academic Press, New York, NY, 1965.

[13] I.M. Cohen, Diffusion of Precursor Ions and Electrons Upstream of a Strong Shock Wave Propagating down a Shock Tube, Phys. Fluids 12 (1969) 361; I.M. Cohen, Transient current overshoot to electrostatic probes in continuum, slightly ionized plasmas, Phys. Fluids 16 (1973) 700.

[14] S.H. Lam, in: Dynamics of Ionized Gases, University of Tokyo Press, Tokyo, 1973.

[15] A. Bergmann, Two-dimensional particle simulation of Langmuir probe sheaths with oblique magnetic field, Phys. Plasmas 1 (1994) 3598.

[16] I.G. Brown, A.B. Campher, W.B. Kunkel, Response of a Langmuir Probe in a Strong Magnetic Field, Phys. Fluids 14 (1971) 1377.

[17] C. Rusteberg, M. Lindmayer, B. Juttner, H. Pursch, On the ion energy distribution of high current arcs in vacuum, IEEE Trans. Plasma Sci. 23 (1995) 909.

[18] I.I. Beilis, R.L. Boxman, S. Goldsmith, V.L. Paperny, Radially expanding plasma parameters in a hot refractory anode arc, J. Appl. Phys. 88 (N11) (2000) 6224−6231; A. Shashurin, I.I. Beilis, R.L. Boxman, Angular distribution of ion current in a vacuum arc with a refractory anode, Plasma Sources Sci. Technol. 17 (2008)015016 (11pp).

[19] G. Bekefi, Radiation Processes in Plasmas, Wiley, New York, NY, 1966.

[20] I.H. Hutchinson, Principles of plasma diagnostics. Cambridge University Press, Cambridge UK, 1987.

[21] M.N. Shneider, R.B. Miles, Microwave diagnostics of small plasma objects, J. Appl. Phys. 98 (2005) 033301.

[22] Z. Zhang, M.N. Shneider, R.B. Miles, Coherent microwave rayleigh scattering from resonance-enhanced multiphoton ionization in argon, Phys. Rev. Lett. 98 (2007) 265005.

[23] A. Shashurin, M.N. Shneider, A. Dogariu, R.B. Miles, M. Keidar, Temporal behavior of cold atmospheric plasma jet, Appl. Phys. Lett. 95 (2009) 231504.

[24] A. Shashurin, M.N. Shneider, A. Dogariu, R.B. Miles, M. Keidar, Temporary-resolved measurement of electron density in small atmospheric plasmas, Appl. Phys. Lett. 96 (2010) 171502.

[25]   P.A. Rizzi, Microwave Engineering: Passive Circuits, Prentice Hall, Englewood Cliffs, NJ, 1988.

[26]   S.Y. Liao, Microwave Devices and Circuits, Prentice Hall, Englewood Cliffs, NJ, 1980.

[27]   W.H. Sutton, Microwave Processing of Ceramic Materials, Ceram. Bull. 68 (1989) 376.

[28]   A. Shashurin, J. Li, T. Zhuang, M. Keidar, I.I. Beilis, Application of electrostatic Langmuir probe to atmospheric arc plasmas producing nanostructures, Phys. Plasmas 18 (2011) 073505.

[29]   L.B. Da Silva, T.W. Barbee, R. Cauble, P. Celliers, D. Ciarlo, S. Libby, et al., Electron density measurements of high density plasmas using soft X-ray laser interferometry, Phys. Rev. Lett. 74 (20) (1995) 3991−3994.

Electrical Discharges

3.1 Electrical breakdown and Paschen law

Electrical breakdown in low-pressure gases is characterized by the short transition time from an insulating interelectrode gap to a relatively high electrical conductivity gap. The transition time varies from ns to μs scale depending on gap size, electrode surface state, etc. (Figure 3.1).

Let us consider phenomena of the electrical breakdown in a low-pressure gas. When a high voltage is applied to the gap, the primary electrons flow in direction from cathode to the anode, multiplying new electrons in an avalanche by the atom ionization. The flux of the electrons Γ_e in any given point x is proportional to the primary flux:

$$d\Gamma_e = \alpha(x)\Gamma_e\,dx \qquad (3.1)$$

Integrating Eq. (3.1) yields the electron flux at point x as $\Gamma_e(x) = \Gamma_e(0)\exp\left[\int_0^x \alpha(x')dx'\right]$, where $\Gamma_e(0)$ is the electron flux at $x = 0$ that is determined at the cathode surface, $\alpha(x)$ is the inverse ionization mean free path, i.e., $\alpha(x) = (1/\lambda_{iz})$ and named first Townsend coefficient.

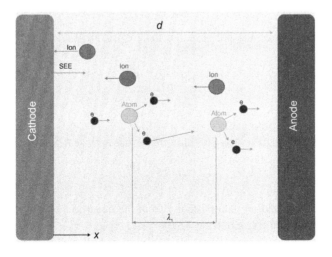

FIGURE 3.1

Schematics of the electron avalanche and gas breakdown.

Recall that if electrical discharge is developed, the current along the interelectrode gap must be constant so that

$$\Gamma_i(0) + \Gamma_e(0) = \Gamma_i(d) + \Gamma_e(0) \exp\left[\int_0^x \alpha(x')\,dx'\right] \qquad (3.2)$$

where Γ_i is the ion flux and d is the interelectrode distance.

Let us determine the condition when an electrical discharge can be self-sustained. The analysis will take into account the secondary electron emission (SEE) by ions. This effect is characterized by coefficient γ_{se} which shows the number of secondary electron emitted by one ion. The secondary electron flux from the cathode surface interacted with ion flux $\Gamma_i(0)$ will be expressed as

$$\Gamma_e(0) = \gamma_{se}\Gamma_i(0) \qquad (3.3)$$

Assuming that the current at the anode is conducted only by electron flux, i.e.,

$$\Gamma_i(d) = 0 \qquad (3.4)$$

one can arrive at the following equality:

$$1 + \frac{1}{\gamma_{se}} = \exp\left[\int_0^x \alpha(x')\,dx'\right] \qquad (3.5)$$

This equality constitutes condition for the self-sustained discharge. In this particular case of constant electron energy, i.e., $\alpha = $ constant, taking the logarithm on both sides of Eq. (3.5) and integrating the Eq. (3.5) from cathode $x = 0$ to anode $x = d$, leads to

$$\alpha d = \ln\left(1 + \frac{1}{\gamma_{se}}\right) \qquad (3.6)$$

The coefficient α depends on the pressure and the electric field and can be expressed as it was proposed in Ref. [1]:

$$\alpha \sim \frac{1}{\lambda_e}\exp\left(-\frac{U_i}{E\lambda_e}\right) \qquad (3.7)$$

where λ_e is the electron mean free path.

Since mean free path depends on the pressure, this equation is usually rewritten as

$$\alpha = Ap\,\exp\left(-\frac{Bp}{E}\right) \qquad (3.8)$$

where A and B are constants and are taken usually from the experiment.

Let us substitute expression for α into Eq. (3.6):

$$Apd\,\exp\left(-\frac{Bpd}{V}\right) = \ln\left(1 + \frac{1}{\gamma_{se}}\right) \qquad (3.9)$$

The above expression determines the electrical breakdown voltage. The electrical breakdown of the gas leads to the self-sustained discharge. Voltage that satisfies the Eq. (3.9) is called breakdown voltage. The Eq. (3.9) can be solved for the breakdown voltage:

$$V_B = \frac{(Bpd)}{\ln(Apd) - \ln[\ln(1 + (1/\gamma_{se}))]} \tag{3.10}$$

One can see that the breakdown voltage depends on the product of pd. The dependence of the breakdown voltage on the parameter pd is called the Paschen law after the nineteenth century German scientist [2]. Paschen curve for various gases is shown in Figure 3.2 indicating a minimum of breakdown voltage.

The physics of the Paschen law could be understood as follows. Larger pressure or larger gap leads to decrease of the mean free path and thus energy that can be gained by electrons between collisions. Since the energy that gained by electrons between collisions should be greater than the ionization potentials, higher overall voltage is required. Smaller pressure or smaller gap leads to increase of the mean free path resulting in decrease of the number of collisions and thus ionization frequency minimizing electron multiplication. Therefore, again, larger overall voltage is required. The value of pd at which the breakdown voltage has minimum can be found by differentiating Eq. (3.10) and setting $(\partial V_B/\partial(pd)) = 0$:

$$(pd)_{min} = \frac{e}{A}\ln\left(1 + \frac{1}{\gamma_{se}}\right) \tag{3.11}$$

By substituting the expression for pd_{min} into Eq. (3.10), one can find the minimum breakdown voltage:

$$V_{BDmin} = \frac{eB}{A}\ln\left(1 + \frac{1}{\gamma_{se}}\right) \tag{3.12}$$

Let us consider an example of the breakdown in air. Using Eqs (3.9)–(3.12), one can calculate that $pd_{min} = 0.83$ Torr cm; $V_{Bmin} = 300$ V at 1 cm gap at the atmospheric pressure.

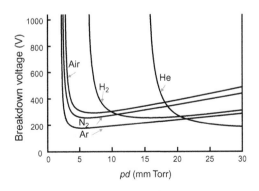

FIGURE 3.2

Pashen curve calculated for various gases.

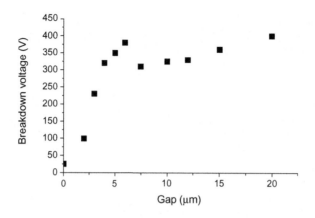

FIGURE 3.3

Electrical breakdown voltage in air as a function of contact gap. After Refs [3–5].

It should be pointed out that experimentally it was found that the breakdown voltage deviates from the Paschen law in the case of a small (micron size) gap between electrodes. In fact, experiments show that the breakdown voltage significantly decreases with the interelectrode gap length decrease contrary to Paschen curve predictions [3,4]. Note that the issue of breakdown at small gaps is very critical in micromechanical systems (MEMS).

Figure 3.3 shows that for gaps greater than 6 μm, the breakdown voltage is in the range 300–400 V. For contact gaps less than 4 μm, the breakdown voltage is a strong function of the contact gap [3–5]. One can see that if the contact gap is greater than 6 μm, the Paschen curve passes through the breakdown data as shown in Figure 3.2. On the other hand, for contact gaps less than about 4 μm, the voltage breakdown points are below the expected values predicted from the Paschen curve and decreases linearly with the gap length.

3.2 Spark discharges and streamer phenomena

Townsend breakdown mechanism relays on importance of ion SEE and as such its timescale is associated with the ion travel time. Experimental evidence suggests that the breakdown can occur very fast that cannot be explained by the Townsend mechanism described above. In addition, it was found that in some instances, breakdown does not depend on the cathode material, which is one of the significant factors in Townsend mechanism. In particular, this is the case of a large gaps and high pressure. The breakdown phenomena in these cases are associated with electron avalanches and streamers.

3.2.1 Electron avalanche

Consider the electron emission from the cathode and let us discuss the electron avalanche in the atmospheric condition as shown in Figure 3.4. Electrons will ionize gas if the ratio of electric field

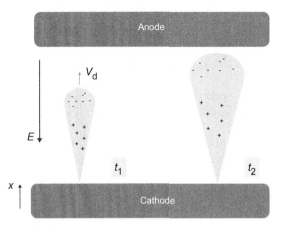

FIGURE 3.4

Schematics of electron avalanche development.

to pressure (E/p) is large. Two effects mainly determine the electron dynamics: ionization leading to electron multiplication and electron attachment (in particular to oxygen) leading to electron losses. Thus the electron avalanche driven by the applied electric field leads to formation of positive and negative ions. One-dimensional approach of the avalanche can be described by the following mathematical formulation.

Electron density distribution can be calculated as

$$\frac{dn_e}{dx} = (\alpha_i - a)n_e \tag{3.13}$$

where α, is the ionization rate and a is the electron attachment rate. Ion density distribution obeys

$$\frac{dn_i^+}{dx} = \alpha_i n_e \tag{3.14}$$

Negative ion density distribution can be calculated as

$$\frac{dn_i^-}{dx} = a n_e \tag{3.15}$$

Integration gives the following result for the electron, ion, and negative ion densities:

$$n_e = n_{eo}\exp[(\alpha_i - a)x] \tag{3.16}$$

$$n_i^+ = \frac{\alpha_i}{\alpha_i - a}(n_e - 1) \tag{3.17}$$

$$n_i^- = \frac{a}{\alpha_i - a}(n_e - 1) \tag{3.18}$$

Drift velocity V_d of electrons is determined by the mobility μ_e and the electric field E as

$$V_d = \mu_e E \tag{3.19}$$

The volumetric charge creates radial electric field. Presence of the radial component of the electric field leads to spreading of the electron cloud as shown schematically in Figure 3.4. Electron cloud spreading with time t is determined by the velocity V_d:

$$x_0 = V_d t \tag{3.20}$$

The ions remain practically fixed during the avalanche due to their low ion mobility. (*Note*: Compare Townsend theory discussed in Section 3.1 in which ion transport is important.) Density of ions is thus accumulated.

Space charge is generated in an avalanche as $\sim \exp(\alpha x)$. The average distance by which electrons are separated from the ions is approximated by the ionization length which is approximately α^{-1}. As a result, space charge forms a dipole with electrons being at the head and ions being behind. When the avalanche reaches the anode, the electrons sink into the metal and only a positive charge will remain. This is so called ion trail; it is shown schematically in Figure 3.5. The ion density is relatively weak except the near-anode region; thus presence of ions does not lead to breakdown. Due to photons created in the strongly ionized positive column, photoelectrons are generated forming secondary avalanches directed toward the column as shown in Figure 3.6. This effect is significant if electric field formed by the volumetric charge is comparable to the external electric field. Ions from the secondary avalanches increase the volumetric positive change that propagates toward the cathode as shown in Figure 3.6. The process develops as a self-propagating streamer.

3.2.2 Streamer mechanism

Streamer is the weakly ionized thin channel formed from primary avalanche in a sufficiently strong electric field. Streamer possesses conductivity, thus can lead to the discharge. Recall that the energetic photons generated in the primary avalanche can also produce secondary avalanches by

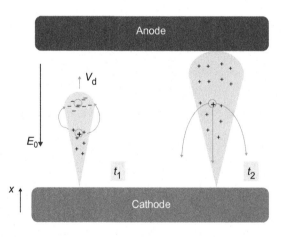

FIGURE 3.5

Schematics of shape and charge in electron avalanche at different moments of time.

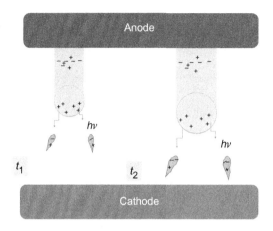

FIGURE 3.6

Cathode-directed streamer at different moments of time.

photoionization as shown schematically in Figure 3.6. The secondary avalanches are pulled into the ion trail due to self-produced electric field. Ions of the secondary avalanches enhance the positive charge at the cathode end leading to the attraction of electrons of the next generation of secondary avalanches. This is the scenario of the streamer growth. *Note*: If the streamer is formed while the avalanche has not yet gone far from the cathode, it will grow toward the anode. This is the anode-directed streamer (or negative streamer).

Streamer can be developed when space charge field at the avalanche head is comparable to the applied field, i.e., $E' \sim E_0$, where E_0 is the applied electric field and E' is the electric field of the space charge of avalanche.

Formation criteria for the streamer is thus can be formulated as

$$E' \approx E_0 \quad \text{and} \quad E' = \frac{e}{R^2} \exp[\alpha(E_0)x] \tag{3.21}$$

where R is the avalanche head radius. Loeb and Meeks [6] showed that in air, one can obtain the following expression for E':

$$E' = 5.27 \cdot 10^{-7} \frac{\alpha}{(x/p)^{0.5}} \exp[\alpha x] \quad (V/cm) \tag{3.22}$$

where p is the pressure. Meek's theory identifies the breakdown with the event of streamer formation. In the case of $d = 1$ cm, calculated parameters for the breakdown are as follows:

$$\alpha d = 18.6; \quad N_{ions} = 3.7 \times 10^{12} \text{ cm}^{-3}; \quad U_B = 32,200 \text{ V} \tag{3.23}$$

The approach resulted in Eq. (3.22) is approximate. In particular, when the gap is larger than 10 cm, the theory underpredicts the breakdown voltage. One reason is that in addition to condition (3.21), the number of changes needed in the streamer head is sufficient for photoionization. Loeb supplementary condition for the ion density in avalanche prior to the streamer formation,

$n_e \sim 10^{13}$ cm^{-3}. This is the necessary density for the photoionization. Raether [7] used more detailed analysis of the volumetric charge to calculate the radius of streamer.

3.3 Glow discharge

Glow discharge is an example of a self-sustained discharge that is supported by SEE from the cold cathode. Typically SEE is caused by the positive ion bombardment. Main feature of the glow discharge is the cathode layer, which is a large positive space charge near the cathode surface having a strong potential drop of about 100−400 V dependent on the gas pressure. Typically a negative space charge layer is formed near the anode. The middle interelectrode region consisting of weakly ionized plasma is called the positive column.

The approximate regions of the glow discharge are shown in Figure 3.7. In the first region immediately near the cathode, electrons emitted from the cathode gain energy in the electric field. No luminescence is observed there and this region is named as Aston dark layer. At the external boundary of Aston layer, the electron energy is sufficient to excite atoms that lead to formation of the cathode glow. The high electron density at the end of the cathode dark space leads to decrease in the electric field. As a result, the ionization rate decreases and radiation intensifies. The next region is called the negative glow, which is sharply separated from the dark cathode space. The dark region (called Faraday dark space) is formed after the negative glow due to fact that electrons lose their energy. As a result, plasma density decreases in the Faraday dark space and electric field increases, thus forming the positive column. Interesting to note that the cathode layer structure does not change when the anode approaches to the cathode, while the positive column length decreases.

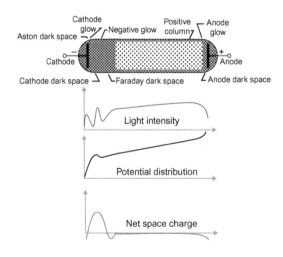

FIGURE 3.7

Characteristic regions of the glow discharge and Axial profiles of various glow discharge properties.

Closer to the anode, there are few distinct regions such as anode glow and anode dark space. The anode repels ions and attracts electrons that creates negative space charge and increases electric field in the near-anode region. Reduction of the electron density explains the dark space, while increase in the electric field explains the anode glow.

One can distinguish between the *normal glow* discharge and the *abnormal glow* discharge. In the case of the *normal glow* discharge, electrons are emitted from only a part of the cathode surface, and both current density and discharge voltage remain constant. An increase in current causes emission to occur from a larger part of the cathode. The abnormal glow discharge is observed at high values of current density. It is characterized by a rapid increase in the voltage between the electrodes as the current is increased.

3.3.1 **Cathode and anode regions**

Let us now consider the regions adjacent to cathode and anode of the glow discharge. In the cathode region, Townsend coefficient, and electric field are not constant. It is known from the above discussion that the secondary electrons created by ion impact are important to maintain the discharge current and sustain the glow discharge. The analysis of the cathode region is therefore similar to the analysis of the Townsend breakdown mechanism described in Section 3.1.

Typically the drift velocity of electrons in the electric field of the positive column is much less than their thermal velocity. In order to provide current continuity, it requires a retarding electric field in the near-anode region to reflect excess of electrons. Thus the interface of the plasma and anode sheath becomes positive with respect to the anode. This is the common situation in many types of discharges and we will describe it in details for the case of a vacuum arc.

3.3.2 **Positive column of the glow discharge**

It is known that the positive column of a glow discharge consists of a low ionized gas. The electrical current density is determined by the electric field, which is relatively small. In order to calculate the current density, one needs to know the plasma density. Thus, one of the main purposes of the modeling is to establish the plasma density distribution in the column. Usually the positive column in a glow is cylindrical symmetric. Therefore, the description can be started with diffusion equation in the cylindrical coordinate system:

$$-\nabla(D_a \nabla n) = \nu_{iz} n \tag{3.24}$$

where D_a is the ambipolar diffusion coefficient and ν_{iz} is the ionization rate. In cylindrical coordinate system, this equation will have the following form:

$$-\frac{d^2 n}{dr^2} = \frac{1}{r}\frac{dn}{dr} + \frac{\nu_{iz}}{D_a} n \tag{3.25}$$

The solution of this equation is the Bessel's function.

$n = n_0 J_0(\beta r)$, if standard boundary condition is used assuming that the plasma density vanishes at the wall, i.e., $n(R) = 0$, then

$$\beta = \sqrt{\frac{\nu_{iz}}{D_a}} = \frac{\chi_{01}}{R} \tag{3.26}$$

where J_o is the zero-order Bessel's function and $\chi_{01} = 2.405$ is the first zero of Bessel's function. The ionization rate depends on the electron temperature (see Chapter 1), i.e.,

$$\nu_{iz} = cp \exp\left(-\frac{U_{iz}}{T_e}\right) \tag{3.27}$$

where c is the coefficient. Substitute expression for the ionization rate (Eq. (3.27)) into Eq. (3.26) leads to the following equation:

$$\sqrt{\frac{cp \exp(-U_{iz}/T_e)}{D_a}} = \frac{\chi_{01}}{R} \tag{3.28}$$

In this equation, only electron temperature is unknown and therefore it can be solved for T_e. Thus, Equation (3.28) can be used to calculate the electron temperature.

Note: The boundary condition used (i.e., $n = 0$) is not self-consistent since finite flux of the particles will require infinite velocity. However, this approximation gives reasonably good estimation for the electron temperature.

Two other parameters of importance are electric field E and the plasma density at the axis n_o. Electric field along the axis of the discharge can be calculated by analyzing the power balance, i.e., power absorbed in the positive column versus power loss:

$$P_{inp} = P_{out} \tag{3.29}$$

$$P_{inp} = 2\pi \int_0^R jEr \, dr \tag{3.30}$$

$$P_{out} = 2\pi R \Gamma_r e W_T \tag{3.31}$$

where Γ_r is the radial particle flux $\Gamma_r = -D_a dn/dr$ and W_T is the total energy carried out per electron–ion pair generated. Assuming a constant electron mobility μ_e, one can calculate the current density as

$$j = en\mu_e E \tag{3.32}$$

The system of equations (3.30) and (3.31) with Eq. (3.32) can be solved for the electric field as follows:

$$E = \sqrt{\frac{\nu_{iz} W_T}{\mu_e}} \tag{3.33}$$

Discharge current can be calculated by integrating the current density:

$$I = \int_0^R 2\pi r e n_0 J_0(\beta r) \mu_e E \, dr \tag{3.34}$$

In this equation, discharge current is known, while electric field is determined by Eq. (3.33). Thus this equation can be solved for plasma density at the axis, n_0.

3.4 Arc discharges

A relatively low cathode potential drop characterizes the arc discharge, which is in the order of 2−3 times of ionization potentials of the atoms of cathode materials. This is what makes arc discharge different from the glow discharge in which cathode fall is in the order of hundreds volts (see Section 3.3). The small cathode fall in the arc is the result of the collective emission phenomena typical for arcs, i.e., thermoionic, field emission, and thermofield emission. Such emission mechanism can supply large electron current.

Typically arc current is large and in the order of 10−100 A in the arc deposition systems and up to 10^5 A in vacuum arc interrupters. Arc discharges are also characterized by strong heating of the electrodes and in particular cathodes due to large power deposition. This leads to significant electrode erosion. In fact, vacuum arc discharge is supported by metal vapor ionization, which is supplied by cathode erosion.

In this chapter, we consider two examples of the arc discharge, namely, atmospheric pressure arc and vacuum arc.

3.4.1 Atmospheric arc

A classical example of the arc discharge is the carbon atmospheric arc. Typically arc is started by separating initially contacting electrodes. If electrodes are arranged horizontally, the arc channel in the interelectrode gap is bended upward due to central part being floated as a result of the buoyancy force. This led to discharge from being dubbed as "arc."

Typical photograph of the arc discharge with the clearly visible plasma−gas interface is shown in Figure 3.8. $V-I$ characteristics of the arc are presented in Figure 3.9. Typically $V-I$ characteristic has V-type shape for all tested conditions. This is the typical $V-I$ dependence that is also predicted in simulations as shown in Figure 3.9.

3.4.1.1 Interelectrode plasma column of the arc discharge

Let us consider a long cylindrical plasma column (interelectrode plasma) in an axial electric field. A steady state arc is burning in a nonmoving gas enclosed in a cylindrical chamber with radius R that is cooled externally. Electric field uniform in both axial and radius directions is considered.

Equilibrium of the plasma column of the arc discharge is determined by the energy balance between the Joule heating and heat conductivity. The Joule heating is the energy source mechanism:

$$W = jE = \sigma E^2 \tag{3.35}$$

where σ is the electrical conductivity. Heat transfer leads to arc cooling, so the balance equation will have the following form:

$$-\frac{1}{r}\frac{d}{dr}(rq) + \sigma(T)E^2 = 0 \tag{3.36}$$

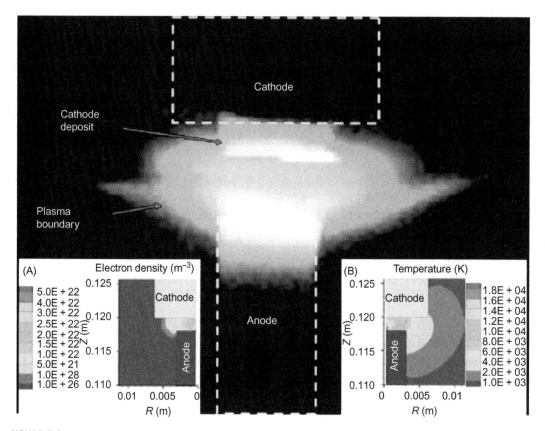

FIGURE 3.8

Photograph of the carbon arc discharge. Insets on the bottom of the figure show simulation results for (A) electron density and (B) temperature distribution from multispecies simulation.

Source: Reprinted with permission from Ref. [8]. Copyright (2010) by American Institute of Physics.

This equation is known as the Elenbass–Heller equation. The heat flux is calculated as

$$q = -\lambda\frac{dT}{dr} \tag{3.37}$$

where λ is the coefficient of heat conductivity. To solve Eq. (3.36), the condition that the discharge is symmetric was employed, i.e., $dT/dr = 0$ at $r = 0$. Additional boundary condition is given at the wall:

$$T = T_w \quad \text{at } r = R$$

where T_w is the wall temperature.

Total discharge current can be calculated as

$$I = E\int_0^R 2\pi\sigma r\,dr \tag{3.38}$$

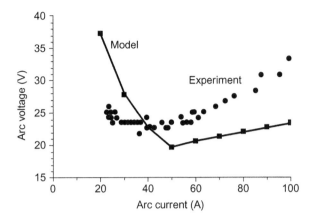

FIGURE 3.9

Calculated VC characteristics of the arc discharge and comparison with experimental data.

Source: Reprinted with permission from Ref. [8]. Copyright (2010) by American Institute of Physics.

Arc is the current-regulated discharge and as such the arc current I is controlled experimentally and thus can be considered as a prescribed known parameter of the problem.

In general, both electrical and thermal conductivities are highly nonlinear functions of the temperature. This makes this problem difficult to solve analytically. On the other hand, one can take into account the fact that the conductivity decreases quickly with temperature decrease, so that it can be expected that the conductivity zone radius is less than the R. In order to simplify the solution, the channel model can be invoked. The arc channel model was proposed by Steenbeck in 1932 [9]. In its essence, the model assumes that current vanishes outside of the channel at the axis and approximately constant in the current channel (conductive zone). In this approach, the total current was calculated using constant σ as

$$I = \sigma_{max} E \pi r_0^2 \tag{3.39}$$

where r_0 is the radius of the conductivity zone.

Two-zone approximation can then be formulated as follows:

$$r_0 < r < R \quad \frac{d}{dr}\left(r\lambda\frac{dT}{dr}\right) = 0 \tag{3.40}$$

$T_0 = T_{max}$ and $T_w = 0$. After manipulations, one can arrive to the following equation for the temperature:

$$\int_0^{T_{max}} \lambda \, dT = \frac{I^2}{\pi r_0^2 \sigma_{max}} \ln\left(\frac{R}{r_0}\right) \tag{3.41}$$

Thus it turns out that we have two equations (3.39) and (3.41) and three unknown (T_{max}, r_0, and E) and this system requires additional equation for closure.

Steenbeck proposed the so-called principle of minimum power, i.e., solution will correspond to $W_{min} = (IE)_{min}$. In other words, for given arc current and chamber radius R, the maximum temperature and the current channel radius must reach such value so that the power W be minimum. This allows for this problem closure.

Note: The principle of minimum is widely used; however, it was never demonstrated that it is followed from the fundamental physics laws. On the other hand, model predictions based on the Steenbeck principles in most cases agree well with experiment.

The problem of the arc column can be solved without invoking the principle of minimum. Such qualitative analysis was proposed by Raizer [11] indicating that the electrical power dissipated in the conductive zone losses by the heat conduction in outside zone. Taking into account this fact, Beilis and Sevalnikov in 1991 [12] developed an analytical model that allows calculating the voltage–current (VC) characteristic and the temperature distribution in the plasma column. The heat conduction equation was solved taking into account the exponential dependence of the electrical conductivity on the plasma temperature and by invoking the patching conditions for the temperature and heat flux at the interface between two zones (radius r_0). The solution depends on parameter $K = a^2/r_0^2$, where $a = 16lT_0^2/(U_i sE)$ is the radius where the heat conduction energy flux equal to electrical power, i.e., $a = r_0$ and $K = 1$. The calculations based on the aforementioned analytical model at $K = 1$ agrees well with the measured VC characteristics and the measured radial temperature distribution.

3.4.2 Vacuum arc

When voltage is applied to an electrode gap in vacuum, an electrical discharge is appeared. Such discharge can support a current from about 1A to few tens kA while the relatively low voltage (~ 20 V) between two electrons is established. This discharge is called the *vacuum arc* [13–15]. The conductive media is ejected in the form of plasma jets from localized small areas—named spots at the negative polarity electrode—named *cathode*. In the case of high-current (i.e., kA) arcs, the spot at the positive polarity electrode named *anode* can also appear. The voltage between the electrodes is not uniformly distributed with the majority of the voltage in the cathode region named *cathode potential drop* and the small fractions in the interelectrode plasma and in the near-anode region known as the *anode potential drop*.

3.4.2.1 Cathode region

3.4.2.1.1 Overview of the experiments

The cathode region consists of few or numerous cathode spots depending on arc current. The cathode spot appears as a very small luminous plasma object with chaotic and high mobility (Figure 3.10). The numerous observations indicate that the cathode spot can disappear and again appear with a characteristic spot lifetime depending on cathode material or cathode structure (i.e., whether cathode material is a bulk or it is covered with a thin film).

In the case of a low arc current ($I < 10$ A), the arc voltage oscillates at a high-frequency (MHz and more) and reaches peak, which is above the cathode potential drop. The minimum arc voltage is considered to be equal to the cathode potential drop U_c. The oscillation amplitude of the single spot decreases with the arc current. The plasma jets produced by the cathode spot consist of energetic ions with the jet velocity exceeding the local sound speed. Vacuum arc at the copper cathode is the most studied case of this type of discharge. Such arc has the following typical cathode spot parameters: cathode potential drop: $U_c = 13–19$ V , threshold arc current (i.e., minimum current at which arc can be sustained): 1–2 A, arc voltage $U_{arc} = 17–30$ V, jet velocity $V_j = 10^6$ cm/s, ion

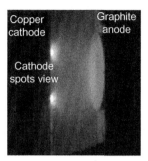

FIGURE 3.10

Photograph of the cathode region by multispot arc operation.

energy per unit charge $30-37$ eV, ion current fraction in the plasma jet $f \approx 0.1$, and erosion coefficient $G_k = 0.05-0.1$ mg/C.

The cathode spots (mainly cathode spots on a copper cathode were studied) were observed mainly by high-speed optical photography and they were classified by different types according to the spot velocity V_s, lifetime t_s, and current per spot I_s. High-speed moving cathode spots ($10-100$ m/s) having a short lifetime ($<10 \mu s$) are associated with type 1 spots. The spot current is relatively small, e.g., $I_s < 10$ A for type 1. The velocity of this spot type (which is of about $10-100$ m/s) can be one order of magnitude higher on the oxide surface in comparison to the clean cathode surface. Slow ($V_s \sim 0.1$ m/s) spots with lifetimes of about $100 \mu s$ and spot currents of about 10 A are defined as type 2 spots that preferably appear when a relatively low gas pressure (mTorr) is present. Several spots can appear together on the cathode surface with distances between them equal or smaller than the overall spot size. This collection of spots is called a "group spot" and appears when arc currents is ≤ 1000 A. The current per group spot in the case of a copper is in the range of $100-200$ A. Separate fragments were observed also in fast moving (type 1) spots. For these spots, the periodic fluctuations of the spot brightness, on a nanosecond timescale, were associated with the lifetime of the spot fragments. A fragment size of about $10 \mu m$ was found [16].

The craters on the cathode surface were observed after arcing. Size of such craters on bulk cathodes and tracks on thin-film cathodes were also studied. Craters with sizes of about $5-30 \mu m$ were observed on copper bulk cathode surfaces as well as on copper thin films on glass with thickness of about $d_f \sim 0.01-0.1 \mu m$. For film Cu cathodes, the following are typical parameters: spot velocity $\sim 10^3 - 10^4$ cm/s, erosion rate $G_f \sim 50-500 \mu g/s$, spot current $I_f \sim 0.1-0.9$ A, and track width $\sim 1-10 \mu m$. The spot current density was estimated from measurements of the optical images of the arc and the crater size to be $10^5 - 10^8$ A/cm^2 on bulk cathodes and about $10^5 - 10^6$ A/cm^2 on thin-film cathodes.

3.4.2.1.2 Principle state of the theory

The cathode spot supports the arc current between the solid material and the near surface cathode material which is in plasma state. Two main theoretical models were developed to explain the cathode spot phenomena: (i) Mesyats explosive model [17] and (ii) Beilis kinetic-vaporization model [18,19].

The first model, i.e., explosive mechanism is based on the premise that cathode spot injects a plasma plume in the interelectrode gap by explosion of a spike or protrusion at the cathode surface. The phenomena occur due to Ohmic overheating when a high-current density is conducted via the irregularities on the cathode surface. In principle, this model reduces to the following basic relation:

$$j_{sp}^2 t_{ex} = C_{ex} \qquad (3.42)$$

where j_{sp} is the current density through the spike, t_{ex} is the characteristic explosion time of the spike (or its part), and C_{ex} is a constant, which is about 1. The experiments show that for metals, the constant is about ~ 1 and the explosion occurs when j_{sp} is about 10^9 A/cm^2 and therefore the time t_{ex} is of order of 10^{-9} s. The mentioned high-current density is provided by field electron emission and can be supported by high electric field of about 10^8 V/cm during the spike. The explosion mechanism is mainly applicable to explain the phenomena during vacuum arc initiation by electrical breakdown of a high-voltage vacuum gap. However the explosive model cannot be applied to the arc after its initiation when the cathode spot phenomena are determined by the energy flux from the near cathode plasma [30].

The kinetic theory [18] consider the local intense cathode heating, the cathode material vaporization, dense plasma generation by the atom ionization near the surface, and the kinetic plasma flow. These processes determine the expanding plasma jet. The main mechanism that supports the current at the cathode–plasma interface is an electron emission from the cathode and ion flux from the plasma to the solid surface. The mathematical formulation of the problem considers three main regions as shown schematically in Figure 3.11:

1. Cathode body, where the energy balance between coming heat flux and energy losses due to cathode three-dimensional heat conduction, evaporation, and electron emission is taken into account. The electron emission is enhanced by a cathode electric field which appears in the electrical sheath at the plasma–cathode surface interface.
2. Kinetic or Knudsen nonequilibrium layer and electron relaxation region where the net of erosion mass is formed and the atoms are ionized.

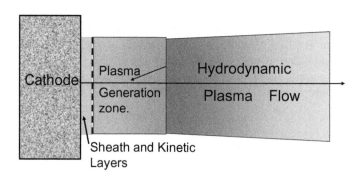

FIGURE 3.11

Schematic presentation of the kinetic-vaporization model.

3. Expansion region where a plasma gas dynamic flow, plasma acceleration, and plasma jet formation occur. The physical phenomena, assumptions, and the complicated system of equations in the near-cathode regions have been described in detail in Refs [18,19]. The phenomena related to the kinetics of material vaporization, sheath, and electron emission was described in Chapter 1. Below an example of calculations is illustrated in order to understand the behavior of plasma and cathode parameters for current characterized different cathode spots.

3.4.2.1.3 Cathode spot: results and calculations

In the framework of the kinetic theory, the system of equations allows description of the cathode spot parameters in a self-consistent manner, i.e., without arbitrary or fitting values which often employed to close the problem formulation. The results of self-consistent calculations are presented here for the case of a copper cathode and constant heat flux to the cathode during the spot life time. The approach with time dependent heat flux was presented in Ref. [30]. The main characteristics that can be calculated according to this theory is the cathode potential drop U_c and the cathode erosion rate per Coulumb G_k. Figure 3.12 shows U_c dependence on spot lifetime t when the spot current I is given as parameter. It can be seen that U_c sharply rise with decreasing the cathode spot lifetime. On the other hand, it decreases with cathode spot lifetime increase and approaches experimental value of about 15 V in the case of 100 A. The last result indicates that the spot can be unstable.

The cathode erosion rate as a function of the cathode spot lifetime is shown in Figure 3.13 with the cathode spot current as parameter. G_k increases with I and it is close to that measured experimentally. The cathode temperature in spot as a function of cathode spot lifetime is presented in Figure 3.14. The cathode temperature increases with arc current and saturates at about 4000 K. When spot lifetime increase from 0.1 μs to 0.1 s, the electron fraction of total spot current is 0.8−0.7, the heavy particle density is $(0.8-3) \times 10^{20}$ cm^{-3}, the degree of the ionization is about 0.9−1 in the case of 10 A and is about 0.55−1 in the case of 100 A, the spot current density is about 2×10^6 A/cm^2, the velocity on the external boundary of Knudsen layer is 0.2 of the ion sound velocity.

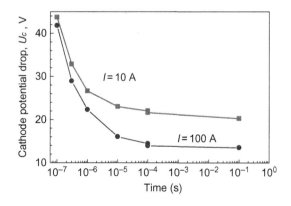

FIGURE 3.12

Cathode potential drop as dependence on cathode spot lifetime.

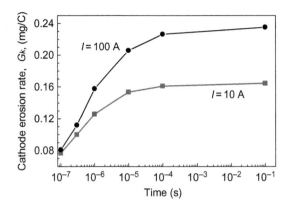

FIGURE 3.13

Cathode erosion rate as dependence on cathode spot lifetime.

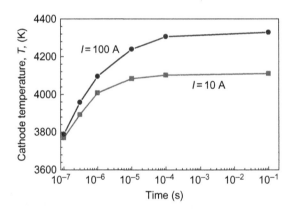

FIGURE 3.14

Cathode temperature as dependence on cathode spot lifetime.

One of the important parameters of the cathode spot is the cathode vaporization fraction, which is the ratio of the cathode erosion rate to the vaporization rate given by Langmuir relation in vacuum (see Chapter 1). This fraction is presented in Figure 3.15 as a function of the cathode spot lifetime with spot current as a parameter. It can be seen that cathode vaporization rate in vacuum arc is about twice lower than that given by Langmuir relation and it increases with the cathode spot lifetime decrease and with the spot current decrease.

3.4.2.2 Anode region

The anode region structure strongly depends on the arc current, the anode geometry, and the interelectrode gap size [20,21].

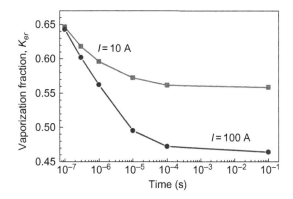

FIGURE 3.15

Cathode vaporization fraction K_{er} as dependence on cathode spot lifetime.

In general, an anode sheath appears if the thermal electron current to the anode from the plasma is different from the local arc current. In the case of a current channel with no anode involvement, the anode potential drop is negative and can be calculated as

$$\Delta U_a = -\frac{kT_e}{e}\ln\left(\frac{en_e(kT_e/2\pi m_e)^{0.5}}{j_a}\right)$$

(3.43)

where j_a is the current density near the anode, T_e is the electron temperature, e is the electron charge, and n_e is the electron density at the plasma−anode sheath interface. When the ratio of the thermal electron current density to the current density decreases, the absolute value anode potential drop decreases until it becomes equal to zero. This effect is also related to the ion starvation condition.

3.4.2.2.1 Conventional vacuum arc: anode plasma observation

Relatively small anode can be significantly heated by the energetic cathode plasma jets even in the case of a moderate arc current. In this case, anodic plasma can be generated by the anode vaporization with following atom ionization. This type of discharges is dubbed the hot anode vacuum arc (HAVA) or hot refractory anode vacuum arc (HRAVA) and can be employed as an efficient source of the metallic plasma. The details of anodic region characteristics can be found elsewhere [20−22].

In a conventional vacuum arc, the entire bulk of the anode is usually cold and it was observed that the anode region can have several distinct modes. The qualitative influence of arc current and gap length on the anode mode appearance is presented in Figure 3.16.

In the case of a relatively low current (<400 A), the anode is a simple collector of electrons provided the arc current is diffuse over the anode, i.e., the *diffuse anode region (mode)*.

When the arc current and the gap increase, the anode begins to play an active role in the discharge and form a *footpoint anode mode*. In this mode, the anode is characterized by a small bright region and is associated with weakly melting anode surface and with anode material evaporation. Furthermore, with arc current increase, the more anode activity leads to formation of an *anode spot*.

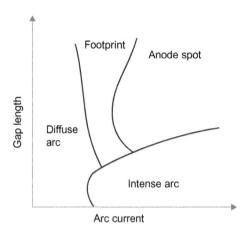

FIGURE 3.16

Anode discharge modes as a function of current and gap length. After Ref. [21].

The anode spot mode was observed in the case of arc current of about 400 A with Cu anode diameter of about 13 mm and gap of about 25 mm [20]. Large anode diameter of 50 mm leads to diffuse arc without the anode spot. Usually the anode spot appears when the threshold current reaches a few kiloampere. This is the typical case in a high-current vacuum arc (HCVA) with planar large copper electrodes. The optically measured Cu anode temperature is in range of 2700−3770 K, the spot mobility is very low, and sometimes a fixed anode spot can occur. The measured electron temperature is about 0.4−0.7 eV in the case of Al anode. The copper anode erosion rate ($I = 0.5 - 3$ kA) can reach relatively large values of about $25 - 120$ µg/C. Another very strong anode activity was observed in the case of so called *intense anode mode*. In this mode, a very bright luminous plasma appears in the interelectrode gap (*interelectrode plasma*). Such mode is typical for a HCVA and it is characterized by strongly anode erosion and intense anode macroparticles emission. Interelectrode plasma can be diffuse or nondiffuse dependent on the gap length and arc current as shown in Figure 3.16 [21].

Experimentally determined arc appearance diagrams, displaying modes of the drawn arc as a function of instantaneous current, and contact gap are shown in Figure 3.17. The lines shown in Figure 3.17 are the boundaries between the regions of the diffuse mode and constricted columnar arcing. The experimental lines indicate the separation between the two arc existence modes in the space for two values of the axial magnetic field (AMF). As the AMF increases, the boundary moves toward higher current. At a given AMF, current levels well to the left of the boundary will produce fully diffuse high-current arcs.

3.4.2.2.2 Anode spot: model and calculation

The mechanism of anode spot existence can be explained by a physical model describing the plasma generation, anode−plasma interaction, and plasma expansion in form of a jet [25]. This

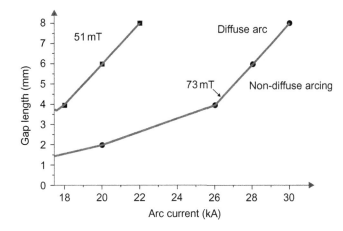

FIGURE 3.17

Lines: Experimental boundaries between regions of diffuse-type arcs and constricted columnar arcs reported in Ref. [23] (arc appearance diagrams). Points: Fitting data for the critical arc current for stable existence of a displaced 100 A jet as a function of the gap, which are used to calculate radii for the main column's cathode arc root from the model.

model includes the kinetic treatment of the anode evaporation and plasma production and takes into account the plasma energy dissipation. The near-anode space sheath, nonequilibrium plasma layer, and a plasma acceleration region are considered. The plasma energy balance includes the Joule energy in the anode plasma jet, the energy dissipation caused by ionization of atoms, energy convection due to the electric current, and the energy required for plasma acceleration in the plasma anode jet. The system of equation allows calculation of the anode spot temperature T_{an}, heavy particle density n_h, degree of ionization, electron temperature T_e, etc. as dependence on anode erosion rate G_{an} in a self-consistent manner. As an example, the calculated anode spot parameters are illustrated for the copper anode spot with the current of about $I = 400$ A and the anode spot lifetime of about 1 ms.

The calculated anode surface temperature T_{an} as functions of the anode erosion rate G_{an} is presented in Figure 3.18. According to the calculations T_{an} increases from 3000 to 4000 K when G_{an} widely varies from 5 to 300 μg/C and it is in range of measured anode spot temperatures (see above). In this case, the density n_h is relatively large of about $10^{19} - 10^{20}$ cm^{-3}, while degree of ionization is significantly low (Figure 3.18) and decreases from 3×10^{-3} to 4×10^{-4} with G_{an} increase. Low ionization degree is the result of the low electron temperature that only slightly varies ($T_e \sim 0.4$ eV) in the considered range of G_{an}. The anode vaporization fraction is also low and increases from 0.01 to 0.06 with G_{an} increase indicating that the erosion mass rate is significantly lower than predicted by the Langmuir evaporation rate. The plasma jet is also generated from the anode spot with relatively large velocity of about 3×10^6 cm/s. In vacuum arcs, the cathode and anode plasma jets formed the interelectrode plasma.

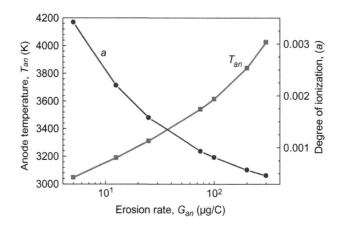

FIGURE 3.18

Anode temperature (T_{an}) and degree of atom ionization in the anode spot plasma as function on anode erosion rate.

3.4.2.3 Interelectrode plasma

In the case of a relatively small arc current (<1 kA), the interelectrode plasma consists mainly of the cathode plasma jets, which interact with the passive anode (collector of current). Primarily the study of the interelectrode plasma focuses on the plasma parameters (e.g., density, temperature) during the plasma jet expansion [26]. The character of the plasma expansion depends on the anode geometries. The free-boundary cathode jet expansion was analyzed using a two-dimensional model for a disk cathode and different anode geometries: disk anode [27], ring anode [28], and small anode relatively to the expanding plasma size [29]. Hydrodynamic model can be employed to study cathode jet expansion and free boundary is calculated as a part of the solution. Detailed analysis of the free-boundary solution will be considered in Chapter 4. The calculation shows that in the case of relatively small jet current (<1000 A), the plasma contracts and the voltage monotonically increases with the AMF strength (as observed experimentally [14]), due to the plasma density and electrical current density increase near the plasma jet axis.

3.4.2.4 HCVA: magnetic field influence

In HCVA, the interelectrode plasma can be controlled by an external AMF. The AMF was used to enhance the ion flux along the arc axis and to reduce the losses in the radial direction. In addition, AMF prevent plasma constriction thus leads to diffuse plasma at the anode surface. When AMF is applied, the arc voltage is monotonically increases with the magnetic field for relatively small arc current (<1 kA). The arc voltage depends nonmonotonically on the magnetic field in the case of a high-current arc (≫1 kA). In the last case, the voltage first decreases as the magnetic field increases from zero, then pass through a minimum value that depends on the arc current and then increases slowly (i.e., voltage dependence on the magnetic field has distinct V-shape). A model of the HCVA is typically based on the hydrodynamic formulation. Simulations are presented in Section 4.2.

Homework problems

1. Gas Breakdown

Calculate the minimum breakdown voltage for Helium, Air and Argon. Take into account that the cathode secondary emission coefficient is 0.12. Plot breakdown voltage versus (p*d).

	Helium	Air	Argon
A	2.8	14.6	13.6
B	34	365	235

2. Compute the breakdown voltage of 1 atm argon as a function of the secondary emission coefficient $(0.05-0.2)$. Consider the interelectrode gap of about 5 mm.

3. **Consider positive column of a DC glow discharge** in the tube having the radius of 1 cm and length of 20 cm. Neon pressure is 100 mTorr. The total energy carried out per electron-ion pair is 40 eV. Diffusion coefficient is $D_a = 10^5$ cm^2/s. Calculate: (a) the electron temperature; (b) the electric field and the potential drop along the column.

  Note: Eq. 1.2.23 can be used to calculate the ionization rate.

4. **Consider the 500 torr carbon arc discharge** with arc current of about 100 A, arc voltage of about 30 V and interelectrode gap of about 5 mm.

  Conductivity of a weakly ionized gas can be calculated as follows:

$$\sigma = \frac{n_e e^2}{m \nu_{en}}$$

  Estimate the radius of the conductivity zone. Following conditions should be considered: temperature: 0.5 eV; electron density: 10^{19} m^{-3}.

5. **Compute the anode sheath potential drop of a 200A cathodic vacuum arc.** Consider the following conditions: cathode material is copper, anode diameter is 1 cm and electron temperature is 2 eV and electron temperature is 10^{19} m^{-3}. The entire anode can be considered as the current collector. Calculate the anode diameter at which anode sheath potential drop vanishes.

References

[1] M.A. Lieberman, A.J. Lichtenberg, Principles of Plasma Discharges and Material Processing, Wiley, NY, 1994.
[2] F. Paschen, Ueber die zum Funkenübergang in Luft, Wasserstoff und Kohlensäure bei verschiedenen Drucken erforderliche Potentialdifferenz, Ann. der Phys. 273 (5) (1889) 69–75.
[3] P.G. Slade, E.D. Taylor, Electrical breakdown in atmospheric air between closely spaced (0.2 µm–40 µm) electrical contacts, IEEE Trans. Comp. Packag. Technol. 25 (3) (2002) 390–396.
[4] T. Lee, H.H. Chung, Y.C. Chiou, Arc erosion behavior of silver contacts in a single arc discharge across a static gap, Proc. Inst. Elect. Eng. 148 (1) (2001) 8–14.

[5] J.M. Torres, R.S. Dhariwal, Electric field breakdown at micrometer separations, Nanotechnology 10 (1999) 102−107.

[6] L.B. Loeb, J.M. Meeks, Mechanism of Electric Spark, Stanford Univ. Press, Stanford, California, 1941.

[7] H. Raether, Electron Avalanches and Breakdown in Gases, Butterworths, London, 1964.

[8] M. Keidar, A. Shashurin, O. Volotskova, Y. Raitses, I.I. Beilis, Mechanism of carbon nanostructure synthesis in arc plasma, Phys. Plasmas 17 (5) (2010) 057101.

[9] A. Engel, M. Steenbeck, Physics and engineering of a discharge of electricity in gases, Properties of Gas Discharges. Engineering Application, vol. 2, Julius Springer Verlag, Berlin, 1934. (in German)

[10] A. von Engel, Ionized Gases, Clarendon Press, Oxford, 1965.

[11] Y.P. Raizer, High Temp. 10 (1972) 1152.

[12] I.I. Beilis, A.Y. Sevalnikov, Column of the atmospheric pressure electric arc, High Temp. 29 (5) (1991) 856−863. (in Russian)

[13] J. Lafferty (Ed.), Vacuum Arcs, Wiley, New York, NY, 1980.

[14] R.L. Boxman, P.J. Martin, D.M. Sanders, Noyes Publications, Park Ridge, NJ, 1995.

[15] I.I. Beilis, Analysis of the cathode spots in a vacuum arc, Sov. Phys. Tech. Phys. 19 (1974) 251−256.

[16] B. Juttner, Cathode spots of electric arcs, J. Phys. D Appl. Phys. 34 (2001) R103.

[17] G.A. Mesyats, Cathode Phenomena in a Vacuum Discharge, Nauka, Moscow, 2000.

[18] I.I. Beilis, Theoretical modelling of cathode spot phenomena, in: R.L. Boxman, P.J. Martin, D.M. Sanders (Eds.), Handbook of Vacuum Arc Science and Technology, Noyes Publications, Park Ridge, NJ, 1995, pp. 208−256.

[19] I.I. Beilis, Vacuum arc cathode spot and plasma jet: physical model and mathematical description, Contrib. Plasma Phys. 43 (3−4) (2003) 224−236.

[20] C.W. Kimblin, Anode voltage drop and anode spot formation in DC vacuum arc, J. Appl. Phys. 40 (1969) 1744−1752.

[21] H.C. Miller, Anode phenomena, in: R.L. Boxman, P.J. Martin, D.M. Sanders (Eds.), Handbook of Vacuum Arc Science and Technology, Noyes Publications, Park Ridge, NJ, 1995, pp. 308−364.

[22] I.I. Beilis, R.L. Boxman, Metallic film deposition using a vacuum arc plasma source with a refractory anode, Surf. Coat. Technol. 204 (6−7) (2009) 865−871.

[23] M.B. Schulman, P.G. Slade, J.V.R. Heberlein, Effect of an axial magnetic field upon the development of the vacuum arc between opening electric contacts, IEEE Trans. Comp. Hybrids Manuf. Technol. 16 (1993) 180−189.

[24] M. Keidar, M.B. Schulman, E.D. Taylor, Model of a diffuse column vacuum arc as cathode jets burning in parallel with a high-current plasma core, IEEE Trans. Plasma Sci. 32 (2) (2004) 783−791.

[25] I.I. Beilis, Anode spot vacuum arc model: graphite anode, IEEE Trans. Comp. Packag. Technol. 23 (2) (2000) 334−340.

[26] M. Keidar, I. Beilis, R.L. Boxman, S. Goldsmith, 2-D expansion of the low-density interelectrode vacuum arc plasma jet in an axial magnetic field, J. Phys. D Appl. Phys. 29 (1996) 1973−1983.

[27] I. Beilis, M. Keidar, R.L. Boxman, S. Goldsmith, Theoretical study of plasma jet expansion in a magnetic field in a disc anode vacuum arc, J. Appl. Phys. 83 (2) (1997) 709−717.

[28] M. Keidar, I. Beilis, R.L. Boxman, S. Goldsmith, Voltage of the vacuum arc with a ring anode in an axial magnetic field, IEEE Trans. Plasma Sci. 25 (1997) 580−585.

[29] I.I. Beilis, M. Keidar, R.L. Boxman, S. Goldsmith, The model of plasma expansion and current flow in a vacuum arc with a small anode, IEEE Trans. Plasma Sci. 27 (1999) 872−876.

[30] I.I. Beilis, Continuous transient cathode spot operation on a microprotrusion: transient cathode potential drop, Part I, IEEE Trans. Plasma Sci. 39 (6) (2011) 1277−1283.

Plasma Dynamics

4.1 Plasma in electric and magnetic field

In this section we will consider charged particle behavior in electric and magnetic fields using a single particle approach. One peculiarity of this approach is that both electric and magnetic fields will be considered given and independent of the plasma properties.

Let us consider firstly the influence of a magnetic field only on a charge q (which could be positive or negative) of a particle with mass m. In this case, equation of motion for a charge particle reads

$$m\frac{d\vec{v}}{dt} = q\vec{v} \times \vec{B} \tag{4.1}$$

If the magnetic field has only component in z-direction (as shown in Figure 4.1), the equations for each component will be

$$m\frac{dv_x}{dt} = qBv_y \tag{4.2}$$

$$m\frac{dv_y}{dt} = -qBv_x \tag{4.3}$$

$$m\frac{dv_z}{dt} = 0 \tag{4.4}$$

Taking derivative we arrive at

$$m\frac{d^2v_x}{dt^2} = qB\frac{dv_y}{dt} = -\left(\frac{qB}{m}\right)^2 v_x \tag{4.5}$$

$$m\frac{d^2v_y}{dt^2} = -qB\frac{dv_x}{dt} = -\left(\frac{qB}{m}\right)^2 v_y \tag{4.6}$$

Equations (4.5) and (4.6) describe a harmonic oscillator with the frequency called the *cyclotron frequency*:

$$\omega_c = \frac{qB}{m} \tag{4.7}$$

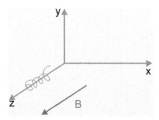

FIGURE 4.1

Schematic representation of charged particle motion in a magnetic field.

The radius of the circular orbit or *Larmor radius* is defined as

$$R_L = \frac{v_\perp}{\omega_c} \qquad (4.8)$$

where $v_\perp$ is the velocity in the plane perpendicular to magnetic field.

The direction of the gyro motion is such that the magnetic field generated by the rotating charged particle is opposite to the externally applied magnetic field, i.e., plasma is said to be diamagnetic. The particle velocity along the magnetic field is not affected by the magnetic field; thus, in general, the trajectory of a charged particle is a helix.

Now we will consider the charged particle motion in a magnetic field when a crossed electric field is present too. In this case, particle motion is the combination of two motions, namely, circular Larmor gyro motion and a drift. The particle motion is described by the following equation:

$$m\frac{d\vec{v}}{dt} = q\vec{E} + q\vec{v} \times \vec{B} \qquad (4.9)$$

Consider components of this equation for electron as an example shown in Figure 4.3. The component along the magnetic field, i.e., in z-direction is

$$m\frac{dv_z}{dt} = -eE_z \qquad (4.10)$$

Equation (4.10) describes the acceleration of electron along the magnetic field. Components of Eq. (4.9) transverse to the magnetic field (x and y) read

$$m\frac{dv_y}{dt} = \omega_c v_x \qquad (4.11)$$

$$m\frac{dv_x}{dt} = -eE_x - \omega_c v_y \qquad (4.12)$$

In order to visually represent the concept of drift motion and to estimate the drift velocity, let us consider the cyclotron circle as shown in Figure 4.2.

Force $F = eE$ acts across the magnetic field during the one half cyclotron motion in the same direction of rotation, while during the second half it acts in opposite direction. As a result, the test particle will be moving faster from up to bottom and slower from bottom to top.

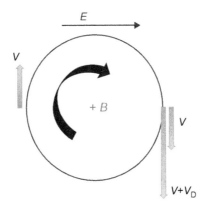

FIGURE 4.2

Schematic representation of drift mechanism. V_D is the drift velocity.

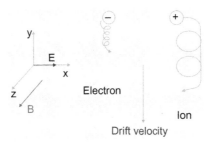

FIGURE 4.3

Schematic representation of the particle drift in crossed electric and magnetic fields.

The difference between these velocities will lead to shift of the cyclotron motion in the direction orthogonal to B and to the force as shown schematically in Figure 4.3. Let us estimate the particle drift velocity taking into account the following acceleration factor as

$$\frac{eE}{m} \qquad (4.13)$$

Thus, during the cyclotron rotation, the particle velocity changes as

$$\Delta v \sim \frac{eE}{m} \cdot \frac{1}{\omega_c} = \frac{eE}{m} \cdot \frac{m}{eB} = \frac{E}{B} = V_D \qquad (4.14)$$

Consider Eq. (4.9) for the case in which electric field has only component in x-direction (see Figure 4.3). In this case, transverse components of velocity

$$\frac{dv_x}{dt} = \frac{q}{m} E_x - \omega_c v_y \qquad (4.15)$$

$$\frac{dv_y}{dt} = \omega_c v_x \tag{4.16}$$

After differentiation (assuming the case of constant electric field, $E_x = E$), one can arrive at

$$\frac{d^2 v_x}{dt^2} = -\omega_c^2 v_x \tag{4.17}$$

$$\frac{d^2 v_y}{dt^2} = \omega_c \left(\frac{q}{m} E - \omega_c v_y\right) = \omega_c^2 \frac{E}{B} - \omega_c^2 v_y \tag{4.18}$$

Since both electric and magnetic fields are constant, one can write

$$\frac{d^2}{dt^2} \left(v_y - \frac{E}{B}\right) = -\omega_c^2 \left(v_y - \frac{E}{B}\right) \tag{4.19}$$

Solution for Eqs (4.17) and (4.19) reads

$$v_x = v_\perp \exp(j\omega_c t) \tag{4.20}$$

$$v_y = v_\perp \exp(j\omega_c t) - \frac{E}{B} \tag{4.21}$$

where $v_\perp$ is the velocity in the $x-y$ plane. These equations describe rotation with the frequency ω_c and superimposed drift in the negative y-direction, or $E \times B$ drift. Drift velocity is the same as it was estimated above (Eq. (4.14)).

The same procedure can be applied for any force transverse to magnetic field, $F_\perp$:

$$V_F = \frac{(F_\perp/q) \times B}{B^2} \tag{4.22}$$

V_F is the drift velocity in the case of applied arbitrary force $F_\perp$.

Let us make several general comments regarding the guiding center motion approximation. Strictly speaking this approximation can be used if motion is collisionless as it was assumed in analysis above, i.e., mean free path for collisions is much larger than the characteristic size of the system. In addition, it is assumed that the magnetic field is spatially uniform on the scale of gyroradius:

$$R_L \ll \frac{1}{(1/B)(\partial B/\partial r)}$$

4.2 Magnetic mirrors

A configuration of magnetic field used to confine charged particles is called a "magnetic mirror." Let us consider electron motion in the magnetic mirror as shown schematically in Figure 4.4. A magnetic field having component along z-axis and whose magnitude varies along this axis is considered.

In addition, radial component of the magnetic field has to be taken into account since magnetic field converges. It will be shown that the particle motion in such configuration gives rise to a force. This force can trap the particle in the magnetic mirror.

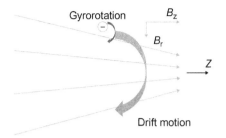

FIGURE 4.4

Schematic representation of the electron motion in a magnetic mirror field.

Assuming the azimuthal symmetry in the cylindrical coordinate system, the equation for the magnetic field (obtained from Maxwell equation) reads

$$\frac{1}{r}\frac{\partial}{\partial r}(rB_r) + \frac{\partial B_z}{\partial z} = 0 \tag{4.23}$$

If we assume that $(\partial B_z/\partial z)$ does not vary with radius, Eq. (4.23) can be integrated:

$$B_r = -\frac{1}{2}r\frac{\partial B_z}{\partial z} \tag{4.24}$$

Therefore, the component of the Lorentz force (Eq. (4.9)) acting on electron in z-direction is

$$F_z = -ev_\theta\frac{r}{2}\frac{\partial B_z}{\partial z} \tag{4.25}$$

where v_θ is the azimuthal velocity. Let us average this force over one gyration consider the guiding center on the axis. In this case, v_θ is constant during the gyration and it is equal to $v_\perp$ and $r = R_L$. Thus the average force is equal to

$$F_z = -ev_\perp\frac{r_\perp}{2}\frac{\partial B_z}{\partial z} = -ev_\perp^2\frac{1}{2\omega_c}\frac{\partial B_z}{\partial z} = -\frac{1}{2}\frac{mv_\perp^2}{B_z}\frac{\partial B_z}{\partial z} \tag{4.26}$$

This force pushes the particle into the region of smaller magnetic field (opposite to magnetic field gradient) and it is independent of charge.

Let us define the *magnetic moment* of gyrating particle as

$$\mu = \frac{(1/2)mv_\perp^2}{B} \tag{4.27}$$

As the particle moves into the region of a strong magnetic field, its kinetic energy $(1/2)mv_\perp^2$ changes; however, the magnetic moment is conserved as will be shown below.

We start with calculating the total energy of the particle that must be conserved:

$$\frac{1}{2}mv_\perp^2 + W_{\text{II}} = 0 \tag{4.28}$$

where W_{II} is the component of the particle energy in the direction along the magnetic field. The change of the W_{II} is due to force F_z:

$$dW_{II} = F_z \, dz = -\frac{W_\perp}{B_z} dB_z \qquad (4.29)$$

where $W_\perp = (1/2)mv_\perp^2$. Using Eq. (4.28), we arrive at

$$dW_\perp = -dW_{II} \qquad (4.30)$$

Thus,

$$\frac{dW_\perp}{W_\perp} = \frac{dB_z}{B_z} \qquad (4.31)$$

it follows from Eq. (4.31) that

$$\frac{W_\perp}{B} = \text{constant} \qquad (4.32)$$

This means that, the magnetic moment is conserved during the particle motion. The invariance of magnetic moment is the basis for one of the primary approaches for plasma confinement called magnetic mirror. Schematically magnetic mirror device is shown in Figure 4.5.

When charge particle moves from a region of a weak magnetic field to the region of a strong magnetic field, its $(1/2)mv_\perp^2$ should increase in order to keep constant μ. Since total energy of particle remains constant, $(1/2)mv_{II}^2$ must decrease. If the magnetic field in the "bottleneck" of the mirror machine is high enough, $(1/2)mv_{II}^2$ becomes zero and the particle will be reflected back to the region of a weak magnetic field.

It should be pointed out that the charged particle trapping in the magnetic mirror machine is not perfect. For instance, if the particle does not have velocity component perpendicular to the magnetic field, i.e., $(1/2)mv_\perp^2 = 0$, it does not have magnetic moment and force F_z does not act on the particle. A particle with a small $v_\perp/v_{II}$ in the region of a weak magnetic field (B_{min}) can also escape if the B_{max} is not strong enough. Let us estimate the condition for particle trapping. Consider a particle in the region of a weak magnetic field having the total velocity v_0 as shown

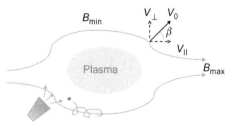

FIGURE 4.5

Schematic of the magnetic mirror machine.

in Figure 4.5. At the turning point $v_{\text{II}} = 0$ and magnetic field is $B_{\max}$. The invariance of the magnetic moment yields

$$m\frac{v_0^2 \sin^2\beta}{2B_{\min}} = m\frac{v_0^2}{2B_{\max}} \tag{4.33}$$

It follows from Eq. (4.33) that

$$\sin^2\beta = \frac{B_{\min}}{B_{\max}} \tag{4.34}$$

where β is the pitch angle in the region of a weak magnetic field. Particles with smaller β will be able to escape and not be trapped in the region of strong magnetic field. In other words, the condition for particle trap is

$$\beta > \beta_{\text{crit}} = \sin^{-1}\sqrt{\frac{B_{\min}}{B_{\max}}} \tag{4.35}$$

Equation (4.35) defines the boundary of the region (of a conical shape) in the velocity space or so-called loss cone. Particles lying within the loss cone (i.e., $\beta < \beta_{\text{crit}}$) are not trapped in the mirror. As a result, plasmas in a magnetic mirror machine are not isotropic. It should be pointed out that discussion so far did not include collisions. Due to collisions, particles can be lost after changing their pitch angle and scattering into the loss cone.

4.3 Remarks on particle drift

In the previous sections, we discussed drift of the charged particle in the uniform magnetic and electric fields. It should be pointed out that the presence of any force in addition to the magnetic field could lead to particle drift as it is reflected in Eq. (4.19). Among the typical forces in addition to electric field that affect particle motions are gravity and centrifugal force. Spatial nonuniformity and temporal variation of the magnetic and the electric fields lead to additional drifts.

For instance, presence of the gravity field leads to the following drift:

$$V_g = \frac{m \cdot g \times B}{qB^2} \tag{4.36}$$

where g is the gravity constant.

Time-varying electric field leads to the so-called polarization drift:

$$V_{E(t)} = \frac{m}{qB^2} \cdot \frac{\partial E}{\partial t} \tag{4.37}$$

It should be noted out that in this case, ions and electrons are drifting in opposite directions. In addition, one can see that the drift velocity is proportional to the particle mass and thus the drift velocity of ions will be larger than that of electrons.

As another example, particle motion in the magnetic field having curvature caused centrifugal force leading to the curvature drift:

$$V_R = \frac{2W_\perp}{qB^2} \cdot \frac{R_c \times B}{R_c^2} \tag{4.38}$$

where R_c is the curvature of the magnetic field.

4.4 The crossed $E \times B$ fields plasma dynamics in plasma devices

The devices based on closed electron drift are currently applied in plasma immersion ion implantation, magnetron sputtering, and electric thrusters for spacecraft propulsion. In these devices, the effect of closed electron drift in a magnetic field is used to maintain large electric field in the quasi-neutral plasma. This electric field is a result of electrostatic coupling of the ion flow and the background charge of the closed electron drift. While ion dynamics in most devices can be relatively easily described, the underlying physics of electron transport is not well understood. As it will be shown in Chapter 5, the main problem lies in description of the electron transport mechanism across a magnetic field, which was found to be largely nonclassical. Several possible mechanisms of anomalous electron transport were proposed and investigated over years and some progress was achieved. The related phenomenon of the electric thrusters based on the electron transport in crossed electric and magnetic fields is described in detail in Section 5.2.

One additional example of $E \times B$ device is the *magnetron*, which is an electrical discharge for sputter deposition of thin films [1,2]. Modern magnetrons employ crossed $E \times B$ fields that cause a Hall current (Hall drift). It is clear that uncontrolled Hall current will lead to electrons escape from the discharge. Electron losses can be limited by closing the drift current, i.e., by the use of such field configuration that will provide a circular pass of the drift current. In this case, the use of closed-drift configurations ensured a significant increase in the plasma density and sputtering rate.

Recently plasma configurations and electron transport in cylindrical magnetron discharges were studied [3]. It was found that two stable plasma configurations are possible around the negatively biased cylindrical target, namely torus and thin disk. Diffuse plasma torus changes the shape with magnetic field to form a thin disk when the target voltage is less than 400 V. Experiments with low-current magnetrons show that the measured electron mobility across the magnetic field scales as $1/B$ and not according to classical scaling as $1/B^2$ [4]. As a result, the concept of anomalous or Bohm diffusion was introduced to describe the electron transport in magnetrons. Recently there was an attempt to analyze the mechanisms of electron cross-field mobility and its impact on the discharge characteristics [5]. The calculated curves for electron energies based on classical and Bohm anomalous diffusion are shown in Figure 4.6. One can see that the classical curve shows too low energy (2 eV versus 4.2 eV or measured/Bohm calculated).

This analysis suggests that the Bohm anomalous diffusion can properly describe the electron transport in cylindrical magnetron. It is interesting to note that in related study of the miniature Hall thruster (that will be discussed in Chapter 5), it was concluded that the Bohm-type diffusion can properly describe the electron cross-field transport [6].

Thus existing experimental evidence and simulations support the idea of the anomalous transport; however, it is not clear what physical phenomena may lead to anomalous electron mobility,

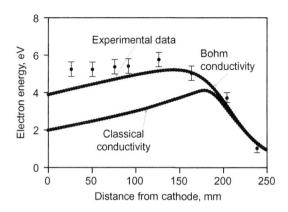

FIGURE 4.6

Dependence of electron energy on distance from cathode for cylindrical magnetron discharge.

and the mechanism of electron transport and plasma turbulence spectra in magnetron plasmas are still an open question.

4.5 Diffusion and transport of plasmas

Understanding the plasma transport phenomena in the presence of gradients of particles and pressure is important for many plasma applications. In this section, we will consider diffusion of plasma in the presence of density gradient. Effect of a magnetic field on the plasma diffusion will also be described.

4.5.1 Basic physics of diffusion

Particle diffusion in a nonuniform plasma arises due to friction term. We start with the steady-state momentum equation:

$$qnE - \nabla p - mn\nu_{coll}v = 0 \tag{4.39}$$

where ν_{coll} is the total momentum transfer frequency. Further analysis will be performed using the following basic assumptions:

Background species are at rest
Collision frequency is constant
Plasma is isothermal, i.e., $\nabla p = kT\nabla n$.

Solving Eq. (4.39) for velocity v, we obtain

$$v = \frac{qE}{m\nu_{coll}} - \frac{kT}{nm\nu_{coll}}\nabla n \tag{4.40}$$

By introducing the flux $\Gamma = nv$, Eq. (4.40) can be rewritten as

$$\Gamma = \frac{qnE}{m\nu_{\text{coll}}} - \frac{kT}{m\nu_{\text{coll}}}\nabla n \tag{4.41}$$

Let us introduce two new parameters:

Mobility coefficient:

$$\mu = \frac{|q|}{m\nu_{\text{coll}}} \quad \left[\frac{m^2}{Vs}\right] \tag{4.42}$$

Diffusion coefficient:

$$D = \frac{kT}{m\nu_{\text{coll}}} \quad \left[\frac{m^2}{s}\right] \tag{4.43}$$

With these coefficients, one can write the equation for flux (Eq. (4.41)) as

$$\Gamma = \pm \mu nE - D\nabla n \tag{4.44}$$

In this equation, positive sign corresponds to ions and negative to electrons. The process of diffusion in the absence of the electric field is called *free diffusion*:

$$\Gamma = -D\nabla n \tag{4.45}$$

Equation (4.45) is called *Fick's law*. Substituting this equation into the continuity equation $((\partial n/\partial t) + \nabla \cdot (nv) = 0)$ without the source and sink terms, one obtains

$$\frac{\partial n}{\partial t} - D\nabla^2 n = 0 \tag{4.46}$$

Note: It should be pointed out that mobility and diffusion coefficients (or transport coefficients) are related by the so-called Einstein relations:

$$\mu = \frac{|q|}{kT}D \tag{4.47}$$

4.5.2 Ambipolar diffusion

In this section we will consider diffusion of plasma, i.e., coupled diffusion of ions and electrons. Typically flux of electrons is initially much larger than that of ions so that electric field will be build up to maintain zero total charge flux. Such diffusion process of electrons and ions without charge buildup will be called *ambipolar diffusion*. In the following analysis, plasma quasi-neutrally is assumed. Thus the basic conditions to be considered are

$$\Gamma_i = \Gamma_e \tag{4.48}$$

$$n_i = n_e = n \tag{4.49}$$

From Eq. (4.48), we obtain that

$$\mu_i nE - D_i \nabla n = -\mu_e nE - D_e \nabla n \tag{4.50}$$

Solving this equation for electric field yields

$$E = \frac{D_i - D_e}{\mu_i + \mu_e} \frac{\nabla n}{n} \tag{4.51}$$

Equation (4.51) can now be substituted into diffusion equation for ions:

$$\Gamma = \mu_i \frac{D_i - D_e}{\mu_i + \mu_e} \nabla n - D_i \nabla n = - \frac{\mu_i D_e + D_i \mu_e}{\mu_i + \mu_e} \nabla n \tag{4.52}$$

This equation has a form of the Fick's law ($\Gamma = -D_a \nabla n$) if we introduce a new coefficient that will be called ambipolar diffusion coefficient:

$$D_a = \frac{\mu_i D_e + D_i \mu_e}{\mu_i + \mu_e} \tag{4.53}$$

Typically in plasma discharges, the electron mobility is much higher than that of ion mobility (i.e., $\mu_e \gg \mu_i$) and thus expression for ambipolar coefficient can be simplified:

$$D_a = \frac{\mu_i D_e + D_i \mu_e}{\mu_i + \mu_e} \approx D_i + \frac{\mu_i}{\mu_e} D_e \tag{4.54}$$

One can see that diffusion is tied to the slower species (ions in this case). Using the Einstein relation, one can obtain that

$$D_a \approx D_i \left(1 + \frac{T_e}{T_i} \right) \tag{4.55}$$

Thus one can see that ambipolar coefficient is proportional to the electron temperature, so that at higher electron temperature, ions and electrons will diffuse at the greater rate that in the case of ion diffusion rate.

4.5.3 Diffusion across a magnetic field

In this section we will discuss effect of the magnetic field on the plasma diffusion. Collision can change particle gyration around the magnetic field. On average collision moves the center of gyration by the Larmor radius. Since such process is random, it is similar to diffusion. In this case the Larmor radius can replace the mean free path for collisions.

In order to derive the diffusion coefficient across the magnetic field, we start with transverse component of the momentum equation:

$$0 = qn(E + v_\perp \times B) - kT\nabla n - mn\nu_{coll} v_\perp \tag{4.56}$$

Schematically the process of diffusion is shown in Figure 4.7. Consider component of Eq. (4.56) in x-direction for electrons:

$$mn\nu_{coll} v_x = - enE_x + nv_y eB - kT_e \frac{\partial n}{\partial x} \tag{4.57}$$

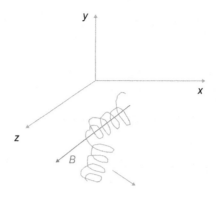

FIGURE 4.7

Schematic representation of particle diffusion across the magnetic field.

and in y-direction:

$$mn\nu_{coll}v_y = -enE_y - nv_xeB - kT_e\frac{\partial n}{\partial y} \tag{4.58}$$

if we use definitions of mobility and diffusion coefficient, these equations can be rewritten as

$$v_x = -\mu E_x + \frac{\omega_c}{\nu_{coll}}v_y - \frac{D}{n}\frac{\partial n}{\partial x} \tag{4.59}$$

$$v_y = -\mu E_y - \frac{\omega_c}{\nu_{coll}}v_x - \frac{D}{n}\frac{\partial n}{\partial y} \tag{4.60}$$

Let us introduce the ratio of the cyclotron frequency to the collisional frequency i.e., β:

$$\beta = \frac{\omega_c}{\nu_{coll}}$$

By substituting expressions for velocities v_x and v_y, Eqs (4.59) and (4.60) can be solved simultaneously for v_x and v_y:

$$v_x = -\frac{\mu}{1+\beta^2}E_x - \frac{D}{n(1+\beta^2)}\frac{\partial n}{\partial x} + \frac{\beta^2}{1+\beta^2}\frac{E_y}{B} - \frac{\beta^2}{1+\beta^2}\frac{kT_e}{n}\frac{\partial n}{\partial y} \tag{4.61}$$

$$v_y = -\frac{\mu}{1+\beta^2}E_y - \frac{D}{n(1+\beta^2)}\frac{\partial n}{\partial y} + \frac{\beta^2}{1+\beta^2}\frac{E_x}{B} + \frac{\beta^2}{1+\beta^2}\frac{kT_e}{n}\frac{\partial n}{\partial x} \tag{4.62}$$

One can introduce transport coefficients across the magnetic field:

$$\mu_\perp = \frac{\mu}{1+\beta^2} \tag{4.63a}$$

$$D_\perp = \frac{D}{1+\beta^2} \tag{4.63b}$$

It should be pointed out that Hall parameter β is an important parameter that indicates the degree of plasma confinement in a magnetic field. In the case of $\beta^2 \gg 1$, the diffusion coefficient across the magnetic field reduces to

$$D_\perp = \frac{1}{\beta^2} \frac{kT}{m\nu_{coll}} = \frac{kT\nu_{coll}}{m\omega_c^2} \tag{4.64}$$

Diffusion of electrons across a magnetic field is much smaller than that of ions due to electron magnetization:

$$D_{\perp ions} > D_{\perp electrons}$$

If plasma flows across the magnetic field, one can introduce the ambipolar diffusion similarly to Eq. (4.54) except the transport coefficients across the magnetic field should be described as follows:

$$D_{\perp a} = \frac{\mu_{\perp i} D_{\perp e} + D_{\perp i} \mu_{\perp e}}{\mu_{\perp i} + \mu_{\perp e}} \tag{4.65}$$

In the magnetic field, the mobility of ions is higher than that of electrons, thus diffusion coefficient across the magnetic field can be approximated as

$$D_{\perp a} \approx D_{\perp e} \left(1 + \frac{T_i}{T_e}\right) \tag{4.66}$$

Recall that the ambipolar diffusion coefficient is determined by the slower species (similarly to the case without a magnetic field), which is the one for electrons in the case across a magnetic field.

Plasmas (and in particular plasmas in a magnetic field) are subject to various instabilities as it was described in Chapter 1. Instabilities tend to destroy magnetic confinement due to development of a larger-amplitude turbulent diffusion. The diffusion has the upper limit of the Bohm diffusion that can be described by coefficient [7]:

$$D_{Bohm} = \frac{1}{16} \cdot \frac{T_e}{B}$$

The scaling of the Bohm diffusion coefficient as $1/B$ (compare to classical diffusion scaling as $1/B^2$, see Eq. (4.64)) makes it very important transport mechanism across the magnetic field as it will be described in Chapter 5.

4.6 Simulation approaches

In this section we will describe basic approaches for plasma modeling and, in particular, approaches that are based on numerical simulations. Two basic approaches were developed in plasma physics and engineering. First one is based on the fluid description of plasma solving numerically magneto-hydrodynamic (MHD) equations. Such approaches capture flow field characteristics, however, lacking the ability to calculate the transport coefficients. The second one is the kinetic model particle techniques that take into account kinetic interactions among particles and electromagnetic fields. This simulation is computationally extensive as it is able to resolve local parameters of the plasma. By taking advantages of both simulation approaches, there are hybrid approaches with various levels of kinetic and fluid combinations. All these simulation approaches have their specific advantages and limitations that will be illustrated in this section.

More rigidly the difference between the fluid and kinetic approaches can be illustrated as follows: fluid approach treats the large-scale properties of plasma involving mass, momentum, and energy transport, while the kinetic approach provides accurate treatment of local processes (collisions, scattering) and transport properties.

We will start with description of the kinetic analysis.

4.7 **Particle-in-cell techniques**

Basic for analytic treatment of the plasma kinetics is the Vlasov equation:

$$\frac{\partial f}{\partial t} + v \cdot \frac{\partial f}{\partial x} + \frac{q}{m}(E + v \times B)\frac{\partial f}{\partial v} = 0 \tag{4.67}$$

where $f(x,v,t)$ is the distribution function, v is the velocity, E is the electric field, and B is the magnetic field.

Note: The Vlasov equation treats the collisionless plasma condition. If collisions between particles are important, one needs to consider a model for Boltzmann equation with the collisional term (collision integral) that will appear in the right-hand side of Eq. (4.67).

Common approach to kinetic modeling is to represent $f(x,v,t)$ by a number of macroparticles (MPs) and compute the particle orbits in a self-consistent E and B. This is equivalent to solving the Vlasov equation by the method of characteristics.

Dawson [8] treated particles as discrete points and computed electric field force assuming explicitly the Coulomb interactions with each other particle. Number of pairs is $N(N-1)/2 \sim N^2$, where N is the number of particles.

Note: Consider 10^7 particle, 10^4 time steps and assume that in order to evaluate the force, we need 10 operations. The total number is $N(N-1)/2 \times 10^4 \times 10 \sim 5 \times 10^{18}$. If each operation will take 10^{-9} s, the total time will be 10^9 s.

Another approach is to use the spatial grid. In this case particles and current densities can be calculated using the interpolation scheme. Fields are solved on the grid while forces are interpolated. This procedure is called particle-in-cell (PIC) method. Number of operation per time step are reduced to $N \log(N)$.

PIC is a technique commonly used to simulate the motion of charged particles or plasma. The calculation flow chart in PIC is shown in Figure 4.8.

For each time step, the equation for the field is solved and the particle motion is implemented.

At $t = 0$ initial conditions are given and particle velocity and positions are prescribed. Particles are loaded with each particle having index (V_i, X_i). Transition from particles to quantities (charge density, current) is made by calculating the charge and current density on the grid. This procedure is called weighting from the grid points and it depends on particle position. After charge density is calculated the electric field can be computed. Similarly once the current is calculated, the self-magnetic field can be computed as well. By calculating the electric field, we interpolate the fields from the grid to particle in order to apply force at the particle by performing again weighting.

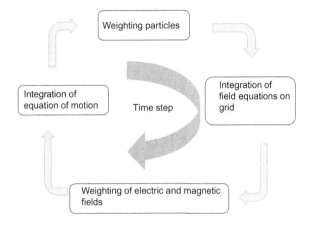

FIGURE 4.8

PIC simulation chart showing the simulation sequence.

The following information for each particle is stored in the computer memory: (V_i, X_i, q_i, m_i). Usually there are more particles than the grid points.

　　Charged particles interact with each other by attracting particles of opposite charge and repelling those with the same charge. This is represented by the Coulomb force and is given by

$$F = \frac{1}{4\pi\varepsilon_0}\frac{q_1 q_2}{r^2} \tag{4.68}$$

where q_1, q_2 are charges and r is the distance between interacting particles.

　　In principle, taking a collection of particles representing the real physical ions and electrons and directly computing this force can simulate plasmas. However, plasma simulations generally require at least 1 million particles in order to reduce numerical errors. Since Coulomb force leads to an n^2 problem, computation of a single time step would require at least 1 trillion operations.

　　Note: In the following, we describe the procedure for PIC simulation following the time chart as shown in Figure 4.8. We will start with integration of the equation of motion.

4.7.1 Equation of motion

In general we have many particles and many time steps, so the method for calculating particle trajectories and electric field should be as fast as possible but should retain some accuracy. Also we want to store the minimum amount of information (not to have previous time steps V_i and X_i for high-order integration). Commonly used integration method is the leapfrog method. It is fast and numerically stable. First, we integrate velocity through the time step. Next, position is updated. The name comes from the fact that times at which velocity and positions are offset from each other by half a time step. As such, the two quantities leap over each other. This idea is shown in Figure 4.9.

　　The equations that must be solved are

$$m\frac{dV}{dt} = F \tag{4.69}$$

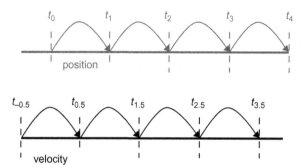

FIGURE 4.9

The leapfrog method. In this approach velocity and positions are offset from each other by half a time step.

$$\frac{dx}{dt} = V \tag{4.70}$$

where F is the force. Finite-difference representation for these equations is

$$m\frac{V_{new} - V_{old}}{\Delta t} = F_{old} \tag{4.71}$$

$$\frac{X_{new} - X_{old}}{\Delta t} = V_{new} \tag{4.72}$$

Positions are given at $t = 0$, while velocities are at $t - \Delta t$ using transformation employing force $F(t = 0)$. If we need to resolve oscillations or waves, we need to preserve

$$\omega_p \Delta t \le 0.3 \tag{4.73}$$

where ω_p is the plasma frequency. If magnetic field is used, the rotation of particle using the following equation should be performed:

$$m\frac{dV}{dt} = qE + q(v \times B) \tag{4.74}$$

After particles are moved to new positions, it is necessary to verify that all particles are still in the computational domain. To this end boundary conditions should be examined. Two boundary interactions are possible. The particles can either exit the domain or collide with solid objects. Computational boundaries are either *open* (or absorbing, allowing particles to leave), *reflective* (elastically returning particles into the domain), or *periodic* (particles are transported to the opposite side of the domain). The reflective boundary is used to identify planes of symmetry.

4.7.2 Integration of the field equations

At this point, charges (ρ) and current densities (J) are assigned to the grid points, so we can calculate E, B using the Maxwell's equations as sources. For instance, electric field can be calculated as

$$\text{div } E = \rho \tag{4.75}$$

Here ρ is the charge density. Electric potential is computed from $E = -\nabla\varphi$. Electric potential can be directly computed from the Poisson equation:

$$\nabla^2\varphi = -\frac{\rho}{\varepsilon_0} \tag{4.76}$$

Here ε_0 is the permittivity of free space. Charge density is calculated as

$$\rho = e(Z_i n_i - n_e) \tag{4.77}$$

The subscripts i and e denote ions and electrons, respectively, and Z_i is the average ion charge number.

Note: Due to very large mass ratio between ions and electrons, there is large difference in the characteristics times for ion and electron motion. Typically in order to resolve the electron motion, one needs to employ extremely small computation time steps (around 10^{-12} s). In other words, in the frame of reference of the ions, electrons move instantaneously. Therefore, if resolving electron motion is not essential for the particular problem of interest, it is possible to simplify the analysis by considering electrons as a fluid.

In this case, the electron density is then given by the Boltzmann relationship

$$n_e = n_0 \exp\left(\frac{\varphi - \varphi_0}{kT_e}\right) \tag{4.78}$$

where n is the reference density and φ_0 is the reference potential.

4.7.3 Particle and force weighting

The following procedure should be conducted to calculate the charge density and forces in the cell:

Calculate charge density on the discrete grid points taking into account particle positions.
Calculate the force on particles from fields on grid points.

This procedure is called weighting (form of interpolation). It is desirable to have the same weighting in both density and force to avoid a self-force (i.e., particle moves by itself).

4.7.3.1 Zero-order weighting

We need to count the number of particles within $\pm(\Delta x/2)$ which is the cell width about the jth grid point (Figure 4.10):

$$n_j = \frac{N(j)}{\Delta x} \tag{4.79}$$

where $N(j)$ is the number of particles.

In this case, the computation is fast with only the one grid lookup. The same force F will be applied for all particles in jth cell. If particle moves into jth cell, density will increase, while if particle moves out of the cell the density will decrease. As a result, motion of the particle in and out of the cell will produce the noise.

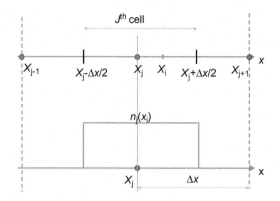

FIGURE 4.10

Illustration of the "rectangular" shape of the particles.

4.7.3.2 First-order weighting

The idea is to smooth out the density and field fluctuations (noise). However, in this case, we need to use two grid points for each particle. This is more computationally expansive but produces better interpolation.

Particles are finite size rigid clouds that can freely pass through each other (the idea was proposed by Birdsall in 1969, called clouds in cells (CIC)) [9].

Let the total cloud charge be q_C. The charge that the particle carries and that is assigned to j is

$$q_j = q_C \left[\frac{\Delta x - (x_i - X_j)}{\Delta x} \right] = q_C \frac{X_{j+1} - x_i}{\Delta x} \tag{4.80}$$

Part of the charge that is assigned to $j + 1$ is

$$q_{j+1} = q_C \frac{x_i - X_j}{\Delta x} \tag{4.81}$$

The net effect is to produce a triangular particle shape $S(x)$ which has width $2\Delta x$. Assignment to nearest grid point is called PIC [9]. As a particle moves through the grid, it contributes to density much more smoothly than the zero-order weight.

The field or force weight is similar. The first-order force comes from linear interpolation:

$$E(x_i) = E_j \frac{X_{j+1} - x_i}{\Delta x} + E_{j+1} \frac{x_i - X_j}{\Delta x} \tag{4.82}$$

4.7.4 Particle generation

In PIC, new particles are generated by sampling sources. Particle generation depends on specifics of the type of particle generation. For instance, particles can be generated at the electrode surface to model emission, or at the spacecraft thruster exit plane to model charge particle ejection. In some instances, all particles will be loaded initially, and the simulation will only compute their final distribution state. In many case of practical interest, particles are generated with the

Maxwellian distribution. According to Birdsall and Langdon [9], the Maxwellian distribution can be approximated as

$$f_M = \sqrt{\frac{M}{12}\left[\sum_{i=1}^{M} R_i - \frac{M}{2}\right]} \tag{4.83}$$

where M is some number and R_i is an ith random number in range [0:1]. The value chosen for M controls the accuracy of this method. It was suggested to use $M = 12$ to prevent entries larger than six times the thermal velocity.

It should be pointed out that often the following simplification can be made. The current carried by the plasma can be assumed to be low and as a result the self-induced magnetic field can be neglected. If no external magnetic field is applied, the set of underlying Maxwell's equations is reduced. This is the *electrostatic* PIC code.

An approach called particle-in-cell−Monte Carlo collisions (PIC-MCC) simulations is widely used to simulate very low-pressure plasmas taking into account rarefied collisions. In this approach the collisions such as ionization, excitation, momentum, and charge exchange between particles are treated with random numbers.

4.7.5 Example of application of PIC simulations

As an example, we consider a case of the plasma flow along the wall leading to formation of the interface-sheath between the plasma and the wall. Let us consider the sheath formation in the presence of a two-dimensional (2D) magnetic field incident to the wall. The analysis is performed using a 2D PIC code. For this example, a simple axisymmetric electrostatic particle-in-cell (ES-PIC) code was used. The code is based on the hybrid approach in which ions are treated as particles, but electrons are represented by a fluid model. As illustrated in Figure 4.11, the computational domain is limited to a small region near the wall. The small size of the domain allows resolving the Debye length and thus directly computes the electric potential in a reasonable amount of time (each simulation takes approximately 20 min). The upper boundary represents the outer wall, while the bottom boundary extends into the quasi-neutral bulk plasma region. Ions are injected into the simulation along the left boundary and leave through the open right and bottom face or by recombining with the upper wall.

To simplify the subsequent computation, we select a simulation mesh in which the radial grid-lines are aligned with the magnetic field. Such a formulation allows us to specify the necessary reference values as a function of the axial grid coordinate only. An example of the simulation is shown in Figure 4.12. Plasma density distribution along the flow and the wall is shown. Plasma density decreases due to acceleration along the flow direction as well toward the wall. Ion densities are shown using the contour plot. Velocity streamlines as well as the sheath boundary are also plotted. The sheath edge is plotted by the solid red line and corresponds to the contour where the radial velocity component (i.e., the component normal to the wall) reaches the Bohm velocity. The sheath forms a short distance from the injection plane and continues to grow as more ions are accelerated from the bulk plasma toward the wall. Plasma density decrease is also influenced by the net increase in ion velocity due to the axial electric field.

The particle-in-cell code is available as a supplemental material to this book and it was provided by Particle-in-Cell Consulting Inc www.particleincellconsulting.com (Dr. Lubos Brieda).

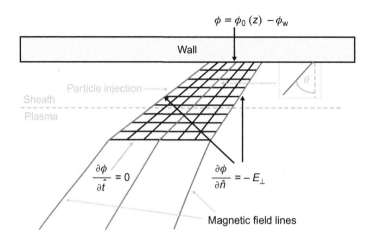

FIGURE 4.11

Schematics of the computational domain. Ion particles are injected from the left. Poisson equation with Boltzmann electrons is used to compute electric potential. Reference potential and density for the Boltzmann relationship vary with the magnetic field line.

Source: *Reprinted with permission from Ref. [39]. Copyright (2012) by American Institute of Physics.*

4.8 Fluid simulations of plasmas: free boundary expansion

In this chapter, we consider fluid simulation of plasmas. In particular, an example of plasma jet expansion will be described. This example will be illustrated on the specific case of the free boundary plasma jet expansion from the cathode spot of the vacuum arc.

4.8.1 Fluid model of vacuum arc plasma jet

A general model for the plasma jet expansion, in particular, plasma jet from the cathode spots of a cathodic arc, should consider the changes in the plasma parameters at distances sufficiently large where plasma density is much lower than that in the region near to the spots.

The theoretical study of vacuum arc plasma jets has been concentrated on the plasma jet formation and magnetic collimation effects. Beilis et al. [10,11] and Hantzsche [12] studied the plasma jet without the presence of an axial magnetic field (AMF) for the various plasma density regions. It may be noted that in the low-density part of the plasma jet, an AMF may have a very significant effect on the jet expansion. 2D free jet plasma expansion in an AMF was analyzed and simulated by Keidar et al. [13]. The problem of the 2D free plasma jet expansion is very complicated and a numerical method is needed in order to obtain the solution in the general case. One of the main difficulties in the jet expansion problem is the formulation of the physical conditions at the plasma−vacuum boundary.

The free boundary conditions at the interface between the plasma and the vacuum in the case of large current density, where the self-magnetic field plays a significant role, were discussed by Huysmans et al. [14]. The plasma boundary was defined by the condition that the magnetic field

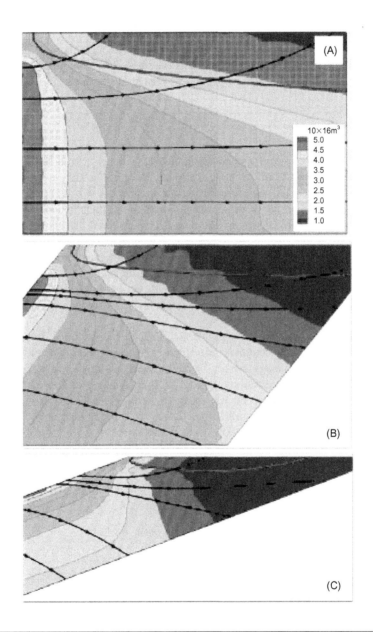

FIGURE 4.12

Simulation results showing the ion density profile for three different magnetic field line angles, 0° (A), 40° (B), and 70° (C), respectively. Streamlines show ion trajectories. The red lines correspond to the sheath edge as computed with the Poisson solver. The sheath edge is plotted by the solid red line and corresponds to the contour where the ion velocity component normal to the wall reaches the Bohm velocity. (For interpretation of the references to colour in this figure legend, the reader is referred to the web version of this book.)

Source: *Reprinted with permission from Ref. [39]. Copyright (2012) by American Institute of Physics.*

lines cannot point out of the plasma (i.e., the normal component of the magnetic field at the plasma boundary must be equal to zero). The jump in the tangential component of the magnetic field across the boundary is defined as the surface current. The boundary condition that the normal component of the plasma velocity to the fixed plasma boundary must be zero was also used by Kress and Riedel [15,16].

4.8.2 Basic model

Let us discuss the plasma jet with characteristic radius R_o (Figure 4.13) at some distance from the cathode surface where the jets from the individual cathode spots in the case of the multicathode-spot vacuum arc are averaged and therefore all plasma parameters will be assumed to be uniform. This plasma is sufficiently rarefied so that inelastic collisions (ionization, recombination, etc.) do not cause changes in the plasma density. In addition, it is considered that the electron and ion temperatures are different but do not change substantially during expansion. In these conditions, the electric field in the interelectrode gap does not significantly influence the expansion of the plasma jet but may play a dominant role in the current collection at the anode.

The anode is assumed not to influence the flow field and only acts as a current collector. The imposed AMF has only an axial component. The geometric position of the free boundary will be determined as part of a self-consistent solution.

The steady-state plasma flow in the vacuum arc is characterized by a directed velocity V of about 10^4 m/s, which is larger than the thermal velocity. The plasma jet of the low-current vacuum arc ($I_{ARC} \sim 100-300$ A) has a plasma density in the rarefied part of the cathode region [10,11] of about 10^{20} m^{-3} and a characteristic radius of about few cm. Taking into account that the electron temperature (T_e) is about $1-2$ eV, we have a mean free path L_c, which is 10^{-3} m. The magnetic diffusion time τ is about 10^{-7} s. This problem is characterized by a timescale of $\tau_f = L/V$, where L is the interelectrode region length. For $L \sim 0.1$ m, τ_f is about 10^{-5} s.

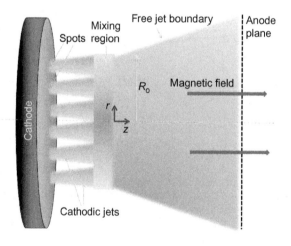

FIGURE 4.13

Sketch of the vacuum arc geometry and model of the vacuum arc plasma expansion.

The current carrying plasma flow in the magnetic field will be studied with the aid of the following assumptions based on the above estimations:

1. The plasma is fully ionized and consists of two species of charged particles.
2. During the expansion, the temperature of each species is not substantially changed and thus can be considered constant.
3. The plasma parameters at the starting plane and in most of the expansion region having a characteristic length L is such that the mean free path for electron–ion collisions (L_c) is much smaller than the L. Thus, the hydrodynamic approximation may be used to describe the flow of the plasma in the magnetic field.
4. The electron component is not inertial and thus $(m_e(\mathbf{V}_e \cdot \nabla)\mathbf{V}_e \approx 0)$.
5. Quasi-neutral plasma expansion is considered, i.e., $|N_e - Z_i N_i|/N_e \ll 1$, where N_e and N_i are the electron and ion densities, respectively, and Z_i is the ionicity.

The behavior of this steady-state plasma may be described by the momentum conservation equations:

$$m_i N_i(\mathbf{V}_i \cdot \nabla)\mathbf{V}_i = Z_i e N_i \mathbf{E} - \nabla P_i - m\, N_i \nu_{ei}(\mathbf{V}_i - \mathbf{V}_e) + Z_i e N_i(\mathbf{V}_i \times \mathbf{B}) \tag{4.84}$$

$$0 = -e N_e \mathbf{E} - \nabla P_e + m\, N_e \nu_{ie}(\mathbf{V}_i - \mathbf{V}_e) - e N_e(\mathbf{V}_e \times \mathbf{B}) \tag{4.85}$$

and the mass conservation equations:

$$\nabla \cdot (\mathbf{V}_i N_i) = 0 \tag{4.86}$$

$$\nabla \cdot (\mathbf{V}_e N_e) = 0 \tag{4.87}$$

where $\mathbf{V}_i$ and $\mathbf{V}_e$ are ion and electron velocities, respectively, m_i and m_e are the ion and electron masses, respectively, $m^* = m_i m_e/(m_i + m_e) \cong m_e$ is the effective mass. P is the partial pressure, $\mathbf{E}$ is the electric field, ν_{ei} is the electron–ion collision frequency:

$$\nu_{ei} = \frac{(\ln\Lambda/10)Z_i N_e \,[\mathrm{m}^{-3}]}{3 \cdot 10^{10} T_e^{3/2}\,[\mathrm{eV}]}$$

where T_e is the electron temperature (in eV) and $\ln\Lambda$ is the Coulomb logarithm. In addition, we use the definition $\nu_{ei} N_e = \nu_{ie} N_i$.

Let us define the electric current density $\mathbf{j} = eN(\mathbf{V}_i - \mathbf{V}_e)$ and the plasma conductivity $\sigma = (e^2 N/m_e \nu_{ei})$. The ions and electrons are assumed to be ideal gases with partial pressures $P_\alpha = eT_\alpha N_\alpha$, where $\alpha = e$, i, and k is Boltzmann's constant. By using the last expressions and condition, $N = N_e = Z_i N_i$, and adding Eq. (4.84) with Eqs (4.85) and (4.86) with Eq. (4.87), Eq. (4.84) may be written in the following form:

$$m_i(\mathbf{V}_i \cdot \nabla)\mathbf{V}_i = -k(Z_i T_e + T_i)\nabla \ln N + \frac{\mathbf{j} \times \mathbf{B}}{N} \tag{4.88}$$

$$\mathbf{j} = \sigma\left\{\mathbf{E} + \frac{kT_e}{e}\nabla\ln N - \frac{\mathbf{j} \times \mathbf{B}}{eN} + (\mathbf{V}_i \times \mathbf{B})\right\} \tag{4.89}$$

$$\nabla \cdot (\mathbf{V}_i N) = 0 \tag{4.90}$$

$$\nabla \cdot \mathbf{j} = 0 \tag{4.91}$$

Let us discuss the situation in which the plasma parameters are such that the electrons are magnetized, while the ions are not, i.e., where $\rho_i \gg L \gg \rho_e$, where ρ_e, ρ_i are the electron and ion Larmor radii, respectively. Equations (4.88)–(4.91) may be written in component form in cylindrical coordinates (see Figure 4.13) by taking into account that the magnetic field has only an axial component (i.e., $\mathbf{B} = i_z B_z$):

$$V_z \frac{\partial V_r}{\partial z} = -V_r \frac{\partial V_r}{\partial r} - C_s^2 \frac{\partial \ln N_r}{\partial r} + \frac{j_\theta B}{N m_i} \tag{4.92}$$

$$V_z \frac{\partial V_z}{\partial z} = -V_r \frac{\partial V_z}{\partial r} - \frac{1}{1 - (V_z^2/C_s^2)} \cdot \left\{ V_z \frac{\partial V_r}{\partial r} + V_r V_z \frac{\partial \ln N}{\partial r} + \frac{V_r V_z}{r} - V_r \frac{\partial V_z}{\partial r} \right\} \tag{4.93}$$

$$(C_s^2 - V_z^2) \frac{\partial N}{\partial z} = N V_z \frac{\partial V_r}{\partial r} + V_r V_z \frac{\partial N}{\partial r} - N V_r \frac{\partial V_z}{\partial r} + \frac{N V_r V_z}{r} \tag{4.94}$$

$$j_z = \sigma \left(E_z + \frac{kT_e}{e} \cdot \frac{\partial \ln N}{\partial z} \right) \tag{4.95}$$

$$j_r = \frac{\sigma}{1 + \beta_e^2} \left(E_r + \frac{kT_e}{e} \cdot \frac{\partial \ln N}{\partial r} \right) \tag{4.96}$$

$$j_\theta = j_r \beta_e \tag{4.97}$$

where V_z and V_r are the axial and radial components of the ion velocity, respectively, j_r, j_z, j_θ are radial, axial, and azimuthal components of the current density, respectively, $Cs^2 = k(Z_i T_e + T_i)/m_i$, $\beta_e = \omega_e/\nu_{ei}$, where $\omega_e = eB/m_e$. In the momentum conservation equation for ions, the term with magnetic field is absent as a result of the ion being unmagnetized.

Let us use Eqs (4.94)–(4.97) in order to express the equation for the electric field in explicit form. Taking into account that the Coulomb logarithm $\ln \Lambda$ has a weak dependence on the density N, we obtain that $\nu_{ei} = a \cdot N$, where $a = $ constant. After some linear transformations in the expressions for $\partial j_r/\partial r$ and $\partial j_z/\partial z$, and using the definition $\mathbf{E} = -\nabla \varphi$, we obtain the following expression for the potential distribution φ:

$$\frac{\partial^2 \varphi}{\partial z^2} - \frac{kT_e}{e} \left\{ \frac{\partial^2 N}{N \partial z^2} - \left(\frac{\partial \ln N}{\partial z} \right)^2 \right\} + \frac{2\beta_e^2 (\partial \ln N/\partial r)}{(1 + \beta_e^2)^2} \left(\frac{\partial \varphi}{\partial r} - \frac{kT_e}{e} \frac{\partial \ln N}{\partial r} \right)$$

$$+ \frac{1}{1 + \beta_e^2} \left\{ \frac{\partial^2 \varphi}{\partial r^2} + \frac{\partial \varphi}{r \partial r} + \frac{kT_e}{e} \left(\frac{\partial \ln N}{\partial r} \right)^2 - \frac{kT_e}{e} \cdot \frac{\partial^2 N}{N \partial r^2} - \frac{kT_e}{e} \cdot \frac{\partial \ln N}{r \partial r} \right\} = 0 \tag{4.98}$$

The boundary conditions for this system of equations will be discussed below for free plasma jet expansion.

4.8.3 Free plasma jet expansion

In this section, the current carrying plasma flow in the magnetic field is analyzed using the free jet expansion approach. We assume that the arc runs to an annular or ring anode, and that the characteristic size of the cathodic plasma (e.g., the cathode radius or the cathode-spot radius) is smaller than the inner radius of the anode and thus a free jet flow occurs.

4.8.4 Boundary condition for free plasma jet expansion

Mathematically, the free boundary problem consists of a system of partial differential equations, together with the necessary boundary conditions. The free boundary is unknown and must be determined as part of the solution. General approach for the solution of the free boundary problem is the trial free boundary method.

Generally, in order to solve the system of equations (4.92)–(4.98), it is necessary to know the geometric position of the plasma boundary, where two kinds of the boundary conditions, (1) hydrodynamic conditions for the plasma density and velocity, and (2) electrostatic conditions for the electric potential and electric field, are required.

If the plasma is in a stable steady state, the shape of the boundary is described by a function $\alpha(z)$, which is the angle of the jet boundary (Figure 4.14).

The main problem then is to develop an equation for this function. Let us introduce the spatial coordinate in the direction n normal to the boundary and the coordinate τ in the direction tangential to the boundary. The plasma velocity direction is tangential to the boundary and therefore the normal component V_n is equal to zero (point A in Figure 4.14). At the point C, the normal n_1 and tangential τ_1 vectors are translated and rotated relative to the normal and tangential vectors at point A. The angle between the tangential axis τ_1 and the old axis τ at C is equal to the jet angle change from point A to point C. According to Figure 4.14, the jet angle change from the point A to the point C in the limit of very small distances between points A and C may be defined as $\alpha(C) - \alpha(A) = \tan^{-1}(V_n/V_\tau)$, where $\alpha(A)$ is the jet angle at point A, $\alpha(C)$ is the jet angle at C, and V_n and V_τ are the normal and tangential components of the velocity vector at C in

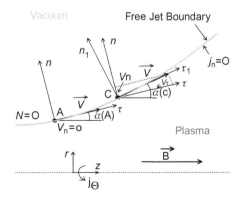

FIGURE 4.14

Schematic representation of the boundary conditions at the free plasma jet boundary.

the coordinate system $\{n,\tau\}$. The change in the jet angle α can be obtained by differentiation of the above expression:

$$\frac{\partial \alpha}{\partial \tau} = \frac{1}{1 + (V_n/V_\tau)^2} \cdot \frac{(\partial V_n/\partial_\tau)V_\tau - (\partial V_\tau/\partial_\tau)V_n}{V_\tau^2} \tag{4.99}$$

The velocity component V_n changes along axis τ from zero at the point A to some nonzero value. Taking into account that $(\partial V_n/\partial \tau) \neq 0$ and $V_n = 0$ at point A, together with Eq. (4.99), yields the final expression for the jet angle change at point A:

$$\frac{\partial \alpha}{\partial \tau} = \frac{\partial V_n}{\partial \tau} \cdot \frac{1}{V_\tau} \tag{4.100}$$

In order to solve this equation, an expression for $\partial V_n/\partial \tau$ and V_τ will be obtained from the n-components of Eq. (4.92) (see Figure 4.14):

$$V_\tau \frac{\partial V_n}{\partial \tau} + V_n \frac{\partial V_n}{\partial n} = -C_s^2 \frac{\partial \ln N}{\partial n} + \frac{[j \times B]_n}{Nm_i} \tag{4.101}$$

where the normal component of the vector $\mathbf{j} \times \mathbf{B}$ is given $[j \times B]_n = j_\theta B \cos(\alpha)$ and α is the jet angle. Taking into account that $V_n = 0$ at the boundary and using Eq. (4.100), we have the following simplified form of Eq. (4.101):

$$V_\tau \frac{\partial V_n}{\partial \tau} = -C_s^2 \frac{\partial \ln N}{\partial n} + \beta_e \omega_i \cos(\alpha) \frac{j_r}{eN} \tag{4.102}$$

where $\omega i = eB/m_i$. V_τ is obtained from the tangential component of Eq. (4.88) using the boundary condition $V_n = 0$:

$$V_\tau \frac{\partial V_\tau}{\partial \tau} = -C_s^2 \frac{\partial \ln N}{\partial \tau} + \beta_e \omega_i \sin(\alpha) \frac{j_r}{eN} \tag{4.103}$$

The velocity components will be obtained from Eqs (4.102) and (4.103) assuming that the plasma density and current distribution are known (a specific boundary condition will be discussed below), and the jet angle change will be calculated.

The generalized equation for the velocities will now be obtained in order to investigate the influence of the boundary conditions. We can solve Eq. (4.102) for j_r and then substitute this expression into Eq. (4.103), we obtain the following expression:

$$V_\tau \frac{\partial V_\tau}{\partial \tau} = -C_s^2 \frac{\partial \ln N}{\partial \tau} + \tan\alpha \left\{ V_\tau \frac{\partial V_n}{\partial \tau} + C_s^2 \frac{\partial \ln N}{\partial n} \right\} \tag{4.104}$$

The main theoretical problem in the plasma expansion is defining the boundary conditions for the density and electric potential at the plasma−vacuum interface. The condition for the density on the external surface of the boundary layer is $N = 0$ in the case of plasma expansion into vacuum, as illustrated in Figure 4.14.

In order to close the problem of the free boundary jet expansion, the boundary conditions for the potential will be formulated. The present model is based on the premise that the current carrying plasma flows from the cathode to the anode, which does not influence the flow field. At the anode plane, we set the anode potential, which is the boundary condition for the potential equation (see Eq. (4.98)). This procedure models the physical conditions presented by a grid anode, and to some extent those presented by a large diameter thin ring anode. At the free plasma boundary, the condition $J_n = 0$ is used and then we have following equation for the electric field at the plasma boundary:

$$E_n + \frac{kT_e}{e} \cdot \frac{\partial \ln N}{\partial n} - \frac{j_\theta B \cos(\alpha)}{eN} = 0 \qquad (4.105)$$

Let us discuss two limiting cases for the above equation. When $B = 0$, it reduces to the simple equation in which the electrical field at the plasma boundary depends on the electron temperature and density gradient. In the general case $(B \neq 0)$, the component of the electric field normal to the plasma boundary is dependent on the magnetic field and for large magnetic fields may change sign, and in the limiting case when $\partial N/\partial n = 0$, it will be determined by the magnetic field and current density.

The system of equations (4.92)–(4.105) with boundary conditions for the density, potential, and current fully determines the free boundary jet expansion problem. The ion velocity, current density, and density distribution are obtained numerically using the implicit second-order accuracy method [17].

4.8.4.1 Example of calculation of free boundary plasma jet expansion

The model that will be described here is related to the single or multicathode-spot vacuum arc plasma jet expansion. The problem was analyzed using dimensionless variables, where all velocities are normalized by C_s, the density by density at the starting boundary N_0, all coordinates by R_0, electrical current density by eN_0C_s, and the electrical potential by the electron temperature (kT_e/e). In this example, plasma parameters were chosen to correspond to the T_i plasma jet.

The calculated plasma jet boundary in the $r-z$ plane is shown in Figure 4.15 for different magnetic fields for the following conditions: arc current $I = 200$ A, initial Mach number $M = 5$, electron temperature $T_e = 2$ eV [10,11], $T_i/T_e = 0.5$, $Z_i = 1$, ion current fraction $f_i = 0.1$, plasma density at the starting plane $N_0 = 10^{20}$ m^{-3}, which corresponds to the conditions in the external part of the cathode jet.

The jet has an approximately conical shape with an angle which changes in the axial direction and depends on the magnetic field. The main jet shape changes occur over axial distances $z \sim (3-4)R_0$. The jet angle and the plasma density decrease did not change substantially at the larger distances $z > 4R_0$. The potential drop U between $(r,z) = (0,0)$ and $(0,4R_0)$ depends on the electron temperature and in the case of $T_e = 2$ eV it was found that $U = 8.4$ V for $B = 0$ and $U = 9$ V for $B = 0.01$ T at $z = 4R_0$. The electric equipotential surfaces were normal to the z-axis (i.e., E_r is small) except near the boundary as shown in Figure 4.16.

The normalized axial current density distribution $J_z(r)$ for different distances from the initial plane and magnetic fields is shown in Figure 4.17, and the radial current density distribution $J_r(r)$ is plotted in Figure 4.18. The radial current density component has its maximum value off-axis, while the maximum $J_z(r)$ is on the axis. This off-axis maximum in $J_r(r)$ is caused by the strong plasma density changes in the region of maximum.

The normalized radial plasma density distribution for $z = 4R_0$ with the magnetic field as parameter is presented in Figure 4.19. For the relatively large magnetic fields of $B > 0.01$ T, the plasma

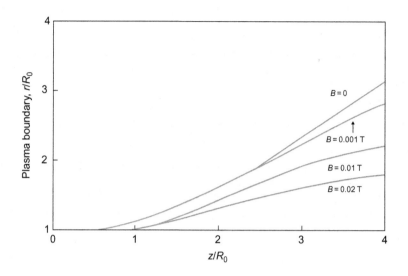

FIGURE 4.15

The plasma free jet boundary in the r–z plane with magnetic field as the parameter. Initial Mach number $M = 5$. Ion temperature to the electron temperature ratio $T_i/T_e = 0.5$, electron temperature $T_e = 2$ eV, arc current $I = 200$ A, initial plasma density $N_o = 10^{20}$ m^{-3}, ion current fraction $f_i = 0.1$.

Source: *Reprinted with permission from Ref. [13]. Copyright (1996) by Institute of Physics.*

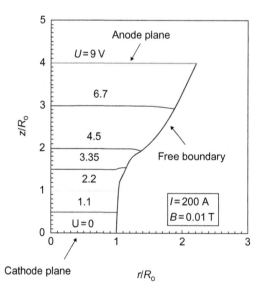

FIGURE 4.16

Voltage distribution in the interelectrode gap.

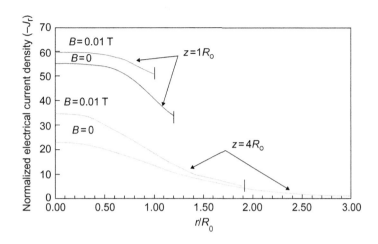

FIGURE 4.17

Normalized axial electrical current density distribution for magnetic fields $B = 0.01$ T and $B = 0$. I marks the plasma boundary.

Source: Reprinted with permission from Ref. [13]. Copyright (1996) by Institute of Physics.

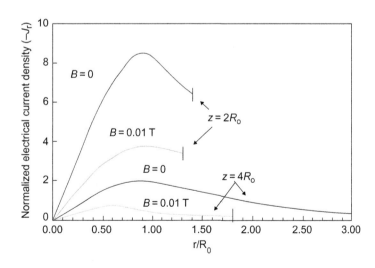

FIGURE 4.18

Normalized radial electrical current density distribution for the magnetic field $B = 0.01$ T and $B = 0$. I marks the plasma boundary.

Source: Reprinted with permission from Ref. [13]. Copyright (1996) by Institute of Physics.

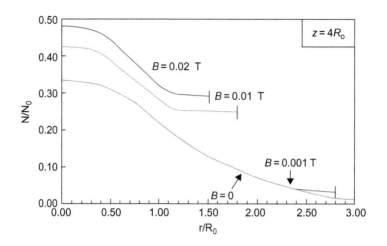

FIGURE 4.19

Normalized radial density distribution with magnetic field as a parameter. I marks the plasma boundary position.

Source: *Reprinted with permission from Ref. [13]. Copyright (1996) by Institute of Physics.*

density distribution near the boundary asymptotically approaches the condition of the radial density gradient equaling zero.

The plasma density distribution for different distances from the initial plane in the case of a 0.01 T magnetic field is plotted in Figure 4.20. The radial plasma density gradient in the boundary region is seen to approach zero with $z > 3R_0$.

The conditions for plasma collimation depend on the initial plasma density at the starting plane. If the plasma density at the starting plane is known, then it is possible to obtain good collimation (jet angle less than 20°) by the proper choice of magnetic field.

The radial distribution of the axial and radial velocity components at $z = 4R_0$ are presented in Figure 4.21 for initial Mach number $M = 5$ with magnetic fields of 0.001 and 0.01 T. The axial velocity in both cases slightly increases due to the pressure gradient. The radial velocity increases approximately linearly starting from zero on the axis and reaching significant values at the plasma boundary as a result of the expansion into the vacuum. In the 0.001 T case, the magnitude of the radial velocity approaches the axial velocity.

The presence of a stronger magnetic field, however, restricts the axial growth of the radial velocity, as can be seen in Figure 4.22, where the radial to axial velocity ratio at the plasma boundary is shown. Based on this results one can conclude that the streamline angle is about 40° for the 0.001 T magnetic field and about 20° for 0.01 T.

The normal component of the electric field at the plasma boundary is shown in Figure 4.23. In the case of $\beta_{eo} \rightarrow 0$ at the starting plane, a space charge layer is formed at the plasma–vacuum transition, in which the potential drop depends on the electron temperature and plasma density gradient. In the more general case with the presence of a magnetic field (values at the starting plane of $\beta_{eo} = 0.01$ or $\beta_{eo} = 0.5$ in Figure 4.23), this layer is also present, but at some point in the jet expansion where $\beta_e \geq 1$ the electric field changes sign. This occurs because the electrons are strongly magnetized while the ions are unmagnetized, and thus the electrons are confined and produce an

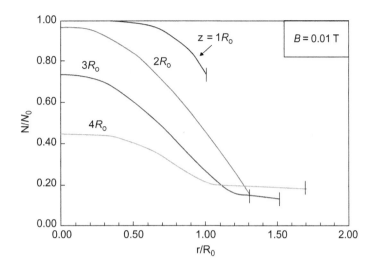

FIGURE 4.20

Normalized radial density distribution for different distances from the initial plane. / marks the plasma boundary position.

Source: *Reprinted with permission from Ref. [13]. Copyright (1996) by Institute of Physics.*

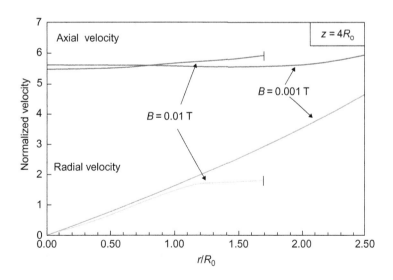

FIGURE 4.21

The radial distribution of the axial and radial velocity components at $z = 4R_0$ with the magnetic field as parameter. I marks the plasma boundary position.

Source: *Reprinted with permission from Ref. [13]. Copyright (1996) by Institute of Physics.*

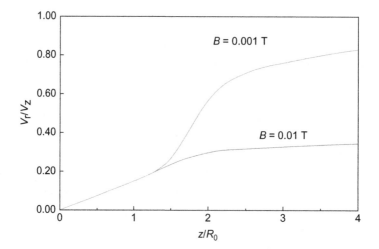

FIGURE 4.22

A plot of the radial to the axial velocity ratio at the plasma boundary with magnetic field as the parameter.

Source: *Reprinted with permission from Ref. [13]. Copyright (1996) by Institute of Physics.*

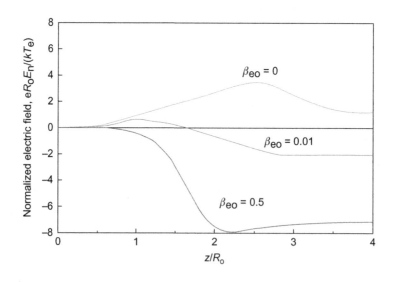

FIGURE 4.23

The distribution of the normal component of the electric field at the plasma boundary with β_{eo} at the starting plane ($z = 0$) as parameter.

Source: *Reprinted with permission from Ref. [13]. Copyright (1996) by Institute of Physics.*

electric field with a polarity that retards ion flow from the plasma. This electric field perturbs the plasma neutrality at the plasma boundary.

The normal component of the electric field at the plasma boundary decreases with axial distance. This is caused by decrease in the plasma density gradient.

4.8.4.2 Example of calculation: hollow anode vacuum arc

It is follows from Section 4.8.4.1 that in order to control the plasma jet parameters far from the cathode spot and to collimate the plasma flux from the plasma source, it is necessary to apply an AMF. Magnetic collimation of the plasma jet is limited by the growth of arc voltage, and the arc becomes increasingly unstable with increasing of the magnetic field component perpendicular to the cathode–anode direction. To this end, in this example, we analyze electrical characteristics such as current collection on the anode and voltage distribution in the interelectrode gap of the vacuum arc with a hollow anode. The problem is shown schematically in Figure 4.24.

Let us focus now on the boundary conditions at the plasma-near anode region interface.

4.8.4.2.1 Near anode region (anode sheath)

The plasma–anode interface consists of a sheath with charge separation. A 1D sheath will be considered because generally the sheath thickness ($\sim L_D$) is smaller than the anode length L_a (Figure 4.25). Let us assume that the electrical current in the sheath is carried mainly by electrons, i.e., $J_i/J_e \ll 1$. The electrons are magnetized in the sheath and drift in the direction parallel to the wall. The electron current in the direction toward the wall is the result of electron collisions even when the collision mean free path is smaller than the sheath thickness. Taking into account strong electron magnetization ($\beta_e \gg 1$), and after neglecting the pressure gradient force in Eq. (4.89) the following expression for the electron velocity can be obtained:

$$V_{er} = \frac{e v_{ei}}{m_e \omega_e^2} \cdot \frac{\partial \varphi}{\partial r} \qquad (4.106)$$

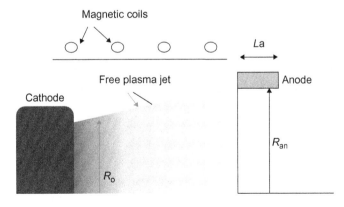

FIGURE 4.24

A sketch of the interelectrode gap geometry with ring anode.

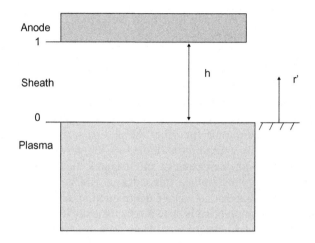

FIGURE 4.25

Geometry of the plasma–anode interface.

In the sheath (where generally $N_e \neq Z_i N_i$), the equation system (4.92)–(4.97) will be supplemented by Poisson's equation:

$$\nabla^2 \varphi = \frac{e}{\varepsilon_o}(N_e - Z_i N_i) \tag{4.107}$$

The solution of Poisson's equation (Eq. (4.107)) for the potential distribution under the above conditions, after taking into account Eq. (4.106), has the following form [40]:

$$\varphi(r') = \varphi_o + \frac{2}{5A}\left(\frac{3}{2}A \cdot r' + \left(\frac{\partial\varphi}{\partial r}\right)_o\right)^{5/3} \tag{4.108}$$

$$A = \frac{e}{\varepsilon_o}B\sqrt{\frac{j_{er}}{am_e}}$$

where J_{er} is the radial current density on the anode, φ_o and $(\partial\varphi/\partial r)_o$ are the potential and the potential gradient at the plasma–sheath boundary, respectively (Figure 4.25, boundary 0).

Taking into account that the thickness of the space charge sheath approximately equal to the Debye length, it is possible to obtain the dependence of the anode sheath drop U_{sh} as a function of the magnetic field. The total arc voltage U (excluding the cathode voltage drop) consists of the voltage drops in the quasi-neutral plasma U_{pl} and sheath voltage drop U_{sh}. The voltage in the plasma U_{pl} depends on the electrical current distribution $j(r,z)$ which is obtained from the 2D calculation of free jet expansion.

The calculated jet voltage U_{pl} in the plasma and the sheath voltage drop U_{sh} are shown in Figure 4.26. The jet voltage U_{pl} was calculated in the quasi-neutral plasma for fixed arc current and interelectrode gap geometry. The sheath voltage drop was calculated using the definition $U_{sh} = \varphi(h) - \varphi(0)$.

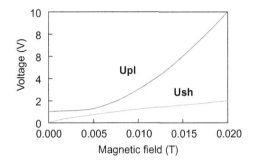

FIGURE 4.26

Voltage drop as a function of magnetic field. T_i, $I = 200$ A, $T_e = 2$ eV, $T_i/T_e = 0.5$, $z = 0.5R_0$, $R_{an}/R_0 = 1$, $r = 0.5$.

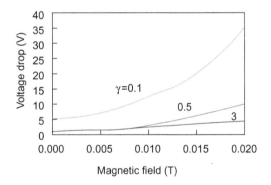

FIGURE 4.27

Total voltage drop U as a function of magnetic field with γ as parameter. $z = 0.5R_0$, $R_{an}/R_0 = 1$.

The voltage drop decreases with anode length L_a as shown in Figure 4.27, where $\gamma = L_a/R_0$. The calculated results were compared with experiments with different anode length to the cathode radius ratio γ and were plotted in Figure 4.28. Note that in experiment, cathode radius R_c was used as a characteristics scale of the plasma jet radius.

Experimental parameters and assumptions taken into account in this calculation are given in Table 4.1. In both the experiments and theory, larger voltage drops are observed with smaller values of γ. One can see that the good qualitative agreement was obtained.

4.8.4.3 Discussion of the free boundary model

The objective of this section is to describe the behavior of the expanding vacuum arc plasma jet in a vacuum environment in the presence of a magnetic field. In the region of the plasma generation (i.e., at the cathode spots), the plasma is so dense that the magnetic field does not influence the plasma expansion, which depends only on the plasma parameters, and the plasma jet has a conical shape. In the rarefied part of the vacuum arc jet, a magnetic field can significantly affect the plasma expansion,

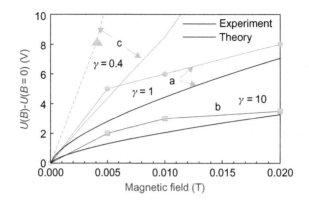

FIGURE 4.28

Voltage drop as a function of magnetic field, experiment, and theory for different γ: (a) ($\gamma = 1$) [18], (b) ($\gamma = 10$) [19], and (c) ($\gamma = 0.4$) [20].

Table 4.1 Experimental and Theoretical Data for Figure 4.28

	Cathode	Current I, A	R_c (m)	L_a (m)	γ	N_o (m^{-3})	Ref.
(a) exp.	Ti	100	0.05	0.08	–		[18]
(b) exp.	Ti	100	0.03	0.28		–	[19]
(c) exp.	Ti	170	0.045	0.020		–	[20]
theory	Ti	100–200	–	–	0.4, 1, 10	10^{20}	

and the jet shape will depend on the magnetic field. At the distance of about 3−4 initial radii, the radial component of the velocity at the plasma boundary becomes comparable to the axial component due to radial plasma expansion. Although the expanding jet becomes rarefied, the plasma remains quasi-neutral in all regions except at the free boundary, where an electrostatic sheath is formed. The electric field at the boundary of the sheath depends on the magnetic field. When the electrons are strongly magnetized and the ions remain unmagnetized, the magnetic field bounds the plasma and can act as a "magnetic wall." The boundary condition on the plasma density, $N = 0$, produces a radial plasma density distribution having a density gradient $\partial N/\partial n \rightarrow 0$ for large magnetic fields.

Nowadays, vacuum arc plasmas are employed in variety of applications ranging from circuit breakers [21], thin film deposition [22], ion implantation [23], and space propulsion [24]. Such wide applications of the vacuum arc are warranted due to outstanding properties of plasma jets (high degree of atom ionization, high directed ion energy, etc.) ejected from high dense cathode spots [25]. However, there are disadvantages such as relatively high cathode erosion and emission of macroscopic droplets MPs, which are critical in some instances [26]. In particular, MP emission is detrimental for thin film.

The simulation of vacuum arc for some applications will be considered in detail in the following section including the high-current vacuum arc interrupters. In the case of a high-current vacuum arc interrupters, the main issue to be considered is the interaction between the individual plasma jets and formation of the plasma channel. This approach helps to explain longstanding problem of the nonlinear behavior of the arc voltage as a function of the magnetic field.

4.8.4.4 Modeling of plasma flow in high-current vacuum arc interrupters

In this section we describe application of the free boundary plasma jet analysis for the case of high-current interrupters. One critical feature of vacuum arcs in the high-current range is columnar arc formation [27]. The vacuum arc in a stationary columnar mode (fixed arc roots) is characterized by high erosion that leads to substantial vapor density in the interelectrode gap. It has been found that the application of an AMF in the gap allows the arc to remain in a diffuse mode up to a much higher current [28]. Without an AMF, a high-current vacuum arc is characterized by a high arc voltage U_{ARC} [29], which may lead to an increase in the heat flux on the electrodes. Imposition of an AMF also changes the arc's current−voltage characteristic [30]. At high currents, U_{ARC} initially decreases as the AMF increases from zero, then reaches a minimum at a critical value of the AMF, and thereafter increases slowly with further increase of the field. For several metals including Cu, the increasing branch of the voltage characteristic was found to be independent of the arc current at a given AMF if the current is above a certain level.

Recall that the problem of a vacuum arc in an AMF was considered elsewhere [31]. It is typically associated with splitting of the overlapping cathode-spot plasma jets into columns, which are separated, on the average. Furthermore, an abrupt transition to connect the falling branch of the voltage-magnetic field dependence to the rising branch is employed.

Using a plasma jet simulation tool outline in this section, it is possible to calculate the arc voltage dependence on the AMF from the unified point of view. Such 2D model explains the physical reasons for the rising and falling branches of the dependence of U_{ARC} on an AMF as well as the smooth transition from one region to the other. In addition, the model predictions will be verified by comparison with experimental data.

Consider an individual expanding plasma jet from a cathode spot having radius R_o as shown in Figure 4.29. As it was shown in Ref. [13], the plasma jet expands radially due to pressure gradients. The radial expansion of the plasma jets makes possible the mixing of adjacent jets and the forming of the common current channel. Therefore, an important feature of the high-current vacuum arc is that in the interelectrode gap two regions can be present along the length of the gap: individual channels up to a distance L' and a common channel, as shown in Figure 4.29. In the case of high current, the self-magnetic field leads to constriction of the plasma mass density and current flow in the common channel, which is accompanied by increasing U_{ARC} [32]. An imposed AMF leads to collimation of the individual plasma jets [13]. Thus the plasma jet mixing occurs at larger distances from the cathode surface as the AMF is increased. Correspondingly, (i) the region of the common current channel becomes shorter and (ii) in the common channel the constriction of the plasma and current flow is limited by the AMF. Both of these lead to an initial decrease in the total U_{ARC} with increasing AMF. In a significantly large AMF, the individual current channels may be completely separated. In this case, U_{ARC} is the potential drop of an individual parallel channel and thus does not depend on the arc current. Consequently, this model explains the smooth transition from the low magnetic field branch of the current−voltage characteristic to the high magnetic field branch.

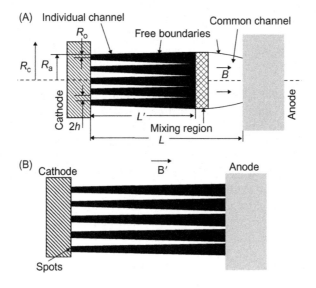

FIGURE 4.29

Schematics of the vacuum arc plasma jet formation: (A) individual cathodic jet mixing and (B) jet separation in a strong magnetic field.

In this consideration we will use the cathode area covered by the cathode spots which is characterized by radius R_a as a parameter. The characteristic distance between spots is $2h$ (see Figure 4.29).

In order to study the plasma expansion and current flow, the model of 2D free-boundary plasma jet expansion in an AMF was used as described above. The model allows us to calculate the geometry of an individual plasma jet as a function of the axial distance from the cathode surface as shown in Figure 4.30. One can see that the plasma jet significantly expands in the radial direction for the case of $B_Z = 0$. Increasing B_Z substantially reduces this expansion effect.

Using this result and having the average distance between adjacent plasma jets, it is possible to calculate the distance from the cathode surface where plasma jet mixing will occur. For some critical magnetic field, this distance is equal to the interelectrode gap. This is the case when the vacuum arc starts to burn as a number of parallel arcs, thus making U_{ARC} independent of the arc current. This case corresponds to the rising branch and the dependence of U_{ARC} on magnetic field. Thus, the condition of the minimum U_{ARC} is connected with individual channel splitting and may be formulated as

$$\Delta r_{jet}(L, B) = h(I_{ARC}, R_a) \tag{4.109}$$

where L is the gap length, $\Delta r_{jet}(L,B)$ is the increase of the plasma jet radius due to the radial expansion, and $h(I_{ARC}, R_a)$ is the distance between two individual adjacent channels.

This model was used to calculate U_{ARC} in the interelectrode gap (except the cathode sheath) for Gundlach's experimental configuration [33] with a cathode radius R_c of 30 mm and electrode separation of 10 mm. The potential drop as a function of AMF is calculated for a case of a 20 kA arc as

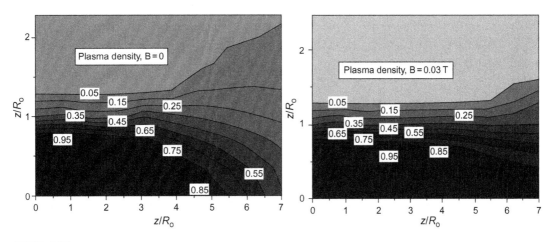

FIGURE 4.30

Plasma density distribution in a single jet with and without magnetic field.

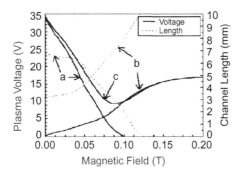

FIGURE 4.31

Voltage (—) and length (—) of the two axial regions of the plasma versus the AMF for $I_{ARC} = 20$ kA:
(a) common channel, toward anode, (b) 100 A individual channels, toward cathode, and (c) total plasma
voltage.

shown in Figure 4.31. In this calculation, we use the area of the cathode surface covered by the
cathode spots (radius R_a) as a parameter. This parameter provides the average distance between
individual cathodic jets, and thus makes it possible to calculate the average distance from the cathode where mixing of two adjacent jets occurs. The individual jet length L' (100 A current channel)
increases with an imposed AMF. Accordingly, the common channel length decreases, since mixing
of individual jets occurs at larger distances from the cathode surface. The total voltage in the interelectrode gap consists of the sum of the potential drop in the region of individual channels and the
potential drop in the common channel. In our model of the configuration of Ref. [33], the length L'
becomes equal to the gap length when $B > 0.1$ T if $R_a = 0.7R_c$.

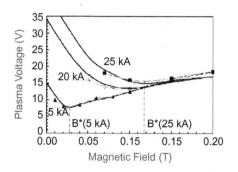

FIGURE 4.32

Voltage in the quasi-neutral plasma versus B_Z for three values of $I_{ARC} = I_{PEAK}$, $R_a = 0.5R_c$ (17 V subtracted for cathode fall).

Source: *Data points from Gundlach [33].*

The total plasma voltage as a function of a magnetic field is shown in Figure 4.32, with arc current as parameters. For comparison, the experimental data of Ref. [33] are also plotted. The model qualitatively predicts the arc voltage behavior with an AMF. One can also see that good agreement with experiment was obtained in the case of $R_a = 0.5R_c$. In particular, the magnetic field that corresponds to the minimum U_{ARC}, $B = 0.1$ T, is also correctly predicted. Overall, these results also show good agreement with experiment.

One can define B^* as the B_Z for which the arc voltage is a minimum for a given I_{ARC}, L, and contact (cathode) diameter R_C (after the transient expansion of the initiated arc is complete [34]). The calculated B^* as a function of I_{ARC} is shown in Figure 4.33 for $L = 10$ mm, with the radius R_a of the area covered by the cathode spots as a parameter. The B^* increases with I_{ARC}, as expected. The data of Refs [33,35] indicate that B^* is probably influenced by an increase in the average contact area covered by cathode spots as I_{ARC} increases.

In summary, it appears that there are two different regions of the plasma along the length of the gap: toward the cathode, the individual cathode-spot plasma jets are separated and noninteracting; toward the anode, the plasma takes the form of a common channel. In our model, the common channel forms at the distance from the cathode where the expanding plasma jets overlap and begin mixing. In the region of the individual jets, the vacuum arc burns as a number of parallel current channels, which makes the voltage drop independent of the arc current. The model predicts that the effect of an AMF on the jet expansion and U_{ARC} depends on the gap length. The model allows calculation of the condition of the transition from the falling branch to the rising branch of the dependence of U_{ARC} on the AMF, and also predicts the smooth transition between these regimes. An AMF has significant influence on the plasma jet expansion. The plasma jet collimation in the AMF leads to increase in the length of the individual channel region and thus to decrease the common channel length. High potential drop occurs in the common channel due to the self-magnetic field effect. Consequently, the plasma voltage decreases with an AMF increase. The rising branch corresponds to the high-AMF regime, when individual jets are well collimated across the gap. In the rising branch, U_{ARC} rises with increasing AMF because of the increasing current density in the individual channels. This results from (i) the decrease in the radial expansion of the jets and (ii) the

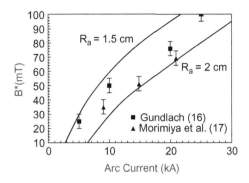

FIGURE 4.33

B^* for minimum arc voltage versus I_{ARC}, with radius of the cathode-spot region $R_a = 1.5$ and 2 cm. Experimental data are shown for Gundlach [33] ($L = 10$ mm, $R_c = 3$ cm) and Morimiya et al. [35] ($L = 10$ mm, $R_c = 2.2$ cm).

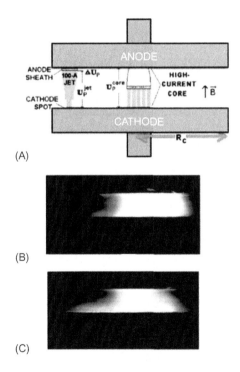

FIGURE 4.34

(A) Sketch of the contact gap in the case of a single cathode plasma jet at a distance from a high-current plasma column ("core"). U is plasma voltage. High-speed photographs of a diffuse column arc in an AMF. (B) Few or no cathode spots outside the central column at 27 kA and 270 mT. (C) Stable cathode spots covering the remaining contact surface at 19 kA and 190 mT.

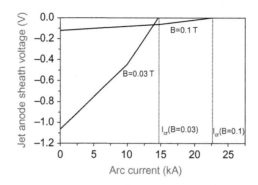

FIGURE 4.35

Calculated anode sheath potential drop for a displaced 100 A jet that is burning in parallel with a main column as a function of *I*, with the AMF as a parameter. Modeled experimental conditions are those of Gundlach [33].

decrease in the axial plasma density gradient as a result of this plasma collimation. The predicted arc voltage behavior for a test case agrees well with experimental data.

4.8.4.5 On the modeling of transition from diffuse to constricted high-current mode

In this section, the high-current diffuse columnar arc in vacuum is described. A separate cathode-spot jet appearing to the side of a high-current plasma column is modeled using the framework described above and described elsewhere [36−38]. Important to note that the plasma expansion and current flow in the jet are affected by the presence of the main column and the applied AMF.

Increasing that the current in the plasma column causes the arc voltage to increase, in turn, affecting the displaced parallel jet. In this case, the anode sheath potential drop in the displaced parallel jet increases from a negative voltage drop of about 1 V toward zero as shown in Figure 4.35 [36−38]. Displaced jet will extinguish when the arc voltage exceeds a critical value defined in Ref. [36] as the arc voltage at which the anode sheath potential drop of the displaced jet changes from negative to positive.

Calculated anode sheath potential drop as a function of arc current with AMF is shown in Figure 4.35. One can note that the increase of arc discharge current leads to decrease of the anode sheath potential drop of the displaced jet and transition from a negative potential drop to zero. Arc current at which anode potential drop of the displaced jet is equal to zero, we shall call a *critical arc current* [36]. Critical arc current is associated with transition from diffuse to collimated arc [36−38].

Homework problems
Section 1

1. **Larmor radius**
   Calculate the Larmor radius for the following cases:
   a. A 20 eV titanium ion in the magnetic field of about 0.001 T

b. A 10 eV electron in a magnetic field of about 0.02 T
c. A 30 keV electron in the Earth's magnetic field of 0.00003 T

2. **Magnetization.**

In collisionless system charge particles are considered magnetized if Larmor radius is smaller than the characteristic size of the plasma. Evaluate whether magnetization condition is fulfilled in the following cases:

a. Hall thruster channel of 0.01 m width and magnetic field of about 0.015 T. Consider 20 eV electrons and xenon ions having velocity of about 1.3×10^4 m/s.
b. Titanium ions in the magnetic field of about 0.01 T and magnetic filter having radius of about 1 cm

3. **Calculate electron drift velocity** in the Hall thruster having magnetic field of about 0.02 T and potential drop of about 200 V across the acceleration region of about 1 cm.

4. **Calculate the magnetic moment** for the following cases:

a. A 10 eV hydrogen atom in a magnetic field of about 0.5 T
b. A 1 keV electron in a magnetic field of about 1 T

5. **Consider mirror machine** with the magnetic field in the center of about 0.01 T and maximum magnetic field is about 0.5 T. Calculate the loss cone angle.

6. A 1 keV helium ion in a mirror device has a pitch angle of 30 degrees at midplane, $B_{min} = 0.5$ T. Calculate its Larmor radius.

7. **Consider partially ionized titanium plasma**. Calculate the ambipolar flux if plasma density gradient is about 10^{21} m^{-4}. Ionization fraction is about 10^{-4}. Consider electron temperature of about 3 eV and electron density of about 10^{18} m^{-3}

Section 2

1. Simulate the plasma-wall transition region (sheath) numerically. Following conditions to consider: wall potential is -5 V with respect to the plasma; electron density is about 10^{17} m^{-3}; electron temperature is 1 eV; Sheath can be considered collisionless. Electrons are distributed according to Boltzmann relation. Use 1D Particle-in-Cell simulation technique.

2. **Free plasma jet expansion.** Estimate the normal component of the electric field at the boundary of the fully ionized plasma jet. The electron temperature is about 2 eV and the plasma density is about 10^{20} m^{-3}.

3. **Anode sheath in a magnetic field.** It can be assumed that the sheath thickness is about the electron Larmor radius. Using this assumption, compute the anode sheath voltage drop as a function of the magnetic field. Consider 100 A arc discharge having the anode surface area of about 1 cm^2.

4. **Vacuum arc plasma jets.** Estimate the distance at which there is mixing of the cathodic plasma jets using results shown in Fig.4.2.24. Cathode diameter is 2 cm and arc discharge current is 5 kA.

References

[1] Z. Wang, S.A. Cohen, Geometrical aspects of a hollow-cathode planar magnetron, Phys. Plasmas 6 (5) (1999) 1655–1666.

[2] I. Levchenko, M. Romanov, M. Keidar, I.I. Beilis, Stable plasma configurations in a cylindrical magnetron discharge, Appl. Phys. Lett. 85 (11) (2004) 2202−2205.

[3] I. Levchenko, M. Romanov, M. Keidar, Investigation of a steady-state cylindrical magnetron discharge for plasma immersion treatment, J. Appl. Phys. 94 (4) (2003) 1408−1413.

[4] S.M. Rossnagel, H.R. Kaufman, Induced drift currents in circular planar magnetrons, J. Vac. Sci. Technol. A5 (1) (1987) 88−91.

[5] I. Levchenko, M. Keidar, K. Ostrikov, Electron transport across magnetic field in low-temperature plasmas: an alternative approach for obtaining evidence of Bohm mechanism, Phys. Lett. A 373 (12−13) (2009) 1140−1143.

[6] A. Smirnov, Y. Raitses, N.J. Fisch, Electron cross-field transport in a low power cylindrical hall thruster, Phys. Plasmas 11 (11) (2004) 4922−4933.

[7] O. Bohm, in: A. Guthrue, R.K. Wakerling (Eds.), The Characteristics of Electrical Discharges in Magnetic Fields, McGraw-Hill, New York, NY, 1949.

[8] J.M. Dawson, One-dimensional plasma model, Phys. Fluids 5 (1962) 445−459.

[9] C.K. Birdsall, A.B. Langdon, Plasma physics via computer simulation, Adam Hilger, New York, (1991).

[10] I.I. Beilis, M.P. Zektser, G.A. Lyubimov, Analysis of the formulation and solution of the problem of the cathode jet of a vacuum arc, Sov. Phys. Tech. Phys. 33 (10) (1988) 1132−1137.

[11] I.I. Beilis, M.P. Zektser, Calculation of the cathode jet parameters of the arc discharge, High Temp. 29 (1991) 501−504.

[12] E. Hantzsche, Theory of the expanding plasma of vacuum arcs, J. Phys. D Appl. Phys. 24 (1991) 1339−1359.

[13] M. Keidar, I. Beilis, R. L. Boxman, S. Goldsmith. 2-D Expansion of the low-density interelectrode vacuum arc plasma jet in an axial magnetic field. J. Phys. D: Appl. Phys., 29(1996) 1973−1983.

[14] G.T.A. Huysmans, J.P. Goedbloed, W. Kerner, Free boundary resistive modes in tokamaks, Phys. Fluids B 5 (5) (1991) 1545−1558.

[15] M.E. Kress, K.S. Riedel, Semi-implicit reduced magnetohydrodynamics, J. Comp. Phys. 83 (1989) 237−239.

[16] J. Crank, Free and Moving Boundary Problem, Clarendon Press, Oxford, 1984.

[17] V.M. Paskonov, A standard program for the solution of boundary-layer problems, in: G.S. Roslyakov, L.A. Chudov (Eds.), Numerical Methods in Gas Dynamics, Israel Program for Scientific Translation, Jerusalem, 1966.

[18] K. Akari, H. Tamagaki, T. Kumakiry, K. Tsuji, E.S. Koh, C.N. Tai, Reduction in macroparticles during the deposition of TiN films prepared by arc ion plating, Surf. Coat. Technol. 43/44 (1990) 312−323.

[19] I.I. Aksenov, V.G. Padalka, V.T. Tolok, V.M. Khoroshikh, Motion of plasma streams from a vacuum arc in a long, straight plasma-optics system, Sov. J. Plasma Phys. 6 (4) (1980) 504−507.

[20] V.N. Zhitomirsky, R.L. Boxman, S. Goldsmith, Unstable arc operation and cathode spot motion in a magnetically filtered vacuum arc deposition system, J. Vac. Sci. Technol. A13 (4) (1995) 2233−2240.

[21] J.M. Lafferty, Vacuum Arcs—Theory and Application, Wiley, New York, NY, 1980.

[22] R.L. Boxman, H. David, P.J. Martin, Vacuum Arc Science and Technology, William Andrew Publishing/Noyes, Park Ridge, NJ, 1995.

[23] A. Anders, Handbook on Plasma Immersion Ion Implantation and Deposition, Wiley, New York, NY, 2000.

[24] M. Keidar, J. Schein, K. Wilson, A. Gerhan, M. Au, B. Tang, et al., Magnetically enhanced vacuum arc thruster, Plasma Sour. Sci. Technol. 14 (2005) 661−669.

[25] I.I. Beilis, State of the theory of vacuum arcs, Plasma Sci. IEEE Trans. 29 (2001) 657−670.

[26] M. Keidar, I.I. Beilis, R.L. Boxman, S. Goldsmith, Transport of macroparticles in magnetized plasma ducts. Plasma Sci. IEEE Trans. 24 (1996) 226−234.

[27] J.V.R. Heberlein, J.G. Gorman, The high current metal vapor arc column between separating electrodes, IEEE Trans. Plasma Sci. PS-8 (1980) 283–288.

[28] C.W. Kimblin, R.E. Voshall, Interruption ability of vacuum interrupters subjected to axial magnetic fields, Proc. IEEE 119 (1972) 1754–1758.

[29] W.G.J. Rondeel, The vacuum arc in an axial magnetic field, J. Phys. D Appl. Phys. 8 (1975) 934–942.

[30] M.S. Agarwal, R. Holmes, Arcing voltage of the metal vapour vacuum arc, J. Phys. D Appl. Phys. 17 (1984) 757–767.

[31] V.A. Nemchinsky, Vacuum arc in axial magnetic field, Proc. XIVth Int. Symp. on Discharges and Electrical Insulation in Vacuum, Santa Fe, 1990, pp. 260–262.

[32] M. Keidar, M.B. Schulman, Modeling the effect of an axial magnetic field on the vacuum arc, IEEE Trans. Plasma Sci. 29 (5) (2001) 684–689.

[33] H.C.W. Gundlach, Interaction between a vacuum arc and an axial magnetic field, Proc. VIIIth Int. Symp. on Discharges and Electrical Insulation in Vacuum, 1978, pp. A2-1–A2-11.

[34] M.B. Schulman, H. Schellekens, Visualization and characterization of high-current diffuse vacuum arcs on axial magnetic field contacts, IEEE Trans. Plasma Sci. 28 (2000).

[35] O. Morimiya, S. Sohma, T. Sugawara, H. Mizutani, High current vacuum arc stabilized by axial magnetic field, IEEE Trans. Power Appar. Syst. 92 (1973) 1723–1732.

[36] M. Keidar, M.B. Schulman, E.D. Taylor, Model of a diffuse column vacuum arc as cathode jets burning in parallel with a high-current plasma core, IEEE Trans. Plasma Sci. 32 (2) (2004) 783–791.

[37] E.D. Taylor, M. Keidar, Transition mode of the vacuum arc in an axial magnetic field: comparison of experimental results and theory, IEEE Trans. Plasma Sci. 33 (5) (2005) 1527–1531.

[38] M. Keidar, E.D. Taylor, A generalized criterion of transition to the diffuse column vacuum arc, IEEE Trans. Plasma Sci. 37 (5) (2009) 693–697.

[39] L. Brieda and M. Keidar, Plasma-wall interaction in Hall thrusters with magnetic lens configuration, J. Appli. Physi., 111(2012) 123302.

[40] M. Keidar, I. Beilis, R. L. Boxman, S. Goldsmith. Voltage of the vacuum arc with a ring anode in an axial magnetic field. IEEE Trans. Plasma Sci., 25(1997), 580–585.

Plasma in Space Propulsion

Plasmas are used in variety of aerospace applications including space propulsion, aero flow control, and plasma-assisted combustion. This section considers specifics of plasma application in space propulsion. Space propulsion is required for satellite motion in the outer space. The displacement of a satellite in space, orbit transfer, and its attitude control are the task of space propulsion, which is carried out by rocket engines. Rocket engines operate according to the basic principle of action and reaction. A force (thrust) F on the spacecraft is formed by ejecting a jet of gas or plasma in the backward direction. In principle, there are two major types of rocket engines which are distinguished by the energy source used to accelerate the gas, namely, chemical and electric rockets. Electric propulsion uses the electric energy form to energize or to accelerate the propellant. The electric propulsion will be considered in which the electrical energy is used to accelerate the propellant in the form of plasma or plasma propulsion [1]. Plasma propulsion utilizes the electric energy to, first, ionize the propellant and, then, deliver energy to the resulting plasma leading to plasma acceleration.

Important characteristic of the plasma thruster is the total propulsive action on a spacecraft which is measured by ΔV. The term ΔV constitutes the sum of all velocity changes by the spacecraft due to thruster operation. In the proximity of large celestial bodies, thrust is needed to compensate various ambient forces, including atmospheric drag and radiation pressure. Such orbital maintenance of the satellite (for instance GEO communication satellite) requires $\Delta V \sim 0.6$ km/s for a 10-year period. The ΔV of deep-space missions is typically of several km/s and ΔV of orbit transfer is about few km/s.

Many types of plasma thruster have been developed over the last 50 years. The variety of these devices can be divided into three main categories depending on the mechanism of acceleration [2]: (i) electrothermal, (ii) electrostatic, and (iii) electromagnetic.

1. *Electrothermal* thruster employs electric energy to heat the gas. In this case, the acceleration mechanism is based on the pressure gradient, which is the same acceleration mechanism as the one employed in chemical thrusters.
2. Electrostatic thruster directly employs electric field to accelerate ions.
3. Electromagnetic thrusters wherein combination of electric and magnetic fields is employed to accelerate the propellant.

In this chapter, the plasma physics and engineering of the thrusters based mainly on the electromagnetic plasma acceleration will be described. Toward the end it will be addressed some basic physics associated with electrostatic acceleration, in particular, ion thruster. Thus this description will not be all inclusive. The selection of these particular cases was motivated by opportunity to demonstrate various physical phenomena present in the chosen mechanisms of plasma acceleration.

Plasma Engineering.

5.1 **Plasma in ablative plasma thrusters**

In this section we describe plasma formation and acceleration in plasma devices based on material ablation. In electrical discharges where plasma generation occurs at the surface such as a vacuum arc or an ablation-controlled discharge (pulsed plasma thruster (PPT), capillary discharge, electroguns, circuit breakers, laser ablation devices, etc.), the plasma—wall interaction plays an extremely important role [2—8]. The accurate description of the plasma—wall interaction is of importance for theoretical investigations of such discharges, because it helps to formulate proper boundary conditions for the overall discharge model. There are several characteristic subregions near the surface in which different physical phenomena play a leading role, namely space-charge sheath, Knudsen layer, presheath, and a region where transition to ionization equilibrium occurs or nonequilibrium ionization region (NEIR). These subregions constitute the entire transition region between the equilibrium plasma and the wall. The role of NEIR in the overall dynamics of the transition region is critical for calculation of the energy and particle fluxes to the wall [9—11].

There are some physical situations in which plasma is accelerated by the external forces in the region near the wall. An example of this is ablative pulsed plasma accelerators [12]. In these accelerators, rapid heating of a thin dielectric surface layer leads to decomposition of the wall material. Ablative products are ionized in the vicinity of the dielectric and are accelerated by an electromagnetic force. Ionization of the ablative products and plasma acceleration take place in the same region and therefore both phenomena are interrelated. These phenomena are interesting from the standpoint of smooth transition between the two ablation modes with small (subsonic) and sonic plasma velocity at the edge of the Knudsen layer that strongly affects ablation [3,13]. Knudsen layer was defined in Chapter 1. The conditions at the Knudsen layer edge are very important for calculation of the flux of returned particles to the surface. For instance, it was shown that the backflux is responsible for surface carbonization in the case of Teflon™ ablation [14]. The effect consists in the flux of returned particles to the surface that is determined by the bulk plasma density and temperature. In principle, two limits are possible. When plasma density is very high, the flux of returned particles is strong. As a result, the velocity at the edge of the Knudsen layer and the corresponding ablation rate is small. The opposite limit corresponds to the case when plasma density is small. The actual limit for the high and small plasma density is determined by specific conditions of the experiment but is close to the equilibrium vapor density. The plasma density and temperature depend on the flow conditions in the plasma bulk. One can expect that there should be smooth continuous transition between these two regimes. In small plasma devices, the spatial extension of the ionization layer may become comparable to the size of the device. An example of such a device is a micro-PPT (μ-PPT) [15].

5.1.1 **Ablation phenomena and the Knudsen layer**

The problem of evaporation under the condition of strong plasma acceleration near the ablated surface is studied. An example of a PPT [12—15] is considered (Figure 5.1). In the PPT, a discharge is struck between two electrodes across the dielectric (propellant). Propellant is heated during the discharge by the plasma and ablates. The ablated vapor is ionized in the vicinity of the propellant and is accelerated by the electromagnetic force.

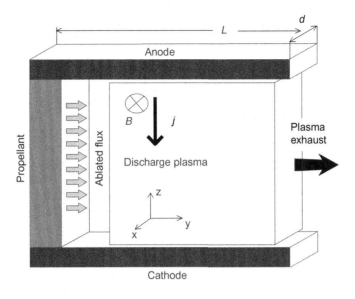

FIGURE 5.1

Schematic of the electromagnetic ablative PPT.

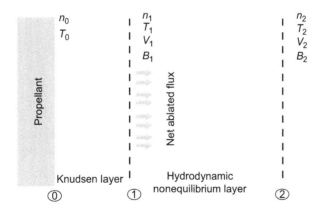

FIGURE 5.2

Schematic of the near surface layers.

Similar to the previous work [13], we consider the multilayer structure of the near surface region (Figure 5.2). One can distinguish two different characteristic regions between the surface and the plasma bulk: (1) a kinetic nonequilibrium layer adjacent to the surface with a thickness of a few mean free paths (the Knudsen layer); boundary 1 corresponds to the Knudsen layer edge, (2) a collision-dominated (hydrodynamic) region; boundary 2 constitutes the edge of the hydrodynamic region. The definitions for the edge of the Knudsen layer and the hydrodynamic region are considered below.

We start from a formulation of the equation set for the hydrodynamic region. For our primary purpose of considering the ablation effects, we simplify the present analyses and assume that ionization equilibrium is reached near the edge of the hydrodynamic region (boundaries 2). Detailed analysis of the ionization phenomena is presented in the Section 5.1.2. Here we assume that the thickness of the NEIR is smaller than the thickness of the hydrodynamic region. The specific conditions under which this assumption is justified are considered in the next section. In this case, the electron density at the boundary 2 will be calculated using Saha equilibrium. In the typical range of parameters (plasma density is about $10^{23} - 10^{24}$ m^{-3} and electron temperature is about 2–4 eV), almost fully ionized plasma is expected according to Saha equilibrium. The plasma parameters (n,T) in the hydrodynamic region depend on the conditions at the boundary 1, which is the Knudsen layer edge. To find the parameters at the edge of the Knudsen layer as a function of velocity at the Knudsen layer edge V_1, we use the mass, momentum, and energy conservation equations in accordance with relations (listed in the Chapter 1, Eq. 1.230) between parameters at the edge of the Knudsen layer and those at the surface presented in Refs [16–18] and compared with Monte Carlo simulation in Ref. [19].

Firstly, the simple model for the heavy-particle (including ions and atoms) flow is considered. Starting from the boundary 1, in the hydrodynamic region, we apply the conservation laws for mass, momentum, and energy. We will consider a simplified, single-fluid MHD approximation taking into account the self-magnetic field. The magnetic field has a primary component in x-direction as shown in Figure 5.1 and decreases between boundaries 1 and 2 due to magnetic diffusion. The mass conservation equation for the heavy particles:

$$\frac{d(nV)}{dy} = 0 \tag{5.1}$$

where n is the plasma density and V is the velocity. The momentum conservation equation for the plasma

$$M(nV)\frac{dV}{dy} = -\frac{d(nkT)}{dy} + j \times B \tag{5.2}$$

where M is the atomic mass, T is the plasma temperature, j is the current density, and B is the magnetic field. The relation between current density and magnetic field can be determined from Faraday's law:

$$j = -\frac{1}{\mu}\frac{dB}{dy} \tag{5.3}$$

Substituting (5.1) into Eq. (5.2) and integrating Eq. (5.2), one can obtain the following relationship between parameters at the boundaries 1 and 2:

$$n_1 kT_1 + Mn_1 V_1^2 + \frac{1}{2}\frac{B_1^2}{\mu} = n_2 kT_2 + Mn_2 V_2^2 + \frac{1}{2}\frac{B_2^2}{\mu} \tag{5.4}$$

The plasma density and velocity distribution in the hydrodynamic region as well as the values of B_1 and B_2 depend on the specific geometry of the accelerator. Let us consider a planar geometry as an example (see Figure 5.1). Further analyses will be based on additional assumptions regarding the plasma acceleration in the electromagnetic thruster. Using the MACH2 computer code, it was demonstrated that the magnetosonic condition exists at the thruster exit plane [20]. The magnetosonic condition is associated with the Alfven critical speed V_c[21]. This velocity is proportional to the ionization potential, $V_c = (2eU_i/M)^{0.5}$, where U_i is the ionization potential. Originally, Alfven [22] put this idea forward in his theory on the origin of the solar system. The idea states that if the relative velocity between plasma and neutral gas in a magnetic field reaches or exceeds a certain critical value, the gas is efficiently ionized. This velocity is determined by the condition that the ion kinetic energy is equal to the ionization energy of the gas. Sometimes this relation is called the Alfven hypothesis and is used in the model of plasma acceleration in a PPT [23]. The Alfven hypothesis was experimentally verified and shown to be correct [24]. Recent simplified model of the plasma acceleration showed that the Alfven sound speed is proportional to the Alfven critical velocity and is about 50% higher [21]. Also recent simulations using MACH2 shows [20] that the flow speed near the exit of the thruster equals the local value of the Alfven wave speed. It was shown that in the framework of the MHD approach, there exists the so-called magnetosonic condition, i.e., plasma accelerates up to the velocity equal to the Alfven sound speed. Based on the above, we will make here an assumption about the existence of a magnetosonic point where the flow velocity equals Alfven's wave speed as shown in Figure 3.2. The assumption about the magnetosonic point allows us to reduce the number of unknowns in the problem. We take the boundary 2 at the distance where the plasma velocity V_2 approaches the Alfven velocity. In other words, this assumption is equivalent to the condition that most of the plasma acceleration takes place in the region between boundaries 1 and 2:

$$V_2 = \frac{B_2}{(\mu n_2 M)^{0.5}} \tag{5.5}$$

where n_2 is the electron density at the boundary 2. To close the system of equations for the hydrodynamic layer, and taking into account that the magnetic field decreases along the channel, we can estimate the magnetic field at the edge of the Knudsen layer (boundary 1) as

$$B_1 = \mu I/d \tag{5.6}$$

where d is the characteristic length of the electrode width (Figure 5.1) and I is the current. Combining Eqs (5.4)–(5.6), one can obtain the following expression for the velocity at the outer boundary of the kinetic (Knudsen) layer:

$$\frac{MV_1^2}{2kT_1} = \frac{(n_1/2) - (T_2 n_2/2T_1) + (1/4)(\mu(I/d)^2/kT_1)}{(3/2)(n_1^2/n_2) - n_1} \tag{5.7}$$

One can see that the velocity at the edge of the Knudsen layer depends upon density at the exit plane (boundary 2), electrode geometry (d), and current. In the case of $I = 0$ (i.e., electrothermal thruster or ablative capillary discharge), this equation reduces to the one obtained previously (Eq. (5.4) in Ref. [13]).

In order to study the general trends of the dependence of the velocity at the Knudsen layer edge V_1 on the current density without details of the flow field along the channel, we consider the plasma density at the magnetosonic plane (boundary 2, Figure 5.2) as a parameter. The calculated velocity V_1 is shown in Figure 5.3 as a function of current density. Parameters (T_1 and n_1) at the Knudsen layer edge (boundary 1) are calculated using Knudsen layer relations (Chapter 1). Heavy-particle density distribution along the acceleration channel depends upon the specific geometry of the channel (for instance, the length of the channel and interelectrode distance as

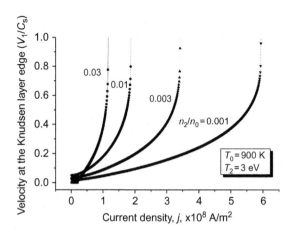

FIGURE 5.3

Velocity at the edge of the Knudsen layer as a function of current density with density at the edge of hydrodynamic layer as a parameter. $L = d = 1$ cm.

Source: *Reprinted with permission from Ref. [251]. Copyright (2004) by American Institute of Physics.*

shown in Figure 5.1). One can see that the solution with a small current density corresponds to small velocity in comparison to the sound speed. The calculation shows that the value V_1 is close to that obtained in the case of an electrothermal plasma accelerator [25] and corresponds to ablation into the dense background plasma [3,13,17]. The velocity V_1 increases with current density and approaches the sound speed limit at some high-current density dependent on the plasma density at the boundary 2. According to the model, the electromagnetic force in the region between boundaries 1 and 2 leads to significant plasma acceleration that in turn affects the plasma parameters (n_1 and T_1) at boundary 1. At some value of current density, the solution approaches the limit of the local sound speed. Note that no solution exists beyond the case when velocity V_1 equals to the sound speed, since no physical mechanism for supersonic acceleration in the Knudsen layer exists. Therefore, density n_2 when $V_1 = C_s$ can be considered as a limiting possible density under the considered conditions. This means that a combination of the current density and the geometry (that determine the flow field, i.e., n_2) uniquely determine the solution. This rather general model (in which three parameters are considered as known) can be coupled with a model of the plasma flow and details of the energy conservation in the particular system. In this case, the number of free parameters will be reduced and a self-consistent solution can be obtained. An example of such a model in the case of an electrothermal device was presented earlier [25].

Plasma acceleration in the hydrodynamic layer causes transition from the ablation mode with small velocity at the Knudsen layer edge and significant backflux to the ablation mode with sonic velocity at the Knudsen layer edge, which is the situation for ablation into vacuum. When the plasma velocity is sufficiently small [3,17,25], it can be assumed that equilibrium ionization (according to Saha equation) takes place. However, when strong plasma acceleration takes place in the hydrodynamic region, this may affect equilibrium conditions. In the next section, we will consider ionization processes under the effects of plasma acceleration.

5.1.2 Ionization in the presence of plasma acceleration in the hydrodynamic region

We start our consideration from the Knudsen layer edge. The reason for this is that in the typical conditions, the mean free path is much larger than the Debye length and therefore the sheath thickness is much smaller than that of the Knudsen layer length. This simplifies the problem significantly since the boundary conditions are known at the Knudsen layer edge. The system of equations reads (using the fluid description of plasma as described in Chapter 4):

$$\frac{d(n_i V_i)}{dy} = \alpha_i n_i n_a - \alpha_r n_i^2 \tag{5.8}$$

$$n_i k(T_h + T_e) + n_a k T_h = P \tag{5.9}$$

$$M n_i V_i \frac{dV_i}{dy} = -\frac{d(P_i + P_e)}{dy} - \frac{1}{2\mu}\frac{d(B^2)}{dy} - M\alpha_i n_i n_a V_i \tag{5.10}$$

$$j = \sigma(E + V \times B) = -\frac{1}{\mu}\frac{dB}{dy} \tag{5.11}$$

where α_i is the ionization rate, α_r is the recombination rate, n_i, n_a are ion and neutral densities, T_e, T_h are electron and heavy-particle (ions, atoms) temperatures, P_i, P_e, P are ion, electron, and total pressure, respectively. We will use the equation of state for an ideal gas (i.e., $P = nkT$). Due to the high electron conductivity, electron temperature variation across the ionization layer is expected to be small and is neglected. We will assume that the ionization rate determines the length of the ionization layer and therefore for simplicity recombination is neglected. As a first approach, we will consider plasma flow with uniform current distribution that approximately corresponds to the slow current layer stage of a PPT [12,26]. This means that we assume that $j = \sigma(E + V \times B) = $ constant.

We use the following dimensionless variables: $\xi = y\alpha_i n_{io}/C_{ia}$; $b = B/B_1$; $n = n_i/n_{io}$; $v = V_i/C_s$, where C_{ia} is the heavy-particle thermal velocity, B_1 is the magnetic field at the edge of the Knudsen layer, n_{io} is the plasma density in the hydrodynamic region at the edge of the NEIR. In the absence of acceleration, the NEIR edge corresponds to full ionization, i.e., $n = 1$. This density is determined by the conditions in the plasma bulk. By NEIR definition, this density can be determined from equilibrium analyses. Taking the above into account, the system reduces to the following:

$$\frac{db}{d\xi} = -\alpha \tag{5.12}$$

$$(v^2 - 1)\frac{dv}{d\xi} = -(v^2 + 1)\varepsilon^{0.5}(1 - n) - \frac{v}{2n}\beta\frac{d(b)^2}{d\xi} \tag{5.13}$$

$$v\frac{dn}{d\xi} = -n\frac{dv}{d\xi} + n\varepsilon^{0.5}(1 - n) \tag{5.14}$$

where $\alpha = C_{ia}/(d\alpha_i n_{io})$; $\beta = (V_a/C_s)^2$; $\varepsilon = (T_i + T_e)/T_i$; and $V_a = (B_1/\mu m n_{io})^{0.5}$, and d is the electrode width. For this system of equations, the following boundary conditions can be used: $n = 0$ and $b = 1$. Another additional boundary condition is the velocity at the edge of the hydrodynamic region. It was shown in the previous section that the velocity at the edge of the Knudsen layer varies in the range from very small $v \ll 1$ up to the local sonic speed. Note that the velocity at the Knudsen layer edge (shown in Figure 5.3) is normalized by the local sound speed determined by the heavy-particle temperature at the Knudsen layer edge. This temperature is close to the surface temperature and much smaller than that of the electrons in the ionization layer. Therefore, a physically reasonable condition for the velocity is $v = 0$ when $\xi = 0$.

5.1.2.1 Limit of small plasma acceleration

Firstly let us consider an asymptotic solution for the case of small plasma acceleration ($v^2 \ll 1$). In this case, Eq. (5.13) reduces to

$$\frac{dv}{d\xi} = \varepsilon^{0.5}(1 - n) + \frac{v}{2n}\beta\frac{d(b)^2}{d\xi} \tag{5.15}$$

By substitution of this equation and Eq. (5.12) into Eq. (5.14), we have

$$\frac{dn}{d\xi} = \alpha\beta(1 - \alpha\xi) \tag{5.16}$$

The last equation has a solution in the form

$$n = \alpha\beta\xi - 0.5\alpha^2\beta\xi^2 \tag{5.17}$$

One can see that the problem is determined by two parameters α and β. Therefore, the ionization region thickness where $n = 1$ (i.e. complete ionization) can be determined as a solution of this equation:

$$\xi^* = \frac{1}{\alpha}\left(1 - \sqrt{1 - \frac{2}{\beta}}\right) \tag{5.18}$$

in the most common case of $\beta \gg 1$, the last expression can be approximated as

$$\xi^* \approx 1/(\beta\alpha) \tag{5.19}$$

where ξ^* is the thickness of the ionization layer.

5.1.2.2 Regular sonic transition

In this section, we describe the asymptotic behavior of the solution in the case of strong acceleration, i.e., $v \to 1$. Let us first calculate the plasma density near the sonic transition place ($v = 1$). The physically meaningful solution with smooth transition requires that the right-hand side of Eq. (5.13) be zero at the sonic plane. In general, using *L'Hospital's* rule, one can find the velocity gradient near the sonic plane, which is finite. This procedure was used by a number of authors [27,28]. The numerator on the right-hand side of Eq. (5.13) has the following form in this case:

$$0 = -(v^2 + 1)\varepsilon^{0.5}(1 - n) - \frac{v}{2n}\beta\frac{d(b)^2}{d\xi} \tag{5.20}$$

Using the above assumption about constant current density, one can find from Eq. (5.12) that $(1/2)(\mathrm{d}(b)^2/\mathrm{d}\xi) = -\alpha(1 - \alpha\xi)$. Taking this into account, the equation for the density reads

$$n^2 - n + \frac{\alpha\beta}{2\varepsilon^{0.5}}(1 - \alpha\xi) = 0 \tag{5.21}$$

Considering that the ionization layer thickness does not change significantly near the sonic transition region (as shown below), we can approximate the thickness as approximately $1/\alpha\beta$ (according to Eq. (5.19). Taking all the above into account, finally the physically meaningful solution that allows smooth transition has the following expression for density in the vicinity of the sonic point:

$$n = 0.5\left(1 - \sqrt{1 - \frac{2\alpha\beta}{\varepsilon^{0.5}}\left(1 - \frac{1}{\beta}\right)}\right) \tag{5.22}$$

This is an expression for density behavior at the sonic plane in the case of regular sonic transition. One can see that a small degree of ionization due to significant convection (plasma acceleration) ($n < 1$) can be expected if the function $2\alpha\beta(1 - 1/\beta)/\varepsilon^{0.5} < 1$ as will be shown in the following section.

5.1.2.3 Numerical examples

In this section, we describe some numerical examples of the solution of the ionization layer problem (Eqs (5.12)−(5.14)) which depends on three parameters α, β, and ε. These calculations are shown in Figures 5.4 and 5.5. Plasma density and velocity in the ionization layer with α as a parameter are shown in Figure 5.4. By definition, the parameter α is inversely proportional to the ionization rate. One can see that when α is large, the ionization layer thickness (determined by condition $n = 1$) becomes smaller. This happens because in the subsonic flow considered here, collisions (ionization in our case) lead to acceleration (which is usual for any one-dimensional (1D) gas flow). When α is small enough ($\alpha < 0.03$), the ionization equilibrium condition cannot be met. This is shown in the bottom of Figure 5.4, where plasma velocity in the ionization layer is plotted. It can be seen that the plasma accelerates up to the sonic velocity when α is small before ionization equilibrium can be reached (Figure 5.4).

Similarly, a dependence on the parameter β is found. Parameter β represents an effect of the electromagnetic acceleration. When β is large, plasma acceleration is small in the subsonic region and as a result ionization equilibrium can be reached.

One can conclude that there is a range of parameters when equilibrium ionization can be achieved, i.e., when $\beta > 60$ and $\alpha < 0.03$. These conditions are necessary conditions for the existence of the ionization equilibrium near the ablated surface under strong acceleration effects. An important conclusion that can be derived is that in order to have significant ionization, the ionization and acceleration regions must be separated. It should also be noted that the above numerical results are obtained considering only the subsonic region and therefore a singularity is found. In reality, this singularity will disappear if the condition for regular sonic transition is taken into account. In this case, the electron density is calculated according to Eq. (5.22).

5.1.3 Example: Application to the carbon−fluorine plasma in a μ-PPT

In this section, we apply the above results to a particular plasma device—a μ-PPT. This is a simplified version of a PPT that was recently developed at the Air Force Research Laboratory (AFRL) for delivery of very small impulse bit [15]. In order to make analyses of this particular device, we

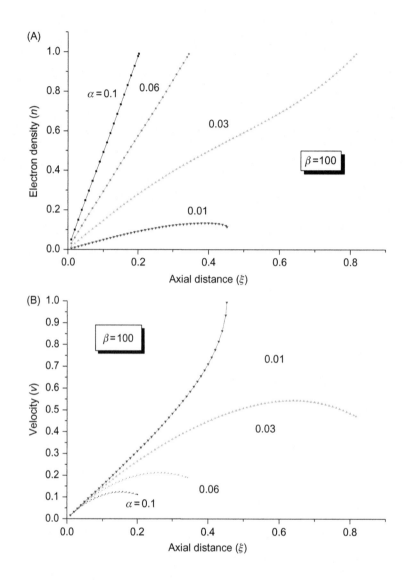

FIGURE 5.4

Plasma density and velocity distribution in the ionization layer with α as a parameter.

Source: *Reprinted with permission from Ref. [251]. Copyright (2004) by American Institute of Physics.*

estimate the parameter β and we calculate the current density distribution in the vicinity of the ablated surface instead of using an assumption about constant current density. The main objective of considering this particular example is to show under which conditions the effect of nonequilibrium ionization due to plasma acceleration is important. The characteristic ionization and recombination times are less than typical time for discharge parameter change in PPT [4,25] and therefore steady-state ionization model developed in previous sections is employed.

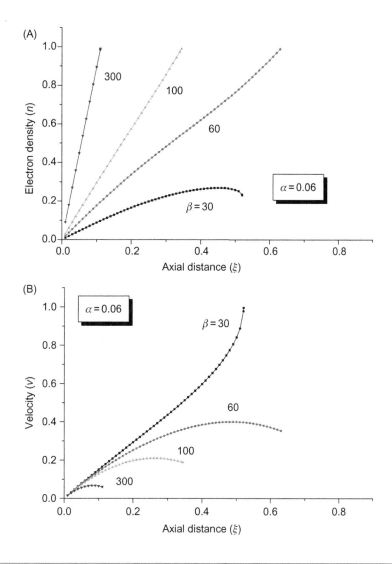

FIGURE 5.5

Plasma density and velocity distribution in the ionization layer with β as a parameter.

Source: *Reprinted with permission from Ref. [251]. Copyright (2004) by American Institute of Physics.*

The schematic of the plasma layer near the ablated surface is shown in Figure 5.6. The plasma is created near the surface and the plasma layer is separated from the surface by the ionization layer described above. In order to calculate plasma properties, we develop a model that includes Joule heating, heat transfer from the plasma to the wall, and dielectric (Teflon) ablation. Mechanisms of the energy transfer from the plasma include heat transfer by particle convection and radiation. It is assumed that within the plasma layer, all parameters vary in the

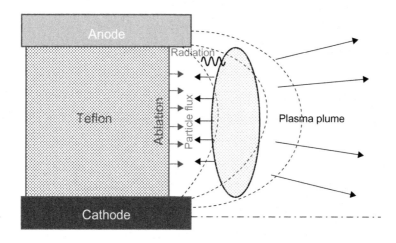

FIGURE 5.6

Schematic of the plasma–surface interaction in a μ-PPT.

Source: *Reprinted with permission from Ref. [251]. Copyright (2004) by American Institute of Physics.*

radial direction as shown in Figure 5.6. The energy balance equation can be written in the following form:

$$\frac{3}{2}n_e dT_p/dt = Q_J - Q_r - Q_F \tag{5.23}$$

where Q_J is the Joule heat, Q_r is the radiation heat, and Q_F is the heat associated with particle flux.

According to Ref. [29], radiation in the continuum from a $C + 2F$ plasma in the considered parameter range provides the main contribution. The radiation energy flux Q_r includes the radiation for a continuum spectrum based on a theoretical model [30,31]. In the expression for $Q_r = AZ_i^2 n_e^2 T^{1/2}(1 + \chi_g)$, the coefficient A is a constant ($1.6 \cdot 10^{-38}$ in SI units) and $\chi_g = E_g/T_p$ with E_g as the energy of the low excited state. The particle convection flux Q_F includes energy associated with electron and ion fluxes to the dielectric wall that leads to plasma cooling. It was shown previously that the energy is carried off mainly by the particle convection [4,25].

The full system of equations is described in detail in Refs [4,25]. The general procedure of calculation of the plasma properties consists of the following. Electron temperature is calculated from the energy balance. For known plasma density and electron temperature, the heat flux to the surface is calculated. In turn, the surface temperature is calculated from the heat transfer equation with boundary conditions that take into account vaporization heat and conductivity. The solution of the heat transfer equation is considered for two limiting cases of small and large ablation rate similar to that described in Ref. [4]. The density at the Teflon surface is calculated using the equilibrium pressure equation for Teflon and the density at the Knudsen layer edge is determined in the framework of the kinetic model. For known pressure and electron temperature, one can calculate the chemical composition of the $C + F$ plasma assuming local thermodynamic equilibrium (LTE) [4,25] or using the nonequilibrium ionization approach developed in Section 5.1.2. The results of calculation according to these two models will be compared.

In order to calculate current density in the near field (as shown in Figure 5.6) we assume that the magnetic field has only an azimuthal component and also neglect the displacement current. The main plasma density gradient developed in the axial direction and therefore gradient mechanism [32] should not affect magnetic transport in this system. The combination of the Maxwell equations and electron momentum conservation gives the following equation for the magnetic field in the case of isothermal flow:

$$\partial \mathbf{B}/\partial t = 1/(\sigma\mu)\nabla^2 \mathbf{B} - \nabla \times (\mathbf{j} \times \mathbf{B}/(en_e)) + \nabla \times (\mathbf{V} \times \mathbf{B}) \qquad (5.24)$$

A scaling analysis shows that the various terms on the right-hand side of Eq. (5.24) may have importance in different regions of the plasma plume and therefore a general end-to-end plasma plume analysis requires keeping all terms in the equation. In the case of the near plume of the μ-PPT with a characteristic scale length of about 1 cm, the magnetic Reynolds number $Re_m \ll 1$ and therefore the last term can be neglected.

In addition, our estimations show that the Hall parameter $(\omega\tau) \ll 1$ if the plasma density near the Teflon surface $n_e > 10^{23}$ m^{-3}. This case is realized in the μ-PPT so the Hall effect is expected to be small for this particular case. Therefore, all results presented below are calculated from simple magnetic diffusion equation without considering the Hall effect, i.e., the second term in the right-hand side of Eq. (5.24) can also be neglected.

The following boundary conditions are employed. We assume that the current is uniform on both electrodes that allow us to estimate the current density on the cathode j_c and on the anode j_a. The magnetic field is assumed to vary as $1/r$ on the upstream boundary between the electrodes. At the lateral boundary, we assume that the normal current $j_n = 0$. The downstream boundary is considered to be far enough away that $B = 0$ can be assumed and finally along the centerline the magnetic field is zero. More details about this model can be found elsewhere [14,33].

From the magnetic transport equation (Eq. (5.24)), the magnetic field and current density distributions can be calculated. In order to calculate the ionization layer properties near the Teflon surface, we will use the magnetic field distribution in the plume instead of assuming $j = $ const as was done in the above.

5.1.4 Ablation-produced plasma: example of Teflon ablation

In order to calculate the ionization rate for the carbon–fluorine plasma, the electron-impact ionization cross sections (available from a database [34]) are used. In these calculations, a specific example is considered [14,15]. The μ-PPT geometry and scale is shown in Figure 5.6. It has a central electrode (cathode) with radius of about 0.4 mm and an outer electrode (anode) with inner radius of about 1.5 mm. The ablative propellant is Teflon.

Firstly, we calculate the ionization layer properties (according to Eqs (5.12)–(5.21)) to understand if ionization equilibrium is achieved. In the Teflon surface temperature–electron temperature $(T_s - T_e)$ plane, these dependencies are shown in Figure 5.7. It can be seen that in general, both regimes with ionization equilibrium and without it can be achieved in the considered range of parameters. One can see that the separation curve depends upon the total current in the discharge. For comparison, we calculate the $T_s - T_e$ dependence from plasma layer model (see previous sections). These data are shown in Figure 5.8. It follows from comparison of Figures 5.7 and 5.8 that during the discharge there are regimes where equilibrium cannot be achieved. Therefore, the plasma

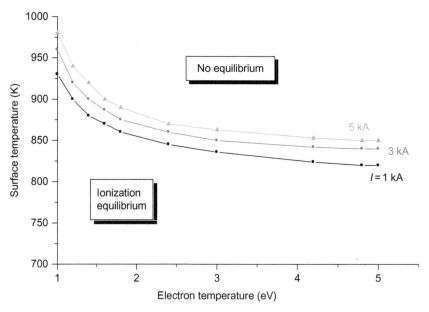

FIGURE 5.7

Separation between two regions with equilibrium and nonequilibrium ionization conditions.

Source: *Reprinted with permission from Ref. [251]. Copyright (2004) by American Institute of Physics.*

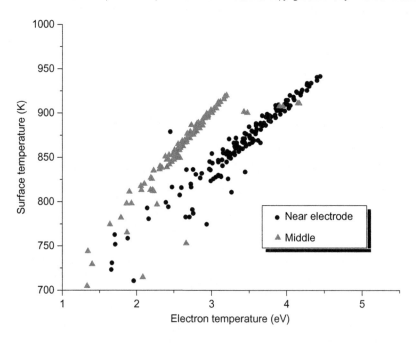

FIGURE 5.8

Calculated dependence of the surface temperature and electron temperature. Sampling was taken at two locations, midway between the electrodes and near the outer electrode.

Source: *Reprinted with permission from Ref. [251]. Copyright (2004) by American Institute of Physics.*

composition in the plasma layer cannot always be calculated based on LTE assumption but rather the more general model (Eqs (5.12)–(5.21)) should be employed.

5.1.5 On the ablation mode

Ablation modes were considered in Refs [13,16,18,19]. It was shown that dependent on the density of the background plasma, there are two solutions with the same ablation rate. One solution corresponds to the case of ablation into dense plasma when backflux is very strong. This condition is realized for instance in an electrothermal PPT [25], an ablation-controlled capillary discharge [13], and in the cathode spot of the vacuum arc [3,17]. On the other hand, in the case of small background plasma density, the ablation mode is close to that realized in the vacuum case. This mode is realized for instance in an electromagnetic PPT and laser ablation [14,16].

The main structure that is studied in this respect is the Knudsen layer, which is the layer attached to the surface in which the distribution function of the emitted particles is transformed from a half-Maxwellian into a drifted Maxwellian at the Knudsen layer edge [16]. This transformation can be viewed as jump conditions in the hydrodynamic description [16,18]. The relations between parameters at the Knudsen layer edge (boundary (1)) and surface (boundary (0)) are as described by the following set of equations [16]:

$$\frac{n_o}{2(\pi d_o)0.5} = n_1 V_1 + \beta \frac{n_1}{2(\pi d_1)0.5} \left\{ \exp(-\alpha^2) - \alpha \pi^{0.5} \, \mathrm{erfc}(a) \right\} \tag{5.25}$$

$$\frac{n_o}{4d_o} = \frac{n_1}{2d_1} \left\{ (1 + 2\alpha^2) - \beta \left[(0.5 + \alpha^2)\mathrm{erfc}(\alpha) - \alpha \exp(-\alpha^2)/\pi^{0.5} \right] \right\} \tag{5.26}$$

$$\frac{n_o}{(\pi d_o)^{1.5}} = \frac{n_1}{(d_1)^{1.5}} \pi^{-1} \left[\alpha(\alpha^2 + 2.5) - \beta/2 \left\{ (2.5 + \alpha^2)\alpha \, \mathrm{erfc}(\alpha) - (2 + \alpha^2)\exp(-\alpha^2)/\pi^{0.5} \right\} \right] \tag{5.27}$$

where $d_o = m/2kT_o$, $d_1 = m/2kT_1$, $\alpha = U_1/(2kT_1/m)^{0.5}$; $\mathrm{erfc}(\alpha) = 1 \, \mathrm{erf}(\alpha)$, $\mathrm{erf}(\alpha)$ is the error function, T_o is the surface temperature, and n_o is the equilibrium density. One can see that in this system of equations, the velocity at the edge of the Knudsen layer remains a free parameter.

Mainly, the ablation mode is determined by the mechanisms of plasma acceleration in the immediate vicinity of the ablated surface. In order to study this effect, the model of the hydrodynamic layer should supplement the Knudsen layer analyses. Using particle and momentum conservation, the velocity at the edge of the Knudsen layer can be calculated. This approach was undertaken in Ref. [13] where plasma acceleration in the hydrodynamic layer without magnetic force effects was investigated. To illustrate the effect of the background plasma parameters on the ablation rate, the contours of the ablation rate in the plasma density–surface temperature plane with electron temperature as a parameter can be calculated. These calculations are shown in Figure 5.9. Comparing ablation rates corresponding to several electron temperatures (T_2), one can see that in the case of larger electron temperature, the ablation rate decreases. Another interesting effect that is shown in Figure 5.9 is that in the low-density regime, electron temperature does not affect the ablation rate. On the other hand, in the high-density regime, there is a very strong effect of T_2 on the ablation rate. This happens because the total ablation rate is the resulting flux between the primary flux and the backflux. In the case of higher T_2, the backflux increases due to the higher electron temperature in the plasma bulk leading to a decrease in the ablation rate.

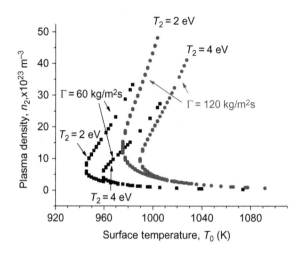

FIGURE 5.9

Ablation contours with electron temperature as parameter.

Source: *Reprinted with permission from Ref. [251]. Copyright (2004) by American Institute of Physics.*

5.1.6 Electrothermal capillary-based PPT

PPTs have the combined advantages of system simplicity, high reliability, low average electric power requirement, and high specific impulse [12]. The PPT is considered as an attractive propulsion option for orbit insertion, drag makeup, and attitude control of small satellites. PPTs, however, have very poor performance characteristics with an efficiency [35] at the level of about 10% leaving an opportunity for substantial improvement. To improve the PPT performance, several directions are being considered, such as elimination of later ablation and choice of the proper current waveform [36].

Currently new PPT devices with electromagnetic acceleration mechanism [37,38] and electrothermal mechanism are under development [39–41] that reflect the increased interest in these types of thrusters. In this study, we concentrate on a PPT device developed at the University of Illinois, the so-called PPT-4 [39,42]. This is an electrothermal device that derives most of its acceleration from the electrothermal or gasdynamic mechanism. This thruster is axially symmetric and a discharge occurs between the annular cathode at the thruster exit plane and the circular metal anode located at the far end of a cylindrical cavity made of Teflon (Figure 5.10). The plasma generated inside this cavity is accelerated in a diverging nozzle that is attached to the downstream end of the cavity. The device has a pulse length of about 10 μs, and the overall specific impulse was measured to be 850 s. The main physical processes in this type of PPT occur in the Teflon cavity in a similar way to an ablation-controlled discharge. Rapid heating of a thin dielectric surface layer leads to decomposition of the material of the wall. As a result of heating, decomposition, and partial ionization of the decomposition products, the total number of particles increases in the cavity. The problem of the ablation-controlled discharge has also a more general interest since it can be used for various applications such as electric fuses, circuit breakers, soft X-ray, and extreme ultraviolet sources [5–8]. In these devices, the discharge energy is principally dissipated by ablation of wall material, which then forms the main component of the discharge plasma.

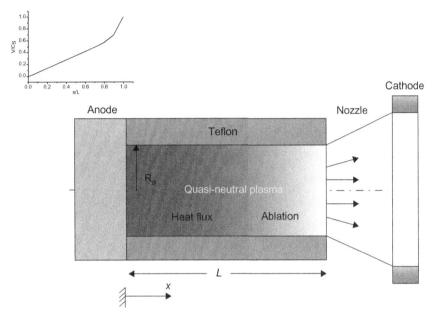

FIGURE 5.10

Schematics of the PPT.

5.1.7 The model of the ablation-controlled discharge

We consider the plasma generation processes (ablation, heating, radiation, ionization, etc.) and plasma acceleration along a Teflon cavity of a pulsed electrical discharge. Figure 5.10 shows some characteristic regions such as the Teflon bulk, the electrical sheath near the dielectric, and the quasi-neutral plasma. Different kinetic and hydrodynamic phenomena determine the main features of the plasma flow including plasma Joule heating, heat transfer to the dielectric and electrothermal acceleration of the plasma up to the sound speed at the cavity exit. Below we discuss the model in different regions and the full system of equations including the final expressions obtained in Chapter 1.

5.1.7.1 Teflon ablation

The Teflon ablation is modeled in the framework of the approximation [43] based on a previously developed kinetic model of metal evaporation into surrounding plasma [44]. In order to understand the mathematical description of the model used, we distinguish two different layers between the surface and the plasma bulk (Figure 5.11): (1) a kinetic nonequilibrium layer adjusted to the surface with a thickness of about one mean free path; and (2) a collision-dominated layer with thermal and ionization nonequilibrium. The plasma–wall transition layer includes also an electrical sheath described in the previous section. This model makes it possible to calculate the plasma parameters (density and temperature) at the interface between the kinetic and hydrodynamic layers (boundary 1 in Figure 5.11) if the velocity at this boundary V_1 is known. The velocity V_1 can be determined

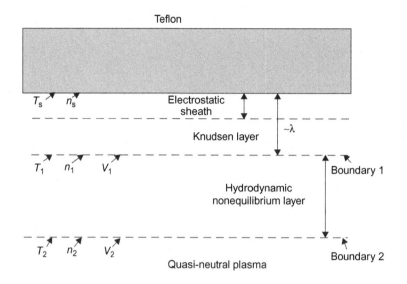

FIGURE 5.11

Schematics of the layer structure near the ablated surface.

by coupling the solution of the hydrodynamic layer and the quasi-neutral plasma. For known velocity and density at this interface, it is possible to calculate the ablation rate. In the hydrodynamic layer, the relation between the velocities, temperatures, and densities at the boundaries 1 and 2 as well as the ablation rate are formulated in the form Eq. 1.234:

$$\Gamma = mV_1 n_1 = n_1[(2kT_1/m)(T_2 n_2/2T_1 - n_1/2)/(n_1 - n_1^2/n_2)]^{0.5} \tag{5.28}$$

The system of equations is closed if the equilibrium vapor pressure can be specified that determines parameters (n_s and T_s) at the Teflon surface. In the case of Teflon, the equilibrium pressure formula is used [12,33]:

$$P_{eq} = P_c \exp(-T_c/T_s) = n_s kT_s \tag{5.29}$$

where $P_c = 1.84 \times 10^{15}$ N/m^2 and $T_c = 20,815$ K are the characteristic pressure and temperature, respectively.

5.1.7.2 Quasi-neutral plasma analyses

The energy transfer from the plasma column to the wall of the Teflon cavity consists of the heat transfer by particle fluxes and radiation heat transfer. Previous models of the ablation-controlled discharge show that the axial pressure and velocity variations are much greater than those in the radial direction [45,46]. Therefore, we will assume that all parameters vary in the axial direction x (see Figure 5.10) but are uniform in the radial direction. The axial component of the mass and momentum conservation equations reads

$$A(\partial \rho/\partial t + \partial(\rho V)/\partial x) = 2\pi R_a \Gamma(t, x) \tag{5.30}$$

$$\rho(\partial V/\partial t + V\partial V/\partial x) = -\partial P/\partial x \tag{5.31}$$

The energy balance equation can be written in the form [18]

$$\frac{3}{2}n_e(\partial T_e/\partial t + V\partial T_p/\partial x) = Q_J - Q_r - Q_F \tag{5.32}$$

The radiation energy flux Q_r includes the radiation for a continuum spectrum based on a theoretical model [30]. According to Ref. [29], the radiation in continuum from a C + 2F plasma in the considered parameter range provides the main contribution. The particle convection flux Q_F includes energy associated with electron and ion fluxes to the dielectric wall that leads to plasma cooling. Previous calculation shows [25] that the electron temperature varies only slightly with axial position and therefore we performed the calculation assuming $\partial T_e/\partial x = 0$.

Radiation and convection heat fluxes from the plasma to the cavity wall (see Figure 5.10) determine the thermal regime of the Teflon. The temperature inside the Teflon wall can be calculated from the heat transfer equation:

$$\partial T/\partial t = a\partial^2 T/\partial r^2 \tag{5.33}$$

where a is the thermal diffusivity. This is the 1D equation in the radial direction. This assumption can be made since the heat layer thickness near the surface is smaller than the Teflon cylinder curvature R_a and also less than the characteristic length of plasma parameter changes in the axial direction. In order to solve this equation, boundary and initial conditions must be specified [25]:

$$\begin{aligned} &-\lambda\partial T/\partial x(x=0) = q(t) - \Delta H \cdot \Gamma - C_p(T_s - T_o)\Gamma \\ &\lambda\partial T/\partial x(x=\infty) = 0 \\ &T(t=0) = T_o \end{aligned} \tag{5.34}$$

where $x = 0$ corresponds to the inner dielectric surface, ΔH is the ablation heat, Γ is the rate of Teflon ablation per unit area, T_o is the initial room temperature, and $q(t)$ is the density of the heat flux, consisting of the radiative and particle convection fluxes (determined by an expression similar to that used in Ref. [25]), and T_s is the Teflon surface temperature. The solution of this equation is considered for two limiting cases of substantial and small ablation rate very similar to that described in Ref. [25].

Having calculated the plasma density and electron temperature, one can calculate the chemical plasma composition considering LTE in the way described previously [25,47,48]. In the considered range of electron temperature (1–3 eV) and plasma density ($10^{21} - 10^{24}$ m^{-3}), we will assume that polyatomic molecules C_2F_4 fully dissociate and we will start our consideration from the point when we have gas containing C and F. The Saha equations for each species (C and F) are supplemented by the conservation of nuclei and quasi-neutrality.

5.1.7.3 Simulations of the plasma in PPT

In this section, we present results of the calculation of the plasma parameter temporary and spatial variation in the Teflon cavity. As working examples, two different configurations of an electrothermal PPT are considered: the so-called PPT-4 [39,42] and PPT-7 [40]. Most of the results are presented for PPT-4 with the following baseline geometry and discharge parameters: anode radius

$R_a = 3$ mm and cavity length $L = 8.3$ mm, pulse duration of about $10\ \mu s$ [39,42]. The current pulse has a peak of about 8 kA and is nonreversive because of the use of a diode across the capacitor. The current reaches a peak at about $3\ \mu s$ (see Refs [39,42] for the current waveform curves used in this model).

The following boundary condition is used for Eqs (5.30) and (5.31) at $x = 0$: $\partial n/\partial x = 0$. The plasma accelerates in the axial direction due to the pressure gradient and achieves the sound speed at the cavity exit plane, $x = L$. The plasma velocity distribution is shown in Figure 5.10 (bottom). One can see that the plasma is significantly accelerated toward the cavity exit. Burton et al. [49] obtained similar flow field development in a liquid-injected capillary discharge when the plasma flow approaches steady-state conditions.

The temporal and axial (along the cavity length) distribution of the velocity at the external boundary of the kinetic layer V_1 (boundary 1, Figure 5.11) is shown in Figure 5.12. In the entire region, the normalized velocity remains smaller than 1. This reflects the fact that the Teflon vapor velocity is subsonic due to the presence of the discharge plasma.

The time evolution of the electron temperature T_e is shown in Figure 5.13. It can be seen that the electron temperature initially increases rapidly and peaks at about 3.5 eV and then decreases to 1 eV toward the pulse end. For comparison, we also plotted experimental results [39,42]. In the experiment, it was measured that the peak electron temperature lies in the range of 2−2.6 eV and decreases down to about 1 eV and varies slightly with axial distance from the

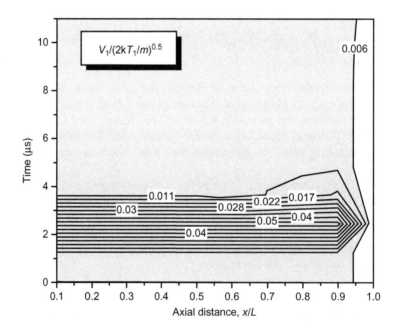

FIGURE 5.12

Velocity at the outer boundary of the Knudsen layer V_1 as a function of axial distance and time.

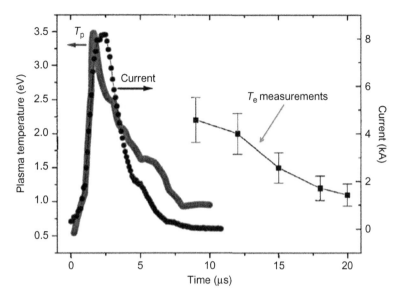

FIGURE 5.13

Temporal variation of the electron temperature in the cavity.

Source: *Experimental data taken from Ref. [39] and experimental current waveform taken from Refs [39,42]. Reprinted from Ref. [25]. Copyright by AIAA.*

thruster exit plane. The experimental data presented here were taken at several probe locations at $10-18$ cm from the thruster exit plane and therefore one can expect a shift of timescale. This shift ($\sim 7\,\mu s$) is approximately equal to the plasma flight time from the cavity to the plane where data were taken according to our simulations [17]. One can see that the model predictions and experimental data have a similar trend that indicates that the model prediction agrees reasonably with the experimental data.

To demonstrate the different ion and neutral species temporal and spatial variation, their distributions are shown in Figure 5.14. Both ion species peak at early times ($\sim 3\,\mu s$), while neutral species peak later. The relative concentration of the species changes along the cavity length and also during the discharge pulse. One can see from Figure 5.14 that the model predicts that initially plasma in the cavity is strongly ionized while after about $3\,\mu s$ the ionization degree decays.

5.1.7.4 Thruster performance
The calculated Teflon mass ablated per pulse is shown in Figure 5.15 as a function of cavity length and radius. Generally, the ablated mass increases with increasing cavity length and decreasing cavity radius. The ablated mass increasing with the cavity length is due to increasing the surface area exposed to the plasma heating. However, when the cavity radius increases, the ablated mass decreases although the ablated surface area increases. This effect is a result of the decreases in the power density (Joule heat) as the cavity radius increases. From comparison with experiment, it can be shown that the model underpredicts the ablation mass by about 25%. However, it should be

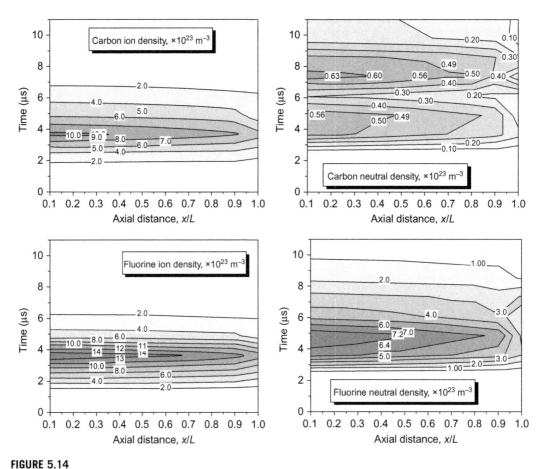

noted that some mass can be ablated in the form of large particulates. This effect for one particular PPT was estimated to be up to 40% of the total ablated mass [50]. The ablation in the particulate phase was not considered in these calculations.

To assess the ability of the model to predict some thruster performance characteristics, we calculated the thrust impulse bit as a function of cavity geometry and compare with experimental data. The gasdynamic thrust impulse is generated due to the pressure force on the anode. The total impulse generated during the pulse is calculated as

$$I_{bit} = \pi R_a^2 \int P \, dt$$

where $P = kT_e(n_a + n_i)$ is the pressure in the cavity, n_a is the neutral atom density, n_i is the ion density. We integrate the pressure during the discharge to calculate the thrust impulse bit produced

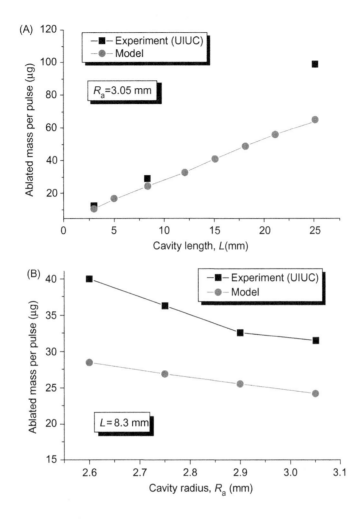

FIGURE 5.15

Ablated mass as a function of cavity geometry and comparison with experiment [39,42]: (A) constant cavity radius and (B) constant cavity length.

Reprinted from Ref. [25]. Copyright by AIAA.

in the Teflon cavity. However, the PPT-4 thruster also has a nozzle with an area ratio of about 100. Such a nozzle may increase the thrust up to a factor of $\beta_n = 1.7$ [51]. The plasma flow in the nozzle is not considered in this section and therefore we will use the nozzle factor as a parameter. The thrust impulse bit dependence on the cavity length is shown in Figure 5.16 with coefficient β_n as a parameter. One can see that the impulse bit increases by a factor of 3 when L increases from 3 up to 25 mm similar to that obtained in the experiment [39,42]. It should be noted that in the entire range of L, the best agreement between the model and experiment is obtained for $\beta_n = 1.3-1.7$. One can conclude that in electrothermal PPT the nozzle has an important effect.

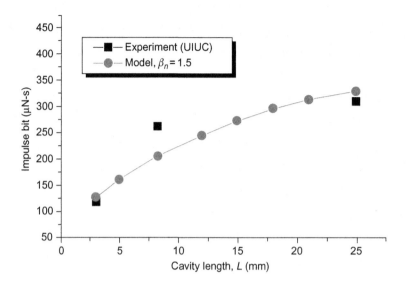

FIGURE 5.16

Thrust impulse bit variation with cavity length with nozzle factor β_n as a parameter and comparison with experiment.

Reprinted from Ref. [25]. Copyright by AIAA.

5.2 Bulk plasma and near-wall phenomena in Hall thruster

Another example from the family of electromagnetic plasma thruster is the so-called stationary plasma thruster (SPT) or Hall thruster. It should be pointed out that this name is not descriptive and it is used just because it was adopted in the West in 1990s [52]. Note that this thruster can also be belonging to electrostatic thrusters family as it will be illustrated below. This particular thruster type uses magnetic field to confine electrons and its physics is based on crossed $E \times B$ field discharge. Discharge in $E \times B$ fields is very interesting unique phenomenon. Not surprisingly such discharges also found various applications that utilize their unique properties i.e., electron drift and high ionization degree. This motivates to start with general description to the physics of $E \times B$ discharge considering some plasma devices having important engineering aspects in their applications.

5.2.1 Plasma acceleration in Hall thrusters

A Hall thruster is currently one of the most advanced and efficient types of electrostatic propulsion devices for spacecraft having efficiency of about 50% [53]. In particular, this configuration is beneficial because the acceleration takes place in a quasi-neutral plasma and thus is not limited by space-charge effects. The electrical discharge in the Hall thruster has an $E \times B$ configuration where the external magnetic field is radial and perpendicular to the axial electric field, which accelerates the ions as shown schematically in Figure 5.17. Passing the electron current across a magnetic field

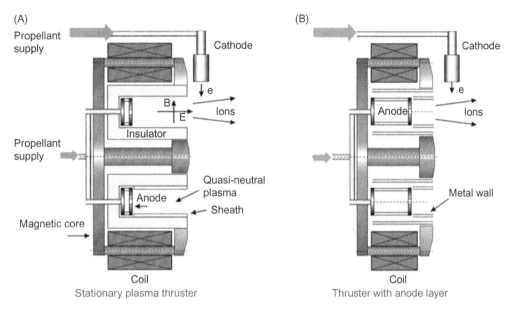

FIGURE 5.17

Schematic of Hall thruster variants. (A) Left image shows Hall thruster with dielectric wall resulting in long acceleration channel. (B) Right image shows thruster with metallic wall that result in a short acceleration zone near the anode, i.e., anode layer.

leads to an electron closed drift or Hall drift. The original idea of ion acceleration in crossed fields by using magnetron-type configuration was first introduced in late 1950s [54,55]. Since initial development of this idea in 1960s [56−62], numerous experimental and theoretical investigations have been conducted. Two different types of Hall thruster were developed: a thruster with closed electron drift and extended acceleration zone, or SPT, and a thruster with short acceleration channel or thruster with anode layer (TAL). Both configurations are shown in Figure 5.17. In an SPT, the plasma interacted at the dielectric wall. Due to this interaction, a secondary electron emission (SEE) from the wall occurs and the electron temperature remains relatively low in comparison to the TAL [63]. In TAL, ion acceleration occurs in a very thin layer near the anode (in electrical sheath) with thickness of about electron Larmor radius that gave the name to this thruster variant. Recently, new mathematical solutions of the anode layer problem were found, i.e., the so called B- and E-layers [64−66]. These solutions are different by the width of ion acceleration zone and by parameters of the cathode plasma. Similarly in magnetically insulated ion diodes, ion acceleration occurs in a thin electric layer with closed electron drift near the anode [67].

The main physical phenomena currently studied in the Hall thruster are the electron transport, propagation and neutralization of the ion beam, plasma interaction with a dielectric wall, and the transition between the quasi-neutral plasma and the sheath. Experimental results suggest that the wall material has substantial effect on the discharge parameters in the Hall thruster [68]. Additionally it was found that use of sectioned electrodes inside the Hall thruster channel has a considerable effect on plasma properties, the discharge current as well as thruster performance [69,70].

Two approaches for modeling plasma flows in Hall thrusters were undertaken in the past: particle simulation and hydrodynamic approach. The first approach consists of fully kinetic simulation and a hybrid model in which ions and neutrals are treated kinetically whereas electrons are treated as a fluid [71−73]. It should be noted that in this numerically expensive approach, very simplified boundary conditions are applied at the walls without considering the plasma−wall transition in detail. In the second approach, the 1D hydrodynamic description for all species is employed [74−77]. However, in 1D analyses, the boundary conditions at the wall are not considered. 2D hydrodynamic model of plasma flow in a Hall thruster suggests that plasma−wall interactions are rather complicated [78−82]. Any state-of-the-art Hall thruster model employs some anomalous cross-field electron mobility in order to reproduce experimental features. Generally there is no clear convincing evidence regarding which one of the possible anomalous transport mechanisms prevails in Hall thruster thus leaving this question unresolved. We just want to point out that very recent experimental and theoretical studies of the Hall thruster with variable channel width shed some light on this revealing that near-wall conductivity (NWC) mechanism may be responsible for electron transport in Hall thruster [83].

Thus the problem associated with closed electron drift configurations (magnetic plasma immersion ion implantation (MPIII), magnetrons, and Hall thruster) can be formulated in a more general manner: what electron transport mechanism sustains an electric field in the partially magnetized quasi-neutral plasma? It was shown that depending on the electron transport across the magnetic field, an electric field is established, resulting in ion acceleration or deceleration [84]. From this point of view, MPIII system and Hall thruster represent two limiting cases with ion deceleration or acceleration conditions across the region with a closed electron drift. Similarly, ion detachment from the magnetized region in the Hall thruster (near the thruster exit plane) and ion flux entrance into the magnetized region in an MPIII are governed by very similar physics. Therefore, despite the fact that the two aforementioned devices (MPIII and Hall thruster) have very different applications, the underlying physics, i.e., closed electron drift, is very similar.

5.2.2 Anomalous electron transport mechanisms

Electron conductivity across magnetic field is one of the long-standing problems of the Hall thrusters. The current continuity, energy balance, and Hall thruster efficiency are determined by the electron conductivity. It was known that classical mechanism of conductivity based on electron collisions could not explain the electron current measured in Hall thrusters. Let us analyze two possible nonclassical mechanisms of electron transport across a magnetic field consisting of plasma oscillation phenomena and NWC effect.

5.2.2.1 Plasma oscillations

The nonclassical or anomalous electron diffusion across a magnetic field has a long history since it was proposed by Bohm to explain high electron transport in magnetic confinement devices [85]. Presence of the turbulent electric field δE results in a random drift across the magnetic field. This leads to the so-called anomalous diffusion coefficient $D_\perp$ to be proportional to $\langle \delta E^2 \rangle$, which is typically larger than the classical diffusion coefficient.

In order to assess a possibility of Bohm-type anomalous transport in specific device such as Hall thruster, we start from review experimental observation of plasma oscillations. Hall thrusters have complex wave structure that goes across wide frequency spectrum [86]. It was established that amplitude and frequency of the plasma oscillations in the Hall thruster depend on mass flow rate, discharge voltage, geometry, facility, magnetic field profile, and cathode operation mode. In addition, it was found that the oscillation spectrum depends on the location inside the Hall thruster channel [87]. The way to characterize the oscillations is to look at the oscillation spectrum as a function of a magnetic field or as a function of discharge voltage [88]. Generally several typical oscillations band were identified in Hall thruster, such as 10−20 kHz discharge oscillations, 5−25 kHz rotating spokes (attributed to ionization process), 20−60 kHz azimuthal modes (drift-type instability associated with gradient of density and magnetic field), 70−500 kHz transient time (ion residence time in the channel), 0.5−5 MHz azimuthal wave. Last oscillations were recently detected by Litvak et al. [89]. Parallel study of the Rayleigh instability [90] suggest that axial density, magnetic field, and electron velocity gradients can drive this type of azimuthal instability. High-frequency instabilities (1−10 MHz) were studied in the Hall-effect thruster [91]. It was found that these instabilities have the highest level near the thruster exit plane. Particle-in-cell (PIC) simulation suggests that high-frequency oscillations with very short wavelength can be developed in the Hall thruster [92]. It was suggested that these oscillations can be responsible for anomalous electron transport. On the other hand, it was argued that axial oscillation (beam-plasma parametric instability type) can promote anomalous electron transport [93]. This type of instability was obtained in the hybrid particle−fluid simulations of the Hall thruster [94]. A correlation between the Buneman instability and low-frequency (1−20 KHz) oscillations was shown theoretically [95]. The oscillations in Hall thruster determine the efficiency of the system and may affect the divergence of the ion beam and electron transport across the magnetic field.

According to the plasma theory, there are several types of instabilities that can lead to anomalous electron transport of the Bohm type. In the low-temperature plasma with parameters typical for the Hall thruster, it was found that the drift-dissipative instability can develop [96]. This instability has the following maximal increment [94]:

$$\gamma \sim \left(\frac{T_e}{B}\right)\frac{1}{n}\frac{dn}{dr} \tag{5.35}$$

where T_e is the electron temperature, B is the magnetic field, n is the plasma density and dn/dr is the plasma density gradient. One can see that the maximal increment is inversely proportional to the magnetic field. In this case, the diffusion coefficient can be estimated as follows:

$$D \sim \lambda_\perp^2 \gamma \tag{5.36}$$

where $\lambda_\perp^2$ is the characteristic size of the plasma turbulence pulsing across the magnetic field, which can be approximated as a wavelength. According to Eqs (5.35) and (5.36), one can see that the maximum increment has the same dependence as empirical coefficient proposed by Bohm [85].

$$D = \frac{1}{16}\frac{T_e}{B} \tag{5.37}$$

It should be pointed out that while this instability was experimentally detected in the afterglow plasma having plasma parameter range somewhat similar to Hall thrusters [96], there is no clear experimental evidence that this type of instability is present in Hall thrusters. Another mechanism that can lead to enhanced electron transport in Hall thrusters, the so-called NWC [60], will be described below.

5.2.2.2 NWC definition

The idea of NWC stems from the fact that, typically in the Hall thruster channel, the mean free path for electron–neutral collisions is about 1 m, while the distance between walls is about 1 cm. Therefore, electron collisions with the wall happen much more often than collisions with neutral particles (the same conclusion is true in comparison to the electron–ion collisions and electron–electron collisions since ionization degree is about 0.01 or less). Without the presence of the axial electric field, electron reflection in the sheath is mirror type and therefore cannot contribute to the conductivity. This makes the axial electric field one of the most important factors in determining the electron transport. In the next section, we will examine the effect of the axial electric field on the electron transport across the magnetic field.

Possible electron trajectories in the sheath near the dielectric are shown schematically in Figure 5.18. The electron trajectories are depended on the initial velocity at the sheath edge. Two electron populations exist dependent on the electron energy distribution function (EDF) and sheath potential drop. Reflected electrons contribute to the low-energy population of the energy distribution having energy smaller than the potential drop in the sheath. The other type are the energetic electrons transited through the sheath and collide with the wall thus leading to SEE as shown schematically in Figure 5.18. For typical Hall thruster conditions (electron temperature 20–30 eV, wall material is boron nitride), the SEE coefficient is about 1. In this case, the sheath reaches the space-charge saturated regime associated with a nonmonotonic potential profile. The sheath voltage drop U_w is relatively small and is about T_e [96]. Under these conditions, the fraction of electron current

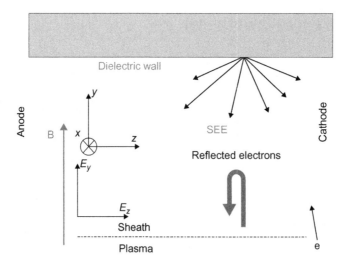

FIGURE 5.18

Schematic of electron interactions in the sheath.

colliding with the wall increases. Therefore, one should consider the SEE effect on the transport across the magnetic field. SEE electrons have an angular distribution that depends on the energy of the primary electrons and angle of incidence. In the presence of the axial electric field in the sheath, the SEE angular distribution could change so that electrons would have some preferable injection in the direction of the electric field. This effect should also contribute to electron transport across the magnetic field. The frequency of electron collisions with walls can be estimated as

$$\nu_{ew} = \sim \langle V_e \rangle / h \tag{5.38}$$

where $\langle V_e \rangle$ is the average electron velocity and h is the distance between walls. Previously, Baranov et al. [97] proposed to take into account that only a fraction of electrons will collide with walls due to reflection in the sheath, i.e.,

$$\nu_{ew} = \frac{\langle V_e \rangle}{h} \exp\left(-\frac{\Delta \varphi_y}{T_e}\right) \tag{5.39}$$

However, the electrons reflected from the sheath boundary can also contribute to the NWC (as will be shown below) and therefore the collision frequency should be close to $\langle V_e \rangle / h$. The next section describes the model of the NWC taking into account details of electron interactions in the sheath.

5.2.2.3 Mathematical model of the NWC

As the mean free path in the typical Hall thruster channel is much larger than the channel width, the electron collisions with the channel wall were determined by the thruster characteristics. Morozov [60] developed a model for calculation of the near-wall current. This method consists of solving the integrals of motion for electrons in a collisionless approach. The initial secondary electron EDF at the wall was chosen to be Maxwellian. The model was further developed over the past decades [98,99]. Recently, a new formulation of the NWC problem based on Morozov's approach was proposed [100] taking into account ion neutralization near the dielectric wall. It was shown that the resulting NWC current contains a correction factor and permits to reach better quantitative agreement with experiment. In addition, recent 2D simulations of the ion dynamics in the Hall thruster demonstrated a significant effect of the ion neutralization near the walls [101].

Phenomenologically, the NWC is the result of electron collisions with walls and consequent cycloidal motion along the magnetic field. Spatially oscillating currents is the essence of the NWC. Electrons reflecting from the wall are not monoenergetic and therefore the resulting current oscillation will rapidly decay with distance from the wall with most electron current concentration in the near-wall region giving the appropriate name for this effect, near-wall conductivity [99].

However, we want to point out an additional important effect that was not considered. Secondary electrons interact with electric fields (both axial and radial) in the sheath and thus their energy distribution function (EDF) is modified. In this section, the NWC problem taking into account this effect is formulated. It should be pointed out that EDF modification due to the radial electric field was very recently considered using a similar formalism [99,102]. We adopt the mathematical description proposed originally by Morozov. The present model [103] is based on the following assumptions: plasma properties are spatially uniform:

a. The EDF of the secondary electrons is Maxwellian with temperature T_w which is different from the bulk electron temperature T_e.
b. Electrons are accelerated in the sheath by potential drops $\Delta\varphi_z$ and $\Delta\varphi_y$ in both axial and radial directions.

Let us describe the NWC mechanism taking into account SEE. The main idea is that the distribution function of the emitted electrons is shifted by the electric field in the sheath. Since both axial and radial electric fields are present, the distribution function is shifted in both directions and the EDF is centered along the direction of the electric field. The electron dynamics can be fully described by the collisionless kinetic equation for distribution function $f(t,r,V)$:

$$\frac{\partial f}{\partial t} + V\frac{\partial f}{\partial r} - \frac{e}{m}(E + V \times B)\frac{\partial f}{\partial v} = 0 \tag{5.40}$$

where E is the electric field and B is the magnetic field. The following distribution function of the emitted electrons from the wall is assumed [99,103]:

$$f(v) = 2n_0\left(\frac{m}{2\pi kT_w}\right)^{3/2}\exp\left(-\frac{mV^2}{2kT_w}\right) \tag{5.41}$$

where n_0 is the electron density. Further consideration is based on the fact that the distribution function is constant along the characteristics, which are determined by the equations of motion for electrons [99]. Assuming constant electric and magnetic fields, the solution of the equation of motion is the electron drift with constant velocity E/B along the x-axis and cyclotron rotation.

We take into account that typically, in Hall thrusters, the electron Larmor radius is much larger than the Debye length. Therefore, we neglect effect of the magnetic field in the sheath. The equations of motion for the electron (characteristics) have the following form:

$$\frac{dV_x}{dt} = -\omega V_z \tag{5.42}$$

$$\frac{dV_y}{dt} = 0 \tag{5.43}$$

$$\frac{dV_z}{dt} = -\frac{eE_z}{m} + \omega V_x \tag{5.44}$$

where ω is the electron cyclotron frequency. Using the above assumptions and conditions, one can integrate the equation for characteristics:

$$V_{ez} = (V_{ex}^o - V_E)\sin(\omega t) + \sqrt{(V_{ez}^o)^2 + \frac{2e\Delta\varphi_z}{m}}\cos(\omega t) \tag{5.45}$$

$$V_{ex} = V_E + (V_{ex}^o - V_E)\cos(\omega t) - \sqrt{(V_{ez}^o)^2 + \frac{2e\Delta\varphi_z}{m}}\sin(\omega t) \tag{5.46}$$

$$V_{ey} = \sqrt{(V_{ey}^0)^2 + \frac{2e\Delta\phi_y}{m}} \tag{5.47}$$

where $V_E = (E_z/B)$, and V_{ex}^o, V_{ez}^o, V_{ey}^o are the velocities at the wall. It was taken into account that electrons are accelerated across the sheath having a potential drop of $\Delta\varphi_y$. It should be noted that Eqs (5.42)–(5.47) describe electron motion in the crossed field outside the sheath, while in the sheath finite jumps of V_{ez} and V_{ey} are considered without considering effect of the magnetic field. The current density (z-component) can be calculated as follows:

$$j_{ez} = \int_{-\infty}^{\infty} \int_{\alpha_y}^{\infty} \int_{-\infty}^{\infty} f(v) V_z dV_x dV_z dV_y \tag{5.48}$$

where $\alpha_y = \sqrt{(2e\Delta\varphi_y/m)}$. Substitute velocity components according to equation of characteristics (Eqs (5.45)–(5.47)). In this case, one can arrive at the following expression for the electron current density:

$$j_{ew} = 2n_0 \frac{E}{B} \left(\frac{m}{2\pi kT_w}\right)^{1/2} \exp\left(\frac{e\Delta\varphi_z}{kT_w}\right) \exp\left(\frac{e\Delta\varphi_y}{kT_w}\right) \int_{\sqrt{(2e\Delta\varphi_y/m)}}^{\infty} \exp\left(-\frac{mV_y^2}{2\pi kT_w}\right) \sin\left(\omega\frac{y}{V_y}\right) dV_y \tag{5.49}$$

Let us introduce the following new variables:

$$\theta = (V_y/\sqrt{(2kT_w/m)}) \text{ and } s = (\omega y/\sqrt{(2kT_w/m)}) = \frac{y}{\rho_{Le}}, \text{ where } \rho_{Le} \text{ is the secondary electrons Larmor}$$

radius. In this case, a new function can be introduced which is the integral in Eq. (5.49):

$$Q(s) = \int_{\sqrt{(e\Delta\varphi_y/kT_w)}}^{\infty} \exp(-\theta^2)\sin\left(\frac{s}{\theta}\right) d\theta \tag{5.50}$$

where s is the nondimensional distance from the wall (y-direction) and the function $Q(s)$ determines the current distribution as a function of that distance. In essence, this function is similar to one introduced originally by Morozov and Savel'ev [99] with one exception, i.e., the potential drop in the sheath is taken into account. In that sense, the present approach is similar to recent work of Barral et al. [102]. The difference is that, in addition, we take into account electron acceleration in the sheath along the axial electric field component. The current density due to NWC can be expressed as follows:

$$j_{ew} = \frac{2}{\sqrt{\pi}} n_0 \frac{E}{B} \exp\left(\frac{e\Delta\varphi_z}{kT_w}\right) \exp\left(\frac{e\Delta\varphi_y}{kT_w}\right) \times Q(s) \tag{5.51}$$

One can see that the function $Q(s)$ provides the dependence of the current density on the distance from the wall. The calculated dependence of $Q(s)$ is shown in Figure 5.19.

One can see that current is concentrated near the wall at a distance of a few Larmor radii in the simplest case of $\Delta\varphi_y = 0$. This case corresponds to Morozov's original solution [99]. Generally, the sheath voltage leads to decrease of the current concentration near the wall and to more uniform current distribution across the channel between the two walls. This was a reason that led some authors to conclude that NWC may be a misnomer [102].

However, it was indicated by some authors that in this formulation, the NWC current underpredicts the measured values [99]. Below we will consider an additional effect associated with electron interactions in the sheath that lead to enhancement of the NWC current density. The main idea is

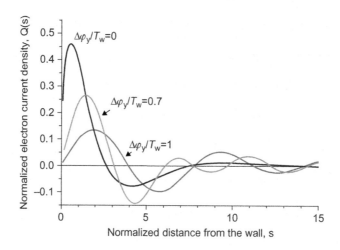

FIGURE 5.19

Calculated function $Q(s)$ as a distance from the wall with sheath potential drop as a parameter.

that the electric field along the wall can affect the near-wall current by producing an additional velocity shift in the axial direction. Typically in the Hall thruster channel, the axial electric field is $E_z = 2-3 \times 10^4$ V/m, which is smaller than the typical radial electric field in the sheath. However, it will be shown that in some cases, the axial electric field can be an important factor contributing to NWC.

The current density increases by a factor of $\exp(\Delta\varphi_z/T_w)$ as can be seen from Eq. (5.51) in which we have to know the potential drop $\Delta\varphi_z$ in the axial direction. In order to estimate the effective potential drop in the axial direction, let us consider in some detail the electron motion in the sheath. It was stated above that typically, in a Hall thruster acceleration channel, the saturated space-charge sheath occurs [70]. In this case, the potential distribution in the sheath has a minimum [96]. In the location of the minimum, the corresponding (y) component of the electric field is zero. On the other hand, there is an axial electric field component which arises from the axial potential distribution in the plasma bulk. The presence of the axial component was discussed previously [104]. It was suggested that the electric field along the dielectric is close to the electric field in the plasma bulk due to high dielectric strength of the wall material. The potential distribution in the axial direction can be obtained from the electron momentum equation as

$$d\varphi_z = dzE_z = E_z^2 \frac{dy}{E_y} \tag{5.52}$$

The total potential drop that electrons are experiencing while being in the sheath can be estimated as follows:

$$\Delta\varphi_z = E_z^2 \int_{y_{min}}^{y_{max}} \frac{dy}{E_y(y)} \approx E_z^2 \frac{L_D}{E_y} \tag{5.53}$$

In the last expression, it was assumed that the sheath thickness is about one Debye length. However, since the space-charge limited sheath is considered, the electric field is not uniform in the y-direction. The largest influence of the axial electric field is near the potential well, since typically the electric field in the y-direction is much larger than that in z, i.e., $E_y/E_z \gg 1$, except near the potential well where $E_y/E_z \leq 1$. The spatial extension of the potential well is about one Debye length L_D [105]. Thus, the potential drop can be estimated with satisfactory accuracy as

$$\Delta\varphi_z \approx E_z L_D \tag{5.54}$$

It should be noted that typically the depth of the potential well is small in comparison to T_w. Thus, secondary electrons reflection back to the wall can be neglected in calculation of the distribution function outside the sheath. Taking Eq. (5.54) into account, one can estimate the NWC current enhancement ($\exp(E_z L_D/T_w)$), which depends on the bulk to wall electron temperature ratio as shown in Figure 5.20. In typical conditions in the Hall thruster channel, we can find that this factor may be about 10.

If we take into account possible enhancement factor due to axial electric field in the sheath, the predicted NWC current will be close to that measured experimentally [99]. Thus it can be concluded that, in general, NWC can explain the high electron mobility in a Hall thruster. However, it is quite interesting to point out that higher NWC current is expected in the case of small T_w as follows from Figure 5.20. On the other hand, according to this model prediction, this is the case in which NWC current is not decaying from the wall as shown in Figure 5.19, thus putting a question mark on the near-wall nature of the current. Bearing this in mind we can conclude that the full picture of electron transport in the Hall thruster is far from completion and further investigation is needed.

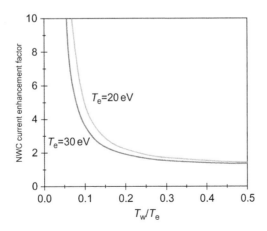

FIGURE 5.20

Enhancement of the NWC by axial electric field effect. The typical conditions are considered: $E_z = 3 \times 10^4$ V/m; $n_o = 10^{17}$ m^{-3}.

5.2.3 Structure of $E \times B$ layer

The idea of using a magnetic field to control the potential distribution in the plasma will be discussed in this section. The region in which magnetic control of the potential distribution is implemented, we shall call the $E \times B$ layer.

The most basic condition is that electrons are magnetized inside the $E \times B$ layer, while ions are not. This means that the magnetic field has to be chosen:

$$r_{Le} \ll L \ll r_{Li}$$

where L is the size of the $E \times B$ layer.

Let us determine the size of the $E \times B$ layer in a typical steady-state condition. This analysis comes from the balance between the electron losses due to transport across the magnetic field and electron generation due to ionization.

Time of electron drift across the $E \times B$ layer can be estimated as

$$\tau_e \approx \frac{L}{V_e} = \frac{L}{\mu_\perp E} = \frac{L^2}{\mu_\perp \phi}$$

where E is the electric field, ϕ is the potential drop, μ is the electric mobility, ν_i is the ionization collision frequency, and V_e is the electron collision frequency. Electron losses to the anode are compensated by the ionization due to electron collisions with neutrals:

$$\nu_i \approx n_a \langle \nu_e \sigma_i \rangle$$

The steady-state condition requires:

$$\tau_e \nu_i = {\sim}1$$

Solving this for the length of the $E \times B$ layer results in the following:

$$L \approx \left(\phi \frac{\mu_\perp}{\nu_i} \right)^{0.5}$$

Taking into account expression for the electron mobility and Larmor radius one can arrive to following:

$$L \approx R_{Le} \left(\frac{\nu_e}{\nu_i} \right)^{0.5}$$

From this equation, one can see that the length of the $E \times B$ layer scales as electron Larmor radius. Further discussion about the scaling of the $E \times B$ layer will be presented in Section 5.2.8.4.

Let us now consider plasma flow focusing in $E \times B$ layer. We will start with considering the Ohm's law:

$$\frac{j}{\sigma} = E + V_e \times B + \frac{1}{en} \nabla P_e$$

Considering this equation along the magnetic field and assuming that the current along the field is small, i.e., $j/\sigma = 0$, one can find

$$0 = E + \frac{1}{en}\nabla P_e$$

or

$$\frac{d\varphi}{ds} = \frac{1}{en}\frac{dP_e}{ds}$$

Using the assumption about the ideal gas ($P_e = n_e kT_e$) after integration, one can find that

$$\varphi(s) = \varphi_0 + \frac{kT_e}{e}\ln\left(\frac{n_e}{n_{e0}}\right) = \text{constant}$$

where φ_0, n_{e0} are the reference potential and electron density.

This concept was introduced by Morozov [106] and it is called the "thermalized" potential. The implication of this result is that the magnetic field can be used to control the electric field distribution in the Hall thruster channel and thus the ion trajectories.

5.2.4 Plasma flow in Hall thruster: example of calculation

In this section, we describe the model and example of calculation of plasma and discharge parameters of the Hall thruster channel. It will be taken in account below that the presheath scale length becomes comparable to the channel width under typical conditions of the Hall thruster plasma flow. The model for the quasi-neutral plasma region is extended up to the sheath edge in order to provide the boundary condition at the plasma–sheath interface as shown in Figure 5.21. In the following sections, a sheath in front of the dielectric wall, a quasi-neutral plasma presheath, and conditions for a smooth transition between these regions are described.

5.2.4.1 Sheath and plasma–sheath transition

In the present calculations, will take into account a nonzero electric field E_o at the presheath–sheath interface that, in combination with the plasma velocity at that interface V_o, will determine the entrance conditions for the sheath. This approach was described in detail in Chapter 1.

Under typical steady-state conditions, the potential drop between the plasma and the wall is negative in order to repel the excess of thermal electrons. However, when the wall has substantial SEE, the floating potential drop may be different from that in the simple sheath. This effect will be considered in this section.

In the considered range of parameters (see below), the electrostatic sheath considered as collisionless and unmagnetized, since the Debye length is much less than the collision mean free path and the Larmor radius. We employ a 1D sheath model that is based on the assumption that the sheath thickness is much smaller than the plasma channel width. SEE from the dielectric wall is taken into account. For simplicity, only singly charge ions are considered. The electrostatic potential distribution in the sheath satisfies the Poisson equation:

$$\nabla^2\varphi = e/\varepsilon_o(n_{e1} - n_i + n_{e2}) \tag{5.55}$$

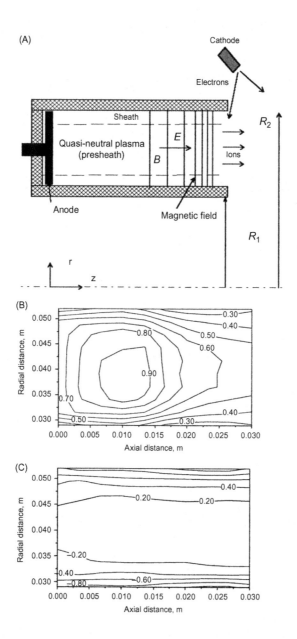

FIGURE 5.21

(A) Schematics of the Hall thruster channel, (B) electron density contours in the channel, and (C) contours of the radial component of ion velocity.

Source: Reprinted with permission from Ref. [78]. Copyright (2001) by American Institute of Physics.

where φ is the potential, n_{e1} is the density of plasma electrons, n_{e2} is the density of secondary electrons, and n_i is the ion density. The plasma electrons are assumed to obey the Boltzmann distribution:

$$n_{e1} = (n_o - n_{e2}(0))\exp(-e\varphi/kT_e) \tag{5.56}$$

where $n_{e2}(0)$ is the density of the secondary electrons at the sheath edge and n_o is the electron density at the sheath edge that is equal to the ion density $n_i(0)$. The ions are assumed to be cold and have at the presheath−sheath interface the energy $U_{io} = 1/2m_iV_o^2$, where V_o is the ion velocity at the sheath edge and m_i is the ion mass. Ions have free motion in the sheath and their density decreases according to

$$n_i = n_o\left(1 + \frac{2e\varphi}{mV_o^2}\right)^{-0.5} \tag{5.57}$$

The current continuity equation can be written in the form

$$j_{e1} + j_i - j_{e2} = 0 \tag{5.58}$$

where j_{e1} is the flux of the primary electrons from the plasma ($j = nV$ is the flux definition for all species), j_i is the ion flux, and j_{e2} is the flux of the secondary electrons. From equation (5.58) one can obtain that

$$n_{e2}V_{e2} = s/(1-s)n_oV_o \tag{5.59}$$

where s is the SEE coefficient determined as $s = j_{e2}/j_{e1}$. We will furthermore assume that the electrons emitted from the surface are monoenergetic and freely move in the sheath. The boundary conditions for the sheath problem are the following:

$$\varphi(0) = 0; \quad d\varphi/dr \ (r' = 0) = E_o; \quad \text{and} \quad V = V_o \tag{5.60}$$

Two parameters are critical for the sheath solution: the electric field and the initial ion velocity. From the current continuity equation (Eq. (5.58)), one can calculate the potential drop across the sheath $\Delta\varphi_w$, as

$$\Delta\varphi_w = kT_e \ln((1-s)/V_o(2\pi m_e/kT_e)^{0.5}) \tag{5.61}$$

This solution is similar to that described first by Hobbs and Wesson [96] except that they *a priory* assumed the Bohm velocity at the sheath edge. One can see that the potential drop across the sheath decreases with the SEE coefficient. It should be noted that in this model, we have assumed that a monotonic potential distribution exists. However, when the SEE coefficient s approaches unity, the solution in form (5.61) breaks down and when s exceeds a critical value, a potential well forms such that a fraction of emitted electrons are returned to the wall. This happens when s is less than 1 and it was obtained that the critical value of the SEE coefficient s^* (based on the condition of monotonic potential distribution) can be calculated as [107]

$$s^* = 1 - 8.3(m_e/m_i)^{0.5} \tag{5.62}$$

From Eq. (5.62), it can be estimated that in the case of BN (the dielectric material usually used in SPT), the critical SEE coefficient is about 0.95 [107].

5.2.4.2 Plasma presheath model

The presheath model is based on the assumption that the quasi-neutral region length is much larger than the Debye radius and therefore we will assume that $Z_i n_i = n_e = n$, where Z_i is the ion mean charge, n_i is the ion density, and n_e is the electron density. For simplicity, only single charge ions are considered in this chapter ($Z_i = 1$). We will consider the plasma flow in a cylindrical channel as shown in Figure 5.21. A magnetic field with only a radial component, $B_r = B$, is imposed. Cylindrical coordinates will be used, as shown in Figure 5.21 with angle θ, radius r, and axial distance from the anode z. The plasma flow starts in the near-anode region and has lateral boundaries near the dielectric wall. The plasma presheath–sheath interface is considered to be the lateral boundary for the plasma flow region. A plasma will be considered with "magnetized" electrons and "unmagnetized" ions, i.e., $\rho_e \ll L \ll \rho_i$, where ρ_e and ρ_i are the Larmor radii for the electrons and ions, respectively, and L is the channel length. We employ a hydrodynamic model assuming (i) the system reaches a steady state and (ii) the electron component is not inertial, i.e., $(\mathbf{V}_e \nabla) \cdot \mathbf{V}_e = 0$. The following system of equations describes the quasi-neutral plasma:

$$nm_i(\mathbf{V}_i \nabla)\mathbf{V}_i = ne\mathbf{E} - \nabla P_i - \beta_i nm_i n_a \mathbf{V}_a \tag{5.63}$$

$$\nabla \cdot (\mathbf{V}_i n) = \beta_i n n_a \tag{5.64}$$

$$\nabla \cdot (\mathbf{V}_a n_a) = -\beta_i n n_a \tag{5.65}$$

$$0 = -en(\mathbf{E} + \mathbf{V} \times \mathbf{B}) - \nabla P_e - n\nu_m m_e \mathbf{V}_e \tag{5.66}$$

$$\frac{3}{2} \partial(j_e T_e)/\partial z = Q_J - Q_w - Q_{ion} \tag{5.67}$$

where n is the plasma density, β_i is the ionization rate, $Q_J = j_e E$ is the Joule heat, E is the axial component of the electric field, j_e is the electron current density, $Q_w = \nu_w n(2kT_e + (1-s)e\Delta\varphi_w)$ represents the wall losses [78], ν_w is the frequency of electron collisions with walls, $Q_{ion} = en_a n U_i \beta(T_e)$ represents ionization losses, U_i is the ionization potential (for xenon, $U_i = 12.1$ eV), and $\beta(T_e)$ is the ionization coefficient (ionization process is described in Chapter 2).

The last term in Eq. (5.66) stands for an effective drag force due to nonelastic collisions similar to that used in Ref. [107]. To simplify the problem without missing the major physical effects, we consider 1D flow of the neutrals. The equations for the heavy particles (ions and neutrals) may be written in component form in cylindrical coordinates by taking into account that the ion temperature is much smaller than the electron temperature (that makes it possible to neglect the ion pressure term in the momentum conservation equation):

$$\frac{\partial(nV_z)}{\partial z} + \frac{\partial(nV_r)}{\partial r} + \frac{nV_r}{r} = \beta_i n_i n_a \tag{5.68}$$

$$V_z \frac{\partial V_z}{\partial z} = -V_r \frac{\partial V_z}{\partial r} + \frac{e}{m_i} E_z - \beta_i V_a n_a \tag{5.69}$$

$$V_z \frac{\partial V_r}{\partial z} = -V_r \frac{\partial V_r}{\partial r} + \frac{e}{m_i} E_r \tag{5.70}$$

$$\frac{\partial(n_a V_a)}{\partial z} = -\beta_i n_i n_a \tag{5.71}$$

In this model, the electron flow will be considered separately along and across magnetic field lines. Due to the configuration of the magnetic field (i.e., only the radial magnetic field component is considered in the model as shown in Figure 5.21), the electron transport is greater in the azimuthal direction ($E \times B$ drift) than in the axial direction (drift diffusion due to collisions). According to Eq. (5.66), the electron transport equation along the magnetic field can be written as a balance between pressure and electric forces assuming that the current component in the radial direction is zero. Assuming that the electron temperature is constant along each magnetic field line, we obtain that

$$\varphi - \frac{kT_e}{e} \ln n = \text{constant} \tag{5.72}$$

The left-hand side of this equation is known as a thermalized potential [106]. This equation makes it possible to reduce the 2D calculation of the electric field to a 1D problem. According to Eq. (5.72), the electric field in the radial direction E_r is determined by the electron pressure gradient in this direction. Calculating the potential distribution along the channel centerline makes it possible to calculate the potential in the entire domain using Eq. (5.72). For known total discharge current and ion current fraction, one can calculate the electron current fraction from the current continuity condition. The equation describing the electron transport across the magnetic field can be obtained from Eq. (5.66) and reads

$$j_{ez} = en \frac{\mu_e}{1 + (\omega_e/\nu_m)^2} \left(E_z + \frac{\partial T_e}{\partial z} + T_3 \frac{\partial \ln n}{\partial z} \right) \tag{5.73}$$

where $\nu_m = \nu_{en} + \nu_{ew} + \nu_B$ is the effective electron collision frequency. In the next section, we will determine different components of the effective electron collision frequency.

5.2.4.3 Electron collisions

For typical conditions of the Hall thruster, the effect of Coulomb collisions appears to be negligibly small [72] and will not be considered here. The total electron collision frequency considered in the present model consists of electron–neutral collisions, electron–wall collisions, and anomalous collisions (Bohm diffusion). The electron–neutral collision frequency may be estimated as follows:

$$\nu_{en} = n_a \sigma_{ea} V_{th}^e \tag{5.74}$$

where n_a is the neutral density, σ_{ea} is the total collision cross section dependent on the electron energy [78] ($\sigma_{ea} \sim (10 \div 40) \times 10^{-20} \, \text{m}^{-2}$ for Xenon in considered electron energy range of $10-30 \, \text{eV}$), and V_{th}^e is the electron thermal velocity.

Inclusion of only the electron–neutral collisions leads to underprediction of the electron mobility observed experimentally in Hall thrusters. This was recognized long ago by many authors [58–61]. Until now, however, there is no consensus about which of the possible mechanisms of electron transport is most significant in the Hall thruster. Recent experimental data support the idea that the second type of collisions prevails especially near the magnetic field peak [108]. In this simulation example, we will account for both the above-mentioned collision mechanisms.

The effective electron collision frequency related to the anomalous turbulent transport (Bohm diffusion) can be estimated as

$$\nu_B = \alpha \omega_e \tag{5.75}$$

where $\alpha \sim 1/16$ is the Bohm empirical parameter [85]. It will be shown below that the exact value of this parameter affects the potential drop across the channel. The best fit with the experimental data [78] on the potential drop corresponds to $\alpha \sim 1/44$ instead of the classical value $\sim 1/16$. It should be noted that the same conclusion derived by different authors was that the best fit with the experimental data on discharge voltage corresponds to $\alpha \sim 1/80 \div 1/100$ [109].

Another possible nonclassical mechanism of the electron transport across a magnetic field is due to collisions with a dielectric wall, the so-called NWC [99,110], that accounts for both elastic and nonelastic electron collisions with a wall. According to [97], the frequency of the electron collisions with a wall can be estimated as

$$\nu_{ew} = (V_{th}^e / h) \exp(-\Delta \varphi_w / T_e) \tag{5.76}$$

where $h = R_2 - R_1$ is the channel width.

5.2.4.4 Plasma flow in Hall thruster channel: simulation results

Computations shown in this section are performed for the geometry of the SPT-100 (which is a SPT with 100 mm outer channel wall diameter) that has a channel length of 3 cm, and inner and outer radii are 3 and 5 cm, respectively. The magnetic field axial profile is parabolic with a magnetic field maximum of $B = 160$ G near the channel exit plane. All results are presented for the fixed discharge current of 4.5 A and mass flow rate of 4 mg/s (xenon). The SEE coefficient will be considered in the range of 0.7–0.95 that correspond to electron energy of 15–30 eV in the case of BN [112]. We have calculated thrust for these conditions, which is on the order of 50 mN that is close to that measured in experiment [113].

At the upstream boundary ($z = 0$), we specify [114] the ion velocity is about $V_o = 2 \times 10^3$ m/s near the anode that corresponds to ion temperature of 3 eV. This upstream condition implies that we are considering only supersonic plasma flow assuming that the transition from subsonic to supersonic flow occurs in the anode vicinity. The atom velocity near the anode is assumed to be $V_{oa} = 2 \times 10^2$ m/s [78]. The atom density at the anode plane depends upon the mass flow rate that will vary from 2 to 5 mg/s. At the downstream boundary (thruster exit plane, $z = L$), we specify an electron temperature of $T_e = 10$ eV [78] that is close to that measured in experiment [115].

The numerical analysis is similar to that described in Chapter 4 and proposed elsewhere [116]. We use the implicit two-layer method to solve the system of equations (5.68)–(5.71). These equations are approximated by a two-layer six-point scheme. The electron temperature distribution is calculated by iteration initially assuming a trial temperature distribution that satisfies the boundary conditions.

Firstly we present an analysis of the sheath solution. As mentioned above, a monotonic potential distribution in the sheath may be obtained if the Bohm condition is fulfilled at the sheath edge. However, if at the same time, the electric field at the sheath edge is finite (nonzero), a monotonic solution can be obtained even when the ion velocity at the sheath edge is smaller than that determined by the Bohm condition. In the present work, the ion velocity in the quasi-neutral plasma presheath up to the presheath edge is calculated from Eqs (5.68)−(5.71). This solution therefore establishes the electric field at the sheath−presheath interface.

The plasma density and radial velocity distribution calculated from Eqs (5.68)−(5.71) is shown in Figure 5.21B. One can see that the peak density that corresponds to the ionization zone is somewhere in the middle of the channel. The radial velocity component distribution (see Figure 5.21C) shows that the region where the plasma develops conditions for the entrance to the sheath is close to the lateral walls of the channel. Ions are accelerated in the direction normal to the wall, which is also the direction of the magnetic field lines. This is not a surprising result as the radial electric field is parallel to the magnetic field lines according to Eq. (5.72). It may be also seen that further downstream the velocity increases at the plasma (presheath) edge. This effect is shown in more detail in Figure 5.22 where the velocity at the plasma edge is displayed as a function of axial position. It can be seen that the boundary velocity increases with axial distance from about $0.7C_s$ up to C_s. The boundary velocity depends also on the SEE coefficient. Smaller SEE coefficient leads to higher velocity as shown in Figure 5.22.

Axial distribution of the electron temperature is shown in Figure 5.23 with SEE coefficient s as a parameter. One can see that the electron temperature peaks at axial distances of about $0.7-0.8\,l$. The peak electron temperature increases with coefficient s and varies from 15 eV up to 30 eV when s decreases from 0.95 down to 0.8. In the first half of the channel, the electron temperature is approximately constant and does not depend on s. It should be noted that the electron temperature predicted by the model is in the range that was measured in experiments [78,115,117,118].

The current−voltage characteristic of the discharge is shown in Figure 5.24 where SEE coefficient s is used as a parameter. It can be seen that the discharge voltage is smaller in the case

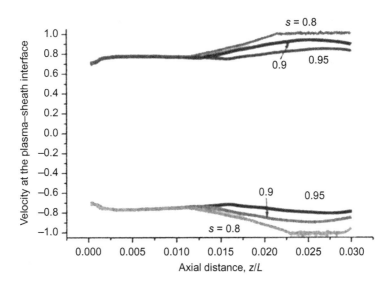

FIGURE 5.22

Velocity (nomalized by Bohm velocity) at the plasma−sheath interface with SEE as a parameter.

Source: *Reprinted with permission from Ref. [78]. Copyright (2001) by American Institute of Physics.*

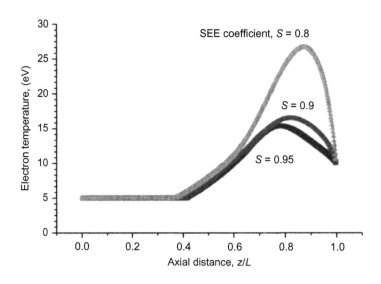

FIGURE 5.23

Electron temperature distribution along the channel with SEE coefficient as a parameter.

Source: *Reprinted with permission from Ref. [78]. Copyright (2001) by American Institute of Physics.*

of the low emissive material for the fixed discharge current. The current–voltage characteristic predicted by the model and its dependence on the SEE coefficient are found to be in agreement with experiment [68]. In that experiment, the current–voltage characteristics of two wall materials (namely BN and glass ceramics (GC)) were measured experimentally and compared. It is expected that the glass ceramics generally has a smaller SEE (note that no data for GC are available, however data on glasses and quartz can be found [119]).

5.2.4.5 Physical interpretation of results

The above sections presented a traditional two-scale analysis of the plasma–wall interface problem. The approach undertaken is however different from the usual presheath formulation since equations for the presheath are formulated in a 2D manner. Therefore, there are no restrictions on the presheath mechanism (geometric, magnetic, collisions) that usually exist in the 1D formulation. In fact an important conclusion that comes out from the model is that the presheath has a 2D nature as shown in Figure 5.21 even though the main dependence in the axial direction is in the near-wall region. Due to the plasma losses on the wall, the radial density gradient is increased. This density gradient leads to ion acceleration in the radial direction that is the main mechanism that provides the conditions at the sheath edge (presheath mechanism).

It was found that the SEE affects the current–voltage characteristic of the $E \times B$ discharge realized in Hall thrusters. It was shown that the discharge voltage is smaller in the case of low emissive material for a fixed discharge current. The reason for this behavior can be understood as follows. Smaller SEE leads to higher electron temperature since less cold secondary electrons are entering the discharge. As a result, the ionization is enhanced and electron density increases.

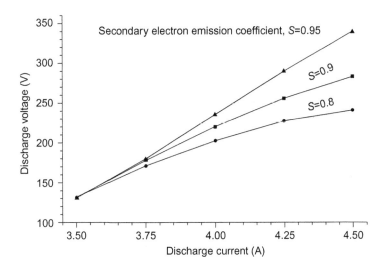

FIGURE 5.24

Current−voltage characteristics of the Hall thruster discharge.

Source: Reprinted with permission from Ref. [78]. Copyright (2001) by American Institute of Physics.

Therefore, both ion and electron current densities increase that eventually leads to total current increase.

In general, two approaches for modeling plasma flows in Hall thrusters were undertaken in the past: particle simulation and hydrodynamic approach. A variation of the first approach is hybrid models in which ions and neutrals are treated as particles whereas electrons are treated as a fluid [72,73,120]. In this numerically expensive approach, however, very simplified boundary conditions are applied at the walls without considering the plasma−wall transition in detail. In the second approach, the 1D hydrodynamic description for all species is employed [75,77,121,122]. However, due to restrictions of 1D analyses, the real boundary conditions at the wall were not considered. 2D hydrodynamic model of plasma flow in a Hall thruster suggests that plasma−wall interactions are rather complicated [78−82]. Any state-of-the-art Hall thruster model employs some anomalous cross-field electron mobility in order to reproduce experimental features. Generally there is no clear convincing evidence regarding which one of the possible anomalous transport mechanisms prevails in Hall thruster thus leaving this question unresolved. The electron conductivity is one of the long-standing problems related to Hall-effect thrusters. This very important problem has various implications on fundamental issues of the Hall thrusters, such as current continuity, energy balance, and ultimately on thruster efficiency. It was known for a long time that classical mechanism of electron collisions could not explain the electron transport experimentally observed in Hall thrusters. Until now, however, there is no consensus about which of the possible mechanisms of electron transport is most significant in Hall thrusters. One idea that was put forward early in Hall thruster history by Morozov relates the anomalous conductivity to near-wall processes (the so-called NWC). This mechanism remains still questionable since current calculated by NWC is less than measured. On the other hand, there is an idea about role of the Bohm-type conductivity across magnetic field due to plasma turbulence. There is some limited indirect experimental evidence of Bohm conductivity, but it was obtained in low-voltage regime of the Hall thruster. On the other hand, most models of Hall thruster relay on either NWC (usually with coefficient to correct for discharge current) or Bohm type. Thus the question about electron conductivity mechanism remains open from both experimental and theoretical points of view. Since all models relay on assumption about electron transport mechanism and relay on some experimental data about integral discharge characteristics (such as discharge current), any modeling predictions are very limited to cases for which experimental data exist. It is obvious that development of a new thruster or predictions regarding existing thruster lifetime cannot be supported by state-of-the art models. Therefore, there is an emerging need to develop high fidelity modeling capabilities.

There is a reliable experimental evidence of the wall material effect on operation of a Hall thruster [123,124]. The existing fluid theories explain this effect due to a strong SEE from the channel walls. In addition to NWC, the SEE is predicted to weaken insulating properties of the near-wall sheaths and, thereby, to cause cooling of plasma electrons. From a practical standpoint, a strong SEE from the channel walls is expected to cause additional inefficiencies due to enhanced power losses in the thruster discharge, and an intensive heating of the channel walls by almost thermal electron fluxes from the plasma. Moreover, because the SEE leads to lower values of the sheath potential drop, ion-induced erosion of the channel walls can also be affected. Although these predictions can be certainly applied for plasmas with electrons, which have a Maxwellian electron velocity distribution function, there is no consensus between the existing fluid and kinetic models on how strong the SEE effects on the thruster plasma are. Recent analytical studies and PIC simulations suggested that the electron velocity distribution function in a Hall thruster plasma is non-Maxwellian and anisotropic [130,131]. The electron average kinetic energy in the direction parallel to walls is several times larger than the electron average kinetic energy in direction normal to the walls. Electrons are stratified into several groups depending on their origin (e.g., plasma discharge or thruster channel walls) and confinement (e.g., lost on the walls or trapped in the plasma). Practical analytical formulas are derived for wall fluxes, secondary electron fluxes, plasma parameters, and conductivity. The calculations based on analytical formulas agree well with the results of numerical simulations. The self-consistent analysis demonstrates that elastic electron scattering on collisions with atoms and ions plays a key role in formation of the electron EDF and plasma—wall interaction. The fluxes of electrons from the plasma bulk are shown to be proportional to the rate of scattering to loss cone, thus collision frequency determines the wall potential and secondary electron fluxes. SEE from the walls is shown to enhance the electron conductivity across the magnetic field, while having almost no effect on insulating properties of the near-wall sheaths. Such a self-consistent decoupling between SEE effects on electron energy losses and electron crossed-field transport is currently not captured by the existing fluid and hybrid models of the Hall thrusters.

Effect of partial thermalization of secondary electrons emitted by the dielectric wall and its influence on the near wall sheath as well as global Hall thruster parameters was considered relatively recently [130−132]. Kinetic simulations revealed that the electron velocity distribution is depleted of the high-energy tail electrons that rapidly leave the plasma along the magnetic field lines and impact the wall. This is particularly strong effect near the space-charge limit where the sheath voltage is small and a large fraction of the electron tail can be lost. The collision frequencies and thermalization rates in the plasma are predicted to be insufficient to re-populate the Maxwellian tail [131]. Penetration of electrons across the plasma and collection at the opposite wall, due to incomplete thermalization of the emitted secondary electrons in the plasma, modifies the space-charge limits and sheath potential. This in turn can affect the electron heat flux to the wall. Partial thermalization of secondary electrons can also explain the absence of the electron temperature saturation as a function of discharge voltage that was observed experimentally [133].

In the next sections, we will describe some advance in understanding various aspects of the Hall thruster physics and engineering.

5.2.5 Peculiarities of plasma flow in Hall thrusters: 2D potential distribution

In this section, we describe the plasma flow in a Hall thruster channel that includes the 2D current conservation effect and relies on some experimental input parameters, such as magnetic field and electron temperature distribution. The discussion and model presented are an attempt to explain the experimentally found nonuniform potential distribution across the thruster channel. This effect is explained by the change of the electron mobility across a magnetic field due to the magnetic field gradient and the electron current along the magnetic field driven by the electron temperature gradient.

According to conventional wisdom, the electric potential in the Hall thruster channel is governed by the magnetic field distribution in that the equipotential contours tend to line up with the magnetic field lines. This stems from the fact that the electric field tends to be zero along magnetic field lines due to high electron mobility in this direction [106]. Indeed, with a correction of a logarithmic factor due to the possible density variation along the magnetic field, this is the case in most conventional Hall thruster channels [53,61,62,78]. In addition, in virtually all models, it is assumed that the secondary electrons emitted from the channel walls are thermalized with the main electron population so that the electron temperature is also constant along the magnetic field line. Note that use of this assumption will be avoided here and will be replaced by measured electron temperature distribution.

However, the nonuniformity of some quantities across the dielectric channel may change the above-mentioned balance and lead to an electron temperature gradient. For instance, placing a segmented electrode with low SEE may create an electron temperature gradient between the dielectric walls [81]. This electron temperature gradient may lead to electron current along the magnetic field and as a result equipotential contours may deviate from the magnetic field lines. An electron temperature gradient may also be induced by other means such as by nonuniformities in the magnetic field [134,135]. It was found in experiments that under certain operating conditions, a distinctive "jet" potential structure developed in the Hall thruster channel [134−136]. In other words, the potential distribution deviates from the magnetic field profile and the peak of the electric field is shifted downstream of the exit plane. This structure may result in a significant divergence of beam ions as they are accelerated out of the thruster. It should be noted that laser-induced fluorescence (LIF) data support the evidence of the potential nonuniformity ("jet") in the radial direction [136]. LIF measurements show that the ion velocity radial profile consistent with the independently measured nonuniform potential distribution. While this abnormal potential distribution was found in several quite different Hall thrusters, it can exist only under certain conditions. For instance, changing from high power to lower power, which results in a uniform electron temperature distribution across the channel width [134,135], or placing two segmented electrodes along the channel wall instead of a single electrode leads to the disappearance of the potential "jet" and results in a radially uniform potential distribution [134−136].

Below the observed potential nonuniformity effect is explained by considering the 2D current conservation problem in the Hall thruster channel.

We will employ the model described in Section 5.2.4. Since only the radial magnetic field component is considered in the model, the electron transport is much greater in the azimuthal direction ($E \times B$ drift) than in the axial direction (drift diffusion due to collisions). Usually in Hall thruster models, an assumption of a constant "thermalized" potential along the magnetic field is used. In this model, however, we will employ a more general 2D model for the electron flow instead of Bolzmann relation along magnetic field. We consider that the electron transport across the magnetic field is due to several collision mechanisms (i.e., electron–neutral and electron–wall collisions) as well as anomalous (Bohm) diffusion with a total effective collision frequency of ν_{ef}. Under the mentioned above conditions, the simplified Ohm's law can be written in component form as

$$j_r = \sigma\left(-\frac{\partial\varphi}{\partial r} + \frac{\partial T_e}{\partial r} + T_e\frac{\partial\ln n}{dr}\right) \tag{5.77}$$

$$j_z = \frac{\sigma}{(1+\beta^2)}\left(-\frac{\partial\varphi}{\partial z} + \frac{\partial T_e}{\partial z} + T_e\frac{\partial\ln n}{\partial z}\right) \tag{5.78}$$

where $\beta = eB/m_e\nu_{ef}$ is the Hall parameter, T_e is the electron temperature, φ is the potential, and $\sigma = e^2 n/m_e\nu_{ef}$ is the classical plasma conductivity. In addition, current conservation implies that

$$\frac{\partial j_r}{\partial r} + \frac{\partial j_z}{\partial z} + \frac{j_r}{r} = 0 \tag{5.79}$$

The numerical analysis is similar to that developed previously [116] and described in Section 5.4. The equation for potential (Eq. (5.79)) is solved numerically by iteration using the successive overrelaxation procedure. Similarly to Ref. [137], instead of employing additional assumptions for the calculation of the electron temperature distribution, we will use an experimentally measured electron temperature distribution. The electron temperature and magnetic field radial distributions are taken from the same experiment [134,135]. In order to simplify the calculations, we will approximate the magnetic field and the measured electron temperature distributions by the following analytical expressions:

$$T_e = T_{eo}\exp\left(-\frac{\alpha(R_{2-r})}{h}\right) \tag{5.80}$$

$$B = B_o\frac{1}{1+\gamma r} \tag{5.81}$$

where h is the channel width, R_2 is the radius of the outer wall, α, γ are numerical coefficients obtained by fitting to experimental data, T_{eo} is the maximum electron temperature, and B_o is the maximum magnetic field strength (radial component) measured in Refs [134–136].

The experimental data for two different thrusters are shown in Figure 5.25B and C. These data were obtained using a similar technique, namely a high-speed probe that induced very small plasma perturbations [134–136]. One can see that in each case, the potential distribution in the channel exhibits a very similar shape having a peak of the potential along the channel centerline. The most

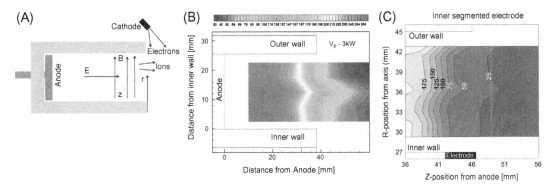

FIGURE 5.25

Schematics of the thruster, model geometry for potential distribution (A) and experimental data on potential distribution in the Hall thruster channel. (B) Experimental data from Ref. [135]. (C) Experimental data from Ref. [81].

Source: *Reprinted with permission from Ref. [80]. Copyright (2004) by American Institute of Physics.*

interesting feature is that significantly different Hall thruster configurations produce very similar and repeatable structures.

In the following, we present the calculated potential distribution in the Hall thruster channel according to Eq. (5.79). These calculations are shown in Figure 5.26. The radial distributions of the experimentally observed ([135], and approximated by Eqs (5.80) and (5.81)) electron temperature and magnetic field are shown in Figure 5.26A. One can see that while the electron temperature peaks near the outer wall, the magnetic field has a maximum near the inner wall [135].

Using the electron temperature and the magnetic field distributions as an input, the potential distribution from the current conservation equation (Eq. (5.79)) and plasma flow model (Section 5.2.4) was calculated. Let us first examine the effect of only a magnetic field gradient in the radial direction (Figure 5.26B). One can see that a nonuniform potential distribution is found with the highest electric field in the region where the magnetic field is high. This is a very much expected result that reflects the fact that the crossed field electron mobility is higher across the magnetic field in the region with a low magnetic field. On the other hand, the presence of the electron temperature gradient leads to additional nonuniformity due to the electron current along the magnetic field. Suppose the electron temperature is higher near the outer wall (see Figure 5.26 A). In this case, the current conservation requirement leads to current density increase in the middle of the channel, since high magnetic field near the inner wall prevents significant current density raise in that region. In this case, the higher current density in the middle of the channel will lead to the higher electric field. Therefore, the equipotential contours create a structure with a potential maximum along the middle of the channel as shown in Figure 5.26C. It can be seen that the model generally predicts the potential structure similar to the one observed experimentally. In addition, similarly to the experiment, this potential structure can be obtained only under certain conditions. These conditions require gradient in both magnetic field and electron temperature gradients.

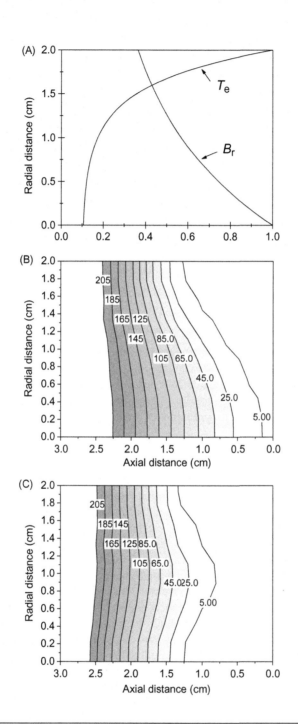

FIGURE 5.26

(A) Electron temperature and magnetic field radial profiles according to experiment [81,134–136].
(B) Potential distribution in the case without electron temperature gradient ($\alpha = 0$). (C) Potential distribution in the case of both electron temperature and magnetic field gradients ($\alpha = 7$, $\gamma = 0.8$).

Source: *Reprinted with permission from Ref. [80]. Copyright (2004) by American Institute of Physics.*

5.2.6 **Anodic plasma in Hall thrusters**

In this section, plasma dynamics and ionization of propellant gas are modeled within the anode holes used for gas injection of a Hall thruster. One of the least studied boundary regions in the Hall thruster channel is the anode. In spite of considerable experimental and theoretical study of the internal structure of the Hall thruster [81,135], general understanding of the near-anode phenomena is very limited. In part this is due to the fact that most of the existing models deal with plasma flow in the channel, plasma–wall interaction, etc. assuming some boundary conditions at the anode. Additionally, it is exceedingly difficult to take measurements very close to or within the anode itself. However, the anode effects potentially play an important role in Hall thruster operation. Previously the near-anode region was studied by Melikov [138,139]. He concluded that high-energy electrons penetrate from the channel into the anode through the holes and that electron-impact ionization begins in the anode cavity and the anode holes. Recently systematic study of the anode phenomena in Hall thrusters was initiated by Dorf et al. [137]. It was found that, dependent on the conditions at the anode, the anode voltage drop can be either positive or negative [137]. While the case with the negative anode fall has been intensively studied [140,141], the positive anode fall is not understood well. Recently, vanishing of the negative anode sheath was considered [142]; however the positive anode sheath formation was not considered. Experimental studies of the positive anode sheath formation and detailed discussion of the possible mechanism was considered by Dorf et al. [143]. The negative anode fall is associated with the sonic transition for the ion flow and due to the fact that the electron thermal current in the near-anode sheath is larger than the discharge current. As a result, there is a requirement for ion flow to the anode. In the case of the positive anode fall, electrons gain energy while moving toward the anode which results in higher power deposition at the anode and therefore possible drop in thruster efficiency.

Anode coating due to dielectric channel sputtering during operation is a common occurrence for Hall thrusters. Dorf et al. [143] demonstrated that anode coating is linked to a "quieting" of the discharge oscillation provides additional incentive for further study. Eventually, understanding of these processes may yield improved thruster designs with less noise in the discharge. Thus this phenomenon yields substantial motivation for detailed study.

In this section, a study of the near-anode region of the Hall thruster in the case of positive anode fall is presented. It was found experimentally [143] that this case is associated with anode coating and current closing through holes in the anode. These holes are used for gas injection into the thruster channel. It was shown that more stable operation of the Hall thruster can be achieved when the anode is in fact coated by a dielectric material. In this case, plasma is formed in the anode holes and creates a jet structure expanding downstream into the channel.

Under conditions of anode dielectric coating (shown in Figure 5.27), these holes behave similarly to hollow anode [144] and hollow cathode [145] devices. Additional ionization can occur within this region. A description of the resulting effects on the current density and internal potential distribution will be attempted. Figure 5.27 shows the various plasma structures near to and within an individual anode hole.

5.2.7 **Model of the hollow anode**

As the electrons stream into the anode hole from the jet, they encounter neutral gas within the hole. The ionization rate and subsequent impact on the electron current density is described by

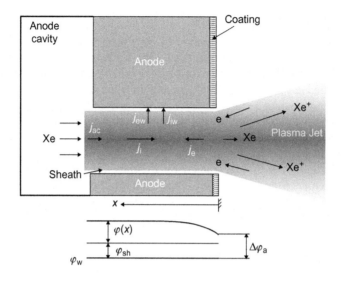

FIGURE 5.27

Schematic of the near-anode plasma structure.

Source: *Reprinted with permission from Ref. [252]. Copyright (2008) by American Institute of Physics.*

$$\frac{dj_e}{dx} = e\beta_i n_e n_a \tag{5.82}$$

where j_e is the electron current density and β_i is the ionization rate. On the other hand, electron-impact ionization leads to neutral density decrease:

$$\frac{d(V_a n_a)}{dx} = -\beta_i n_e n_a \tag{5.83}$$

Within the anode hole, current conservation conditions read as $\nabla \cdot j = 0$. A current conservation equation means that current density is invariant inside the hole:

$$\frac{dj_i}{dx} + \frac{dj_e}{dx} = -\frac{2}{R_h}(j_{ew} - j_{iw}) \tag{5.84}$$

where j_e is the electron current density, j_i is the ion current density, j_{ew} is the electron current density to the wall, and j_{iw} is the ion current density to the wall. The right-hand side of Eq. (5.84) corresponds to the current collected by the anode hole wall. Consequently, the current along the anode hole decreases due to current conservation condition. Schematically, the current balance is shown in Figure 5.27. The electron current density to the wall can be calculated as follows:

$$j_{ew} = e n_s V_{th} \exp\left(-\frac{\Delta\varphi}{T_e}\right) \tag{5.85}$$

where $\Delta\varphi = \varphi(x) - \varphi_w$ is the voltage drop across the sheath, φ_w is the wall potential (see Figure 5.27). Ion current density at the wall can be calculated using the Bohm condition:

$$j_{iw} = en_s\sqrt{\frac{T_e}{m_i}} = en_s C_s \tag{5.86}$$

where n_s is the electron density at the sheath−presheath interface [146]. In the diffusion approximation [147] assuming that $T_e \gg T_i$, the axial component of the electron and ion velocities can be calculated using corresponding expressions for electron and ion mobilities:

$$V_e = \mu_e\left(\frac{d\varphi}{dx} - T_e\frac{1}{n_e}\frac{dn}{dx}\right) \tag{5.87}$$

$$V_i = -\mu_i\frac{d\varphi}{dx} \tag{5.88}$$

The electron energy distribution in the anode hole is determined by the balance between the Joule heating, wall losses, and ionization losses.

$$\frac{3}{2}\frac{dT_e}{dx} = Q_j - Q_{ion} - Q_w \tag{5.89}$$

Combining Eqs (5.83)−(5.89), one can derive the system of equations for potential and electron density distributions inside the anode hole:

$$\frac{d^2\varphi}{dx^2} = \frac{\beta_i n_a}{\mu_i} + \frac{2n_s/n_e}{R_h\mu_i}(V_{the}\exp(-\varphi_w/T_e) - C_s) - \frac{d\varphi}{dx}\frac{d\ln(n_e)}{dx} \tag{5.90}$$

$$\frac{d^2 n_e}{dx^2} - \frac{dn_e}{dx}\frac{1}{T_e}\frac{d\varphi}{dx} - \frac{n_e}{T_e}\frac{d^2\varphi}{dx^2} = -\frac{\beta_i n_e n_a}{\mu_e T_e} \tag{5.91}$$

Electrons can gain some energy in the plasma jet emanating from the gas injection holes. In the plasma jet, the electron density increases toward the hole. In order to calculate electron density variation in the anodic jet, we developed the model of the plasma jet shown schematically in Figure 5.27.

5.2.7.1 Analytical solution

Let us start with analysis of the system of equations (5.90) and (5.91) with an aim to find an analytical solution for some limited case. If we substitute second derivative of potential in Eq. (5.91) with Eq. (5.90), we arrive at the following expression after some modifications:

$$\frac{d^2 n_e}{dx^2} - n_e\left[\frac{2n_{es}/n_e}{T_e R_h\mu_i}(V_{te}\exp(-\varphi_w/T_e) - C_s) + \frac{\beta_i n_a}{\mu_e T_e}\right] = 0 \tag{5.92}$$

We further assume that the coefficient in the right-hand side of Eq. (5.92) is constant, i.e.,

$$\alpha = \left[\frac{2n_{es}/n_e}{T_e R_h\mu_i}(V_{te}\exp(-\varphi_w/T_e) - C_s) + \frac{\beta_i n_a}{\mu_e T_e}\right] = \text{constant}$$

In this case, Eq. (5.92) has an analytical solution that reads as

$$n_e = \exp\{x\alpha\} \tag{5.93}$$

This solution is applicable in the case of $\alpha = $ constant, which can be realized if for instance electron temperature variation is small (this condition will be analyzed below) and neutral density change is also negligible. According to Eq. (5.93), electron density increases exponentially inside the anode hole due to significant impact ionization.

5.2.7.2 Model of the anodic plasma jet

Plasma expansion and electron heating near the hole is calculated from the 2D model elsewhere [116]. Briefly, the model is based on 2D, axisymmetric, two-fluid hydrodynamics that include the mass and momentum conservation for ions, neutrals, and electrons. The following system of equations completely describes the plasma flow and current distribution:

$$m_i(\vec{V}_i \cdot \vec{\nabla})\vec{V}_i = -k(Z_i T_e + T_i) \cdot \vec{\nabla} \ln(n_e) + \frac{\vec{j} \times \vec{B}}{n_e} - \nu_{ia} m(\vec{V}_i - \vec{V}_a) \tag{5.94}$$

$$\vec{j} = \sigma \left\{ \vec{E} + (kT_e/e)\vec{\nabla} \ln(n_e) - \frac{\vec{j} \times \vec{B}}{e \cdot n_e} \right\} \tag{5.95}$$

$$\vec{\nabla} \cdot (\vec{V}_\alpha\, n_\alpha) = 0 \tag{5.96}$$

$$\vec{\nabla} \cdot \vec{j} = 0 \tag{5.97}$$

where $\vec{V}_i$ is the ion velocity, $\vec{E}$ is the electric field, $\vec{j}$ is the current density, and $\vec{B}$ is the magnetic field (self or external). In this particular study, a constant neutral velocity along the jet is assumed and the magnetic field effects are neglected.

The formulation of the boundary conditions was described elsewhere [116]. The *free plasma boundary* is defined as the surface where the plasma jet velocity has only a tangential component, i.e., the normal component V_n is zero. The plasma density at the free boundary of the plasma jet is $n = 0$, and normal current density is $j_n = 0$.

5.2.7.3 Calculations of the anodic plasma

We consider a typical case of middle power, SPT-100-type Hall thruster. Discharge current of 5 A, mass flow rate of 5 mg/s, and a neutral xenon gas temperature of 700 K [148] are considered. The anode hole length of about 1 mm is considered. The electron density in the vicinity of the anode is taken to be $(2-10) \times 10^{16}\,\mathrm{m}^{-3}$ with an electron temperature of about 5 eV [63]. The solution strategy implemented here is based on known electron density at the anode hole exit plane. A condition that the current is collected along the anode hole or inside the anode cavity is employed. Wall potential, φ_w, is found as a part of the solution based on the condition that the current collected within the anode hole and the current inside the anode cavity equal to the total discharge current.

Anodic jet density structure is shown in Figure 5.28. Plasma is generated inside the anode hole and expands. The voltage drop across the plasma jet is about few volts. The fast plasma jet expansion leads to plasma density decrease by more than order of magnitude at a distance of a few

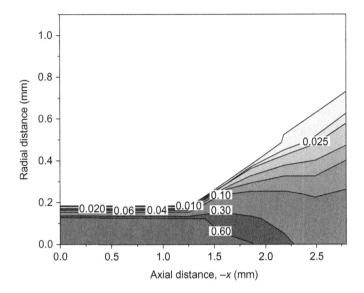

FIGURE 5.28

Plasma density distribution in the jet emanating from a single anode hole.

Source: *Reprinted with permission from Ref. [252]. Copyright (2008) by American Institute of Physics.*

millimeters from the hole. Thus one might expect that the plasma density at the anode hole entrance is in the range of 10^{17} m^{-3} based on plasma density measurements at the distance of about few millimeters from the anode [63].

Electrons entering the anode hole produce significant ionization of the gas. As a result, the electron density inside the hole increases. The characteristic length for electron density gain, $1/\alpha$, can be estimated analytically according to Eq. (5.93). The calculated characteristic length for electron density gain is shown in Figure 5.29. One can see that depending on the number of anode holes and the anode hole radius, the characteristic length, $1/\alpha$, is comparable or smaller than the anode hole length. Thus, based on the analytical estimations, one can expect significant ionization to occur inside the hole.

The numerically calculated potential and electron density distributions along the anode hole are shown in Figure 5.30 with anode hole radius as a parameter. Indeed, this model predicts significant increase of the electron density in the case of a small anode hole radius by few order of magnitude. The potential drop along the anode hole is about 10−30 V dependent on the anode hole radius as shown in Figure 5.30. The potential drop increases with anode radius decreases. The total potential drop in the near-anode region consists of the potential drop along the anode and the anode sheath as will be shown below.

The current density distributions are shown in Figure 5.31. As is clearly shown, the electron current density dominates near the anode hole exit plane which is expected given their higher mobility. Ion current density increases and near the hole inner end becomes very large due to increase of the electric field. According to these results, current continuity condition would require significant current collection inside the anode cavity as will be shown below.

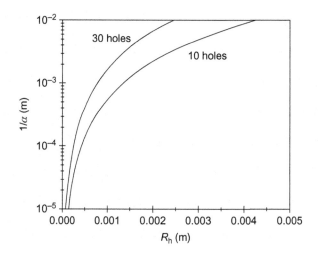

FIGURE 5.29

Characteristic ionization length as a function of the anode hole radius with number of anode holes as a parameter.

Source: *Reprinted with permission from Ref. [252]. Copyright (2008) by American Institute of Physics.*

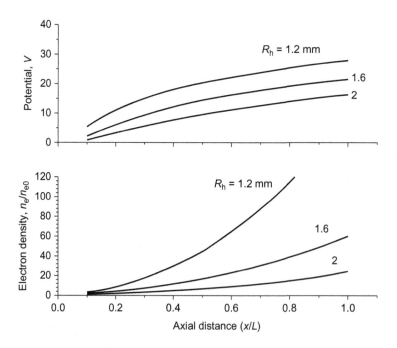

FIGURE 5.30

Potential and electron density distribution along the anode hole with anode hole radius as a parameter.

Source: *Reprinted with permission from Ref. [252]. Copyright (2008) by American Institute of Physics.*

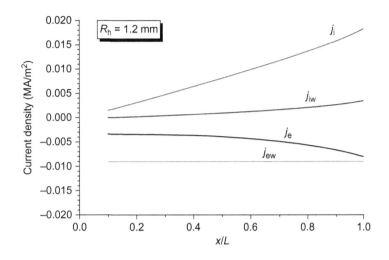

FIGURE 5.31

Components of the current density distribution along the anode hole.

Source: *Reprinted with permission from Ref. [252]. Copyright (2008) by American Institute of Physics.*

The wall potential drop in the anode hole is negative due to excess of electron thermal current as shown in Figure 5.32. The wall potential drop increases with anode hole radius increase. On the other hand, potential drop along the anode hole decreases so that the total potential drop in the anode region becomes negative. Thus, this model predicts that a "double layer" potential structure can exist in the near and interior anode region with the total voltage drop being either positive or negative. This allows significant gain of electron energy and consequently strong ionization in the region of gas injection while reducing the detrimental efficiency effects of the positive anode fall.

In this model, we have implemented the condition of current conservation inside the anode, i.e., the total current at the inner end of the anode hole and the anode hole wall current is equal to the discharge current. As a result, presence of the electric field at the inner end of the anode hole leads to ion current. Figure 5.33 shows the dependence of the ion current at the inner end of the hole on the anode hole radius. One can see that solution indeed requires significant ion current from the anode cavity, i.e., significant ionization inside the cavity. This current decreases as anode radius approaches approximately 1.5 mm due to density decrease. However, larger anode hole radius leads to increase of the sheath voltage (see Figure 5.33) and as a result current collected by the anode hole decreases. This leads to increase of the current fraction inside the anode cavity. Overall, the model predicts that the discharge cannot be supported without significant ionization inside the anode cavity.

5.2.8 Thruster with anode layer (TAL)

Among Hall thruster technologies, the TAL has much wider technical capabilities especially in the high-power regime of operation [79]. While in an SPT, the interaction of the plasma with the

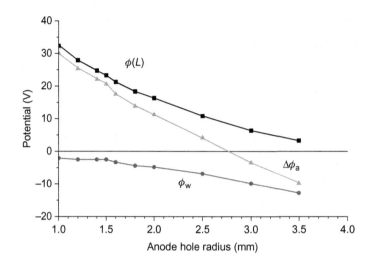

FIGURE 5.32

Dependence of the wall potential and total potential drop across the near-anode region on the anode hole radius.

Source: *Reprinted with permission from Ref. [252]. Copyright (2008) by American Institute of Physics.*

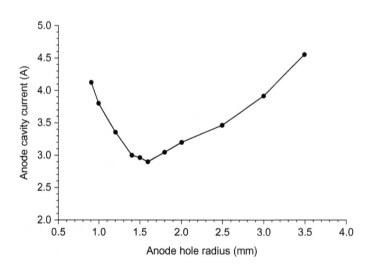

FIGURE 5.33

Dependence of the current fraction inside the anode cavity on the anode hole radius.

Source: *Reprinted with permission from Ref. [252]. Copyright (2008) by American Institute of Physics.*

dielectric wall plays an important role, in TAL plasma—wall interaction is less critical. Due to the collisions of the electrons with the wall and SEE, the electron temperature remains relatively low in comparison to the TAL. As a result, the ion acceleration occurs over a more extended region [63]. Naturally that having acceleration channel, walls made from conductive material would significantly affects Hall thruster operation. In fact, in a TAL, the ion acceleration takes place over a very short length of about the electron Larmor radius near the anode. This is why the term "anode layer" was attributed to the Hall thruster with metal walls. Similarly, in magnetically insulated ion diodes, ion acceleration occurs in a thin layer having thickness of about the electron Larmor radius [149]. Despite many theoretical efforts, the complicated physical processes in the Hall thruster channel are far from being completely understood. Mainly the physics of the electron transport, plasma interaction with the wall, and the transition between the quasi-neutral plasma, and the sheath have not been investigated in satisfactory detail.

In the TAL variant, an electric discharge in the crossed magnetic and electric fields is created in the gap between magnetic poles where closed electron drift takes place. In this gap, the neutral atoms are ionized and accelerated. Some fraction of the accelerated ion stream is directed toward the wall that limits the discharge in the radial direction and protects the magnetic poles from erosion. This leads to wear of these walls due to ion bombardment and results in a shorten lifetime of such accelerators.

Recently great emphasize was put forward into development of high-power Hall thrusters [150]. Among Hall thruster technologies, the TAL configuration seems to have much wider technical capabilities and range of parameters [151]. Since reduced erosion of various thruster components, such as electrodes, insulators, and screens, is critical for the long-term operation of the thruster, the TAL variant of Hall thruster technology seems to be beneficial since it has a very small acceleration region and therefore small area with contact of ions with materials where possible erosion occurs. In order to meet high-power requirements, a TAL was developed and an experimental model was presented. TAL using Bismuth as a propellant demonstrated specific impulse in the range of 2000—5000 s at power levels of 10—34 kW.

The idealized picture of the Hall thruster in which ions are accelerated in the channel is not exact and in reality in TALs as well as in some SPTs, significant ion acceleration occurs outside of the channel in the fringing magnetic field. This may be considered as one possible mechanism of high plasma beam divergence in Hall thrusters [152].

TALs can operate in both single- and two-stage regimes; in the two-stage regime, ionization and acceleration take place in two separate discharges, so that at the exit plane of the first stage, the plasma is highly ionized and additional ionization in the second stage is insignificant [63]. Usually, the two-stage regime of operation has many advantages, for instance, the possibility to achieve much higher specific impulse [153]. When a two-stage thruster is considered, the most important region in terms of long time operation is the second acceleration stage in which a very large voltage drop is applied. The second stage channel geometry is shown schematically in Figure 5.34.

An important aspect of high-power TAL is the sheath formation near the channel wall. Since the channel walls have potential, which is equal to the cathode potential, a significant potential drop can exist between the wall and the plasma. As a result, a high-voltage space-charge sheath is formed and the sheath thickness can be comparable to the channel width. Therefore, a transition

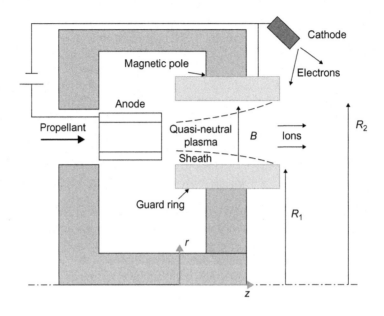

FIGURE 5.34

Schematics of TAL.

between quasi-neutral and space-charge regimes can occur in the high-power TAL channel. This transition can have a significant effect on the TAL operation and therefore extremely important for practical applications. In this section, this effect will be considered in detail.

5.2.8.1 Plasma boundary issues in TAL

The plasma—wall transition region in the TAL channel determines the particle and energy fluxes from the plasma to the wall. Recently, we presented a model of plasma wall transition in the SPT that accounts for SEE [78]. In order to develop a self-consistent model, the boundary parameters at the sheath edge (ion velocity and electric field) are obtained from a 2D plasma bulk model. In the considered condition, i.e., ion temperature much smaller than that of electrons and significant ion acceleration in the axial direction, the presheath scale length becomes comparable to the channel width so that the plasma channel becomes an effective presheath. It was shown that a plasma—sheath matching approach proposed previously could be used [154]. In this approach, the electric field that develops in the presheath can serve as a boundary condition for the sheath. At the same time, the model predicts that the quasi-neutrality assumption at the presheath edge is still valid in typical cases. It was shown [78,79] that the presheath scale length becomes comparable to the channel width under typical conditions of the Hall thruster channel. Thus, the model for the quasi-neutral plasma region is extended up to the sheath edge in order to provide the boundary condition at the plasma—sheath interface. The density N_s and velocity V_s at the presheath—sheath interface serve as a boundary condition for the quasi-neutral plasma model that is described in the next section.

5.2.8.2 Plasma flow of the quasi-neutral plasma

The initial model is based on the assumption that the quasi-neutral region length (i.e., channel width, see Figure 5.34) is much larger than the Debye radius and therefore we will assume that $Z_i n_i = n_e = n$, where Z_i is the ion mean charge, n_i is the ion density, and n_e is the electron density. For simplicity, only singly charged ions are considered in this chapter ($Z_i = 1$). We will consider the plasma flow in an annular channel as shown in Figure 5.34. A magnetic field with only a radial component, $B_r = B$, is imposed. Cylindrical coordinates will be used with angle θ, radius r, and axial distance from the anode z. The plasma presheath−sheath interface is considered to be the lateral boundary for the plasma flow region. A plasma will be considered with "magnetized" electrons and "unmagnetized" ions, i.e., $\rho_e \ll L \ll \rho_i$, where ρ_e and ρ_i are the Larmor radii for the electrons and ions, respectively, and L is the channel length. We employ a hydrodynamic model assuming: (i) the system reaches a steady state and (ii) the electron component is not inertial, i.e., $(\mathbf{V}_e \nabla) \cdot \mathbf{V}_e = 0$.

Generally two regimes are possible in a TAL, the so-called vacuum regime (in which the electron density is much higher than that of ions) and the quasi-neutral plasma regime [155]. Below, we briefly formulate a model for the quasi-neutral regime. A hydrodynamic model is employed in a 2D domain assuming that the system reaches a steady state. The same model as the one described in Section 5.2.4 can be employed. In the component form, the system of equation reads [79]

$$\frac{\partial(nV_z)}{\partial z} + \frac{\partial(nV_r)}{\partial r} + \frac{nV_r}{r} = \beta nn_a \tag{5.98}$$

$$V_z\frac{\partial V_z}{\partial z} = -V_r\frac{\partial V_z}{\partial r} + \frac{e}{m_i}E_z - \beta(V_z - V_a)n_a \tag{5.99}$$

$$V_z\frac{\partial V_r}{\partial z} = -V_r\frac{\partial V_r}{\partial r} + \frac{e}{m_i}E_r \tag{5.100}$$

$$\frac{\partial(n_a V_a)}{\partial z} = -\beta nn_a \tag{5.101}$$

In the system of equations (5.98)−(5.101), the subscript for ions was omitted, i.e., $V = V_i$.

Due to the configuration of the magnetic field (i.e., only the radial magnetic field component is considered in the model), the electron transport is greater in the azimuthal direction ($E \times B$ drift) than in the axial direction (drift diffusion due to collisions).

The equation describing the electron transport across the magnetic field reads

$$j_{ez} = en\frac{(e/m\nu_{ef})}{1 + (\omega_e/\nu_{ef})^2}\left(E_z + \frac{\partial T_e}{\partial z} + T_e\frac{\partial \ln n}{\partial z}\right) \tag{5.102}$$

where $\nu_{ef} = \nu_{en} + \nu_{ew} + \nu_B$ is the effective electron collision frequency. The different electron collision mechanisms are described below. Electron collisions treatment is the same as the one described in Section 5.2.4.

5.2.8.3 Example of TAL simulation: high-power Bismuth TAL

In this section, we describe the simulation of a particular TAL design that uses Bismuth as the propellant. The Bismuth thruster was originally developed in the 1960s in the former USSR. It was demonstrated that specific impulse in the range of 2000–5000 s at power levels of 10–34 kW can be achieved. This thruster has a two-stage configuration in order to separate the ion production and acceleration zones. In this analysis, we will consider only the second stage. In particular, plasma density is used as a parameter. The simulations correspond to the following case: mass flow rate is 20 mg/s, magnetic field is uniform and equals 0.2 T. In this case, the calculated discharge current is about 6 A. The particular design is considered that has following geometry: $R_1 = 6$ cm, $R_2 = 10$ cm, and $L = 5$ mm. The 2D hydrodynamic model described in the previous section is used to calculate plasma flow in the second stage channel.

The computational domain is shown schematically in Figure 5.34. The electron temperature distribution along the channel is shown in Figure 5.35 in the case of $\varphi_a = 3$ kV. One can see that the electron temperature peaks at about 120 eV near the channel exit plane and then decreases toward the anode. In order to validate the model, some integral characteristics of the thruster are calculated and compared with experiment.

The steady-state thrust can be calculated at the thruster exit plane as follows: $T = 2\pi m_i \int_{R_1}^{R_2} n V_z^2 r \, dr$. The calculated thrust increases linearly with the current (ion beam current) in agreement with experiment as shown in Figure 5.36. It can be seen that the model predicts thrust levels close to those measured experimentally over the entire range of the beam current (or equivalently mass flow rate).

5.2.8.4 Example of calculation: TAL—analysis of the space-charge sheath near the channel wall

In the channel of the second stage, the guard ring has the cathode potential as shown schematically in Figure 5.34. When high voltage across the acceleration channel is considered, one should take into account the importance of the sheath development near the screening walls (guard ring) of the channel. When a negative voltage is applied to a surface immersed in a plasma, electrons are repelled from the surface, leading to sheath formation. Electrons drift away from the surface due to

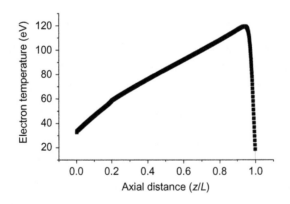

FIGURE 5.35

Electron temperature along the channel.

Source: *Reprinted with permission from Ref. [79]. Copyright (2004) by American Institute of Physics.*

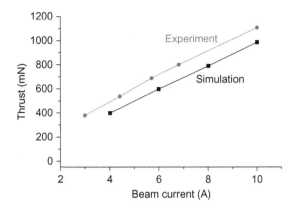

FIGURE 5.36

Thrust as a function of a beam current.

Source: *Reprinted with permission from Ref. [79]. Copyright (2004) by American Institute of Physics.*

the presence of the high electric field. In the steady state, the ions are then accelerated toward the surface by the electric field of the sheath. In the 1D steady-state case, the sheath thickness can be estimated according to the Child−Langmuir law [156,157]:

$$L_s = \left(\frac{4}{9}\varepsilon\right)^{1/2} \left(\frac{2e}{m_i}\right)^{1/4} \frac{U^{3/4}}{(eZ_iN_sV_s)^{1/2}} \tag{5.103}$$

where V_s is the ion velocity at the sheath edge, U is the voltage across the sheath, L_s is the sheath thickness, ε is the permittivity of vacuum, N_s is the plasma density at the sheath edge, and m_i is the ion mass. In a TAL, the channel wall has a potential equal to the cathode potential and therefore the voltage across the sheath, U, is equal to the plasma potential and varies along the channel. Similarly, the steady-state sheath thickness in a partially magnetized plasma was calculated in a plasma immersion ion implantation system [158]. One can see that the steady-state sheath thickness is determined by the plasma density and ion velocity at the sheath edge for a given bias voltage (Equation 5.103).

Based on the 2D hydrodynamic model, the plasma parameters at the plasma−sheath interface (N_s and V_s) as well as ion flux to the wall are calculated. It should be noted that the model used for sheath thickness calculation is one-way coupled, since the 2D plasma flow model does not account for the sheath thickness, while account for the voltage drop in the plasma and the sheath. The sheath thickness is several centimeters in typical high-power TAL conditions and occupies a significant portion of the channel as shown in Figure 5.37. The sheath thickness varies significantly along the channel wall due to variation of the potential drop between the plasma and the wall, which has cathode potential. Note that in typical high-power TAL design, the channel width is about several centimeters. It can be seen that when the discharge voltage is about 10 kV, almost the entire channel becomes nonquasi-neutral (sheath) and the quasi-neutral plasma is confined in the middle of the channel. Therefore, one can conclude that the discharge voltage limits the regime of quasi-neutral acceleration in a TAL with given geometry. Considering the interaction of the highly energetic

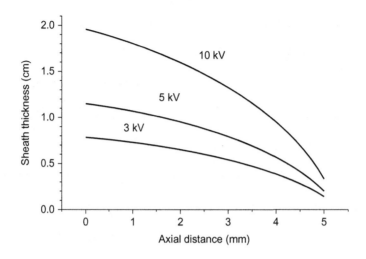

FIGURE 5.37

Sheath thickness along the channel with the voltage drop as a parameter.

Source: *Reprinted with permission from Ref. [79]. Copyright (2004) by American Institute of Physics.*

(due to acceleration in the sheath toward the wall) ion flux with wall materials, one can expect significant erosion.

It was shown above that the effect of the sheath expansion near the acceleration channel wall is very significant in the case of a high-power TAL. It affects mainly the cross-sectional area of the quasi-neutral plasma as shown in Figure 5.39. In turn, the change of the cross-sectional area of the quasi-neutral plasma affects the current density in the axial direction. It is natural to study this effect using a 1D approach since the axial current density (Eq. 5.102) is calculated in a 1D manner along the channel centerline. To this end, we developed a simplified quasi-1D model of the discharge in crossed $E \times B$ fields. The original model of this discharge was developed by Zarinov and Popov [59]. Recently, a modified version of that model that took into account plasma−wall interactions by introducing a more detailed electron energy equation was developed by Choueiri [63]. We present here a model [79] in which we adopt the formulation of Refs [59,63] except that sheath expansion near the channel wall in the acceleration region is taken into account as it occurs in a high-power TAL. In order to compare with previously published results [59,63], we adopt here a rectangular coordinate system shown in Figure 5.38. The z-axis is along the channel axis and the y-axis is along the thruster radius, i.e., along the magnetic field. The x-axis corresponds to the azimuthal direction in the Hall thruster. In this case, the current density in the quasi-neutral region will vary along the channel (z-axis). The characteristic time of sheath formation near the channel wall is about the plasma frequency and therefore a steady-state sheath will be considered. In the case considered, the cross-sectional area $A(z)$ of the quasi-neutral plasma is varied along the channel. We start with the following set of equations for the quasi-neutral plasma:

$$A\frac{\mathrm{d}j_z}{\mathrm{d}z} + j_z\frac{\mathrm{d}A}{\mathrm{d}z} = e\nu_{iz}n_eA \qquad (5.104)$$

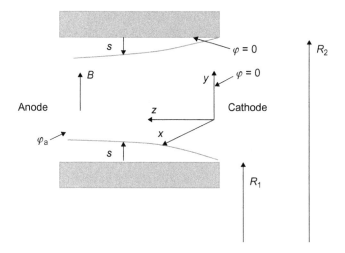

FIGURE 5.38

Schematics of the channel in TAL.

$$j_z = \mu_\perp \left(e n_e \frac{d\varphi}{dz} - \frac{d(n_e T_e)}{dz} \right) \tag{5.105}$$

$$n_e = n_i = \frac{j_i(z)}{e\sqrt{(2e(\varphi_a - \varphi)/M)}} \tag{5.106}$$

$$\mu_\perp = \frac{e}{m_e \nu_e} \frac{1}{1 + (\omega_e/\nu_e)^2} \tag{5.107}$$

where $\mu_\perp$ is the classical cross-field mobility, $j_i(z)$ is the ion current dependent on the cross section, j_z is the electron current, ν_{iz} is the ionization frequency, ν_e is the electron collision frequency, and ω_e is the electron cyclotron frequency. The change of the cross-sectional area of the quasi-neutral plasma due to sheath expansion can be calculated as follows:

$$A(z) = \pi((R_2 - s)^2 - (R_1 + s)^2) \tag{5.108}$$

where R_1 and R_2 are the inner and outer radii of the channel and s is the sheath thickness (dependent on z) that can be calculated from Eq. (5.103). Since the wall has cathode potential (set to 0 for simplicity), the potential drop between the plasma and the wall is simply equal to the plasma potential φ. The channel cross section at the exit plane $A_o = \pi(R_2^2 - R_1^2)$ corresponds to the case of zero sheath thickness, since the wall potential is equal to the exit plane potential and therefore cathode potential.

In addition, following Ref. [59,63,81], a simplified energy equation can be used: $T_e = \beta\varphi$. It should be noted that similar linear behavior of electron temperature with discharge voltage near the channel exit plane was measured very recently [159]. On the other hand, the electron temperature profile in the entire channel is much more complicated and is a result of the balance between electron Joule heating and cooling due to interaction with the channel walls and ionization.

For example, a typical profile is shown in Figure 5.35. Therefore, linear behavior of the electron temperature with plasma potential is generally only valid near the channel exit. In this particular 1D model, we are interested primarily in the effect associated with near-wall sheath expansion. Therefore, in order to simplify the problem, we will consider electron temperature as a parameter, which is constant along the channel.

Let us normalize the system of equations (5.104)–(5.107): $\phi = \varphi/\varphi_a$, $\xi = z/l^*$, $\bar{n} = n_e/n^*$, $\bar{j} = je/j^*$, $\theta = T_e/\varphi_a$. The characteristic quantities can be defined as follows:

$$l^* = \sqrt{\frac{e\varphi_a}{(\nu_{iz}/\nu_e)m_e\omega_e^2}};\ \ n^* = \frac{j_{io}}{e\sqrt{(2e/M)\varphi_a}};\ \ j^* = e\nu_{iz}n^*l^*;\ \ \bar{A} = \frac{A(z)}{A_0}$$

The derivative of the channel cross section $A(z)$ can be calculated as follows:

$$\frac{1}{A}\frac{dA}{dz} = -\frac{2(ds/dz)}{((R_2^2 - R_1^2/R_2 + R_1) - 2s)} \tag{5.109}$$

Finally, the following normalized system of equations is obtained:

$$\frac{d\bar{j}}{d\xi} = \bar{n} - \bar{j}\frac{1}{\bar{A}}\cdot\frac{d\bar{A}}{d\xi} \tag{5.110}$$

$$\bar{j} = \bar{n}\frac{d\phi}{d\xi} - \theta\frac{d\bar{n}}{d\xi} \tag{5.111}$$

$$\bar{n} = \frac{1}{\bar{A}\sqrt{1-\phi}} \tag{5.112}$$

This system of equations is similar to the original formulation of Zharinov and Popov [59] with one exception that change of the cross-sectional area of the quasi-neutral channel $A(z)$ is taken into account.

Below, the following particular case of a high-power TAL is considered: mass flow rate of 10 mg/s, Bismuth as the propellant (i.e., $R_1 = 6$ cm and $R_2 = 10$ cm). As was mentioned above, in the calculation presented here, we assume a constant electron temperature close to the peak electron temperature ($T_e = 100$ eV, see Figure 5.35).

The effect of the near-wall sheath expansion on the anode layer potential profile is shown in Figure 5.39. One can see that the sheath expansion strongly affects the potential profile and the anode layer thickness decreases from 0.7 down to 0.15 when the sheath expansion effect is taken into account. This happens because the quasi-neutral plasma region is confined in the middle of the channel and as a result the current density increases. In turn, this leads to a higher axial electric field and therefore a smaller anode layer thickness.

The anode layer (acceleration channel) thickness dependence on the discharge voltage is shown in Figure 5.40. One can see that the anode layer thickness significantly decreases with the discharge voltage increasing due to the sheath expansion effect. It is interesting to note that shrinking of the anode layer thickness is important as it leads to a smaller area of contact of the plasma with the walls and therefore smaller total erosion of the channel walls.

It should be noted that the normalized length, l^*, is proportional to $\sqrt{\varphi_a}$, which leads to a non-monotonic behavior of the anode layer thickness with the discharge voltage in considered range of

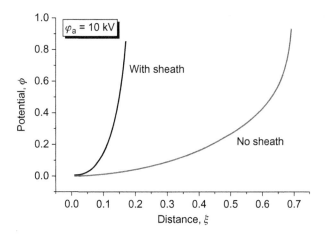

FIGURE 5.39

Plasma potential profile in the quasi-neutral region with and without sheath expansion. Discharge voltage is 10 kV.

Source: *Reprinted with permission from Ref. [79]. Copyright (2004) by American Institute of Physics.*

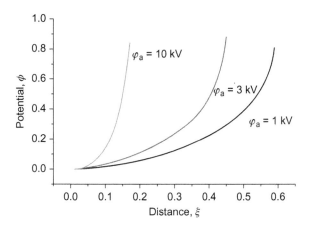

FIGURE 5.40

Potential profile in the quasi-neutral region with discharge voltage as a parameter.

Source: *Reprinted with permission from Ref. [79]. Copyright (2004) by American Institute of Physics.*

discharge voltages. This happens as a result of the near-wall sheath expansion effect. Without considering the near-wall sheath, the dimensional anode layer thickness will monotonically increase with the discharge voltage. Near-wall sheath expansion in the thruster channel leads to the current density increase (see above) and reverses the dependence of the dimensional anode layer thickness on the discharge voltage. The dimensional anode layer thickness will monotonically decrease with the discharge voltage in the large discharge voltage range (>10 kV).

5.2.9 **Present state of the art: multiscale analysis of Hall thrusters**

Recently progress was achieved in two directions: 1D kinetic model of the plasma slab between two walls was developed and analysis of the thruster was performed using 2D hybrid code. The 1D model allows analysis of the electron velocity distribution function, flux, and energy losses to the dielectric walls as well as cross-field conductivity. The model is based on the kinetic approach in which electrons are treated as simulation particles, and the local plasma parameters such as the Larmor radius and Debye length are resolved directly [160,161]. The simulation domain consists of a single magnetic field line bound by the dielectric walls. Magnetic field curvature is accounted for and sheath drop is computed self-consistently. Collisions are treated using the Monte Carlo method, and momentum transfer, Coulomb force, ionization, recombination, and excitation processes are included. Electron transport is computed using two methods. First, bulk mobility is computed from the average axial velocity and the average normal component of the electric field. Then, radial variation in mobility is obtained using newly developed method based on guiding center shift. This method is used to characterize the significance of NWC. The input parameters for the kinetic model include the magnetic field profile, axial electric field, and heavy-particle density distribution. These parameters can be computed by the 2D model. In particular, a hybrid PIC code can be used to obtain global plasma parameters. In this code, ions and neutrals are modeled as particles, but 1D conservation equations are used for the electron density. Radial variation in potential is obtained from the thermalized model.

The primary objective of the multiscale approach is to develop a tool capable of self-consistently determining electron mobility and the thruster plasma properties of interest. To accomplish this goal, we divide the problem into the following three spatial scales as shown in Figure 5.41:

• *Magnetic field line*: On the spatial scale of a magnetic field line, dynamics are driven by the cyclotron motion of electrons. Electrons are magnetized, and individual field lines can be

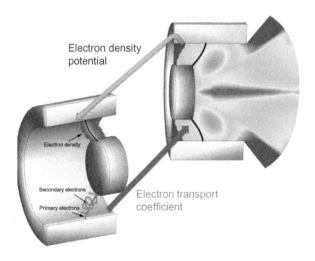

Electron density
potential

Electron density

Secondary electrons

Primary electrons

Electron transport
coefficient

FIGURE 5.41

Illustration of the multiscale model.

considered independent of each other. Heavy particle and properties normal to the field lines are assumed frozen. This approach allows us to rapidly simulate electrons and recover mobility self-consistently. Leveraging modern multicore architectures via multithreading allows us to study multiple field lines simultaneously.

- *Thruster channel*: On the spatial scale of the thruster, plasma is assumed to be quasi-neutral, and electron density can be obtained from kinetic ions. Electron temperature and plasma potential is obtained by solving the quasi-1D equations.
- *Plasma plume*: Outside the thruster exit, the magnetic field plays a negligible role. The plume is quasi-neutral except in low-density sheath regions around the spacecraft. Of interest here is the formation of charge-exchange (CEX) ions and their impact on spacecraft components. Electron density is obtained from Boltzmann relationship, and potential can be solved by direct inversion or by solving Poisson's equation.

Figure 5.42 compares the mobility computed using multiscale approach to that used obtained by hybrid code. Although the background values of mobility are similar quantitatively, we can see stark differences between the two versions. The kinetic solution contains two distinct regions of high mobility, which are not seen in the analytical model. High production of secondary electrons was predicted by the code for the field lines at the left band. NWC may also explain the oscillatory nature of transport, which seems to be related to the high number of SEE seen on the magnetic field lines in this region. On the other hand, the high mobility in the right band may be due to a strong electric field. The right field line corresponds to the location where hybrid predicts drop in potential corresponding the start of the acceleration zone. We also see reduced mobility near the innerpole. This region is dominated by increased magnetic pressure which reduces flux of electrons to this region. It should be noted that the results shown here are statistically accurate.

5.3 **Micropropulsion**

Recent trends in space exploration associate with the paradigm shift toward the small and efficient satellites or micro- and nanosatellites. There are many near-future space missions

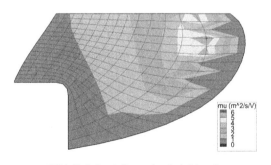
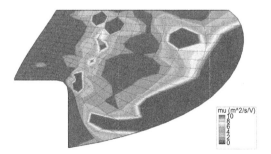

(A) Initial simulation using hybrid code   (B) Kinetic code

FIGURE 5.42

Comparison of the analytical mobility of electrons used by hybrid (A) to the kinetically determined mobility (B).

involving science, military, and commercial payloads utilizing micro- and nanosatellite platforms that require very small thrust levels including very fine attitude control for high-resolution Earth imaging and astronomy, and very fine positioning requirements of spacecraft formation flying that is at the core of many interferometry missions. Nowadays, many basic components of spacecraft are being miniaturized so that micro- and nanosatellites are being designed and built. To satisfy the needs of both the low-thrust missions and the small-scale spacecraft, miniaturized propulsion systems are required.

While, micropropulsion devices producing micro-Newton level thrust or micro-Newton-second impulse bit are currently under development in many academic, government, and industrial laboratories, the field of micropropulsion is still in its infancy.

As it was described in Introduction to Chapter 5, traditionally plasma thrusters are divided into three categories based on the acceleration mechanism: electrothermal (arcjet, resistojet), elecrostatic (ion thruster, FEEP), and electromagnetic (PPT, Hall thruster, MPD thruster). The most important issue for microthrusters is the propellant management and therefore issues related to propellant feeding system become critical. In this section, we describe various propulsion concepts and we adopt natural division based on the propellant mechanism.

5.3.1 Microablative thrusters

Ablative thrusters are most simple devices in terms of functionality and in fact were the first propulsion concept flown in space. Ablative PPTs have achieved a high degree of maturity over several decades of research and development, test, and flight applications. The first flight of a PPT took place on Russian Zond 2 spacecraft in 1964. Subsequent flights took place in United States on the geosynchronous MIT Lincoln Lab LES-6 satellite in 1968 and 1974 and US Navy TIP/NOVA satellite in 1981. Finally, PPT experiment on Earth Observing 1 (EO-1, 2001) spacecraft has demonstrated the capability of new generation PPT to perform spacecraft attitude control. Ablative PPTs are relatively simple, compact devices featuring no moving parts. PPTs are able to provide minute impulse bits making them suitable for fine attitude control. Due to simplicity, PPTs are ready for miniaturization. The success of relatively recent EO-1 PPT enables the new generation of PPT technology to be considered for future missions with negligible risk.

5.3.1.1 Ablative PPT

PPTs are currently considered as an attractive propulsion option for mass and power limited satellites that require µN-s to mN-s impulse bits [162–167]. One of the most extensively analyzed propulsion functions has been attitude control. For missions that require precision pointing, the PPT offers a unique advantage over other technologies as it delivers a small impulse "bit" with high specific impulse (exhaust velocity). A second major functional area is stationkeeping, which includes drag compensation and formation flying. These functions require less than 1 mN of thrust. For instance, interferometer missions [168] such as ST-3 require submillimeter relative positioning precision. In particular, the US Air Force has a growing interest in highly maneuverable microsatellites to perform various missions, such as space-based surveillance, on-orbit servicing, inspection, and space control [162,163]. Recently, an electromagnetic PPT was successfully operated for pitch axis control on the EO-1 spacecraft [168,169]. It was shown that PPT can be easily scaled down in power and size. A µ-PPT is the miniature version of the traditional PPT has been designed at the

AFRL for delivery of very small impulse bit [170,171]. AFRL μ-PPT was developed for a demonstration mission on TechSat21 [172] and it was recently employed at FalconSatIII. The μ-PPT can deliver a thrust in the 10 μN range to provide attitude control and stationkeeping for microsatellites.

In μ-PPT, the discharge across the propellant surface ablates a portion of the propellant, ionizes it, and then accelerates it predominantly electromagnetically to generate the thrust (schematically this thruster is shown in Figure 5.43) [25,173]. It is expected that the use of electromagnetic acceleration to create thrust will also lead to relatively high specific impulse.

It was demonstrated [174] that both discharge energy (peak current) and thruster size affect significantly the discharge uniformity (azimuthal or radial). Azimuthal nonuniformity relates to the current constriction and anode spot formation phenomena. This happens when the discharge current or thruster size exceeds some critical value. Discharge nonuniformity leads to a much higher ablation rate and causes degradation of the specific impulse. On the other hand, small discharge current leads to strong Teflon surface carbonization (charring) and radial nonuniformity, which in turn leads to thruster failure. The primary mechanism of the charring formation was identified and it is related to carbon backflux. Thruster size and discharge energy can be optimized by trading between two conflicting requirements of a large pulse energy (to prevent charring) and a small discharge energy (to prevent current constriction) [174]. Because the Teflon ablation rate grows nonlinearly with the surface temperature, the model predicts a lower rate of ablation in the areas where the surface temperature has a minimum. By taking this into account, the effect of the temperature distribution may be related to the preferential charring of the Teflon surface observed experimentally, as shown in Fig. 5.43 (b). It is interesting to note that comparison of the calculated temperature field and ablation rate with the photograph of the Teflon surface (Fig. 5.43 (b)) shows that the area with surface temperature and ablation rate minimum corresponds to the charred area in the case of the 3.6-mm-diam thruster. Because the charring phenomenon is completely intolerable and leads to thruster failure, the optimal discharge energy should be chosen somewhere near the spot formation limit.

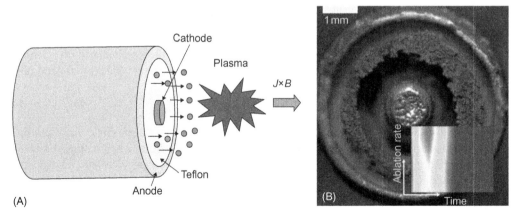

FIGURE 5.43

(A) Schematic of the coaxial μ-PPT. (B) Teflon surface photograph and the ablation rate in the case of a micro-PPT.

5.3.1.2 Microlaser plasma thruster

Another example of the ablative type of micropropulsion system is the micro-laser plasma thruster (μ-LPT). Most notably, μ-LPTs have been developed by Phipps and Luke [175] and Gonzalez and Baker [176]. In Ref. [176], a Q-switched microchip laser, pumped by a CW diode laser, was used to ablate an aluminum target generating thrust in the range of 0.3 nN to 3 μN with power consumption of about 5 W and pulse frequency in the range of 1 Hz to 10 kHz. The wide dynamic range of thrust levels provided by these devices is one of their most attractive features.

The μ-LPT developed by Phipps and Luke [175] uses a $1-10$ W, high-brightness diode laser irradiating various absorbing material and substrate combinations (e.g., black ink on paper, black PVC on Kapton™). Laser coupling coefficients on the order of 60 μN/W and specific impulses on the order of 500 s with a 1 W laser are achieved. One of the major advantages of the μ-LPT is its large dynamic range of impulse bit that can be varied between 0.4 nN-s to 16 μN-s by simply increasing the laser pulse duration. In addition, selection of absorber and substrate materials allows the specific impulse and the laser characteristics to be tailored for specific mission requirements.

The μ-LPT can be operated in two different modes. In reflection mode (R), the laser is incident on the target and the ablated material "reflects" from the surface. This mode has the potential problem of leading to deposition of plume effluent on the laser optics. In transmission mode (T), the laser passes through a transparent substrate film from the back. The substrate is coated on the other side with an absorbing material that is ablated. This approach circumvents the problem of optics contamination found with the R-mode. However, the dynamic range of impulse bit available in T-mode is more restrictive. Coating of laser optics by plume deposition is one of the major lifetime limitations of the μ-LPT in R-mode. Therefore, there is a certain preference for development of T-mode operation. Preliminary computational results for R-mode μ-LPT were presented [177]. A lens focuses the laser diode output on a 25 μm diameter spot on the transparent side of a fuel tape. The beam heats an absorbing coating to high temperature, producing a miniature ablation jet. The material that is ablated is usually PVC or Kapton. Typical parameters of operation are power of $2-14$ W, pulse duration of $3-10$ ms. The fuel tape thickness is about 185 μm, composed of 125 μm of transparent backing (usually cellulose acetate) and about 60 μm of absorbing coating. Typically $Q*$ (energy of laser light required to ablate 1 kg of target material) is about 2×10^7 J/kg and the momentum coupling coefficient is about $C_w = 60-100$ μN/W. Recent computational work on T-mode μ-LPT was also developed and compared with experiment [178]. Generally LPT models satisfactorily describe the main features of the thruster such as ablated mass and plume expansion.

5.3.1.3 Microvacuum arc thruster

One possibility to improve some aspects of PPTs without compromising its efficiency is to use metal as a propellant. Success in development vacuum arc technology attracted much attention to this technology [179]. Metal propellant will have the following benefits: lower energy consumption per mass ionized (due to the low ionization potential of metals), high ionization degree, operation with higher repetition rates since the metal melting temperature is higher than that of polymer dielectrics used in some current thruster, and high pulse-to-pulse stability. For this reason, a vacuum arc source was developed and used for an ion thruster [180]. This work led to development of a coaxial vacuum arc plasma thruster [181]. Schematically this thruster is shown in Figure 5.44. However, the measured efficiency was reported to be only approximately 1.6% which is strongly different from the

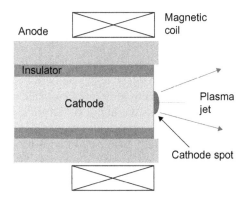

FIGURE 5.44

Schematics of the VAT.

estimations, leaving ample room for improvement. On the other hand, several problems related to this technology were reported, including lifetime and ion current degradation with cathode recession [182]. Therefore, understanding of these problems and the efficiency limitations of the vacuum arc thruster (VAT) technology is an important issue. In addition, basic aspects of VAT such as thrust mechanism, effect of the magnetic field of the thruster operation and its plume, droplet generation were not studied in detail. It was shown that using an axial magnetic field (the so-called magnetically enhanced VAT) helps dramatically to improve thruster characteristics, such as specific impulse (plasma velocity) and to decrease plasma plume divergence [183,184].

In VAT, the thrust is created in the minute spot on the cathode surface called cathode spots (having size of ∼10−100 μm) [185]. The high-pressure plasma in the cathode spot creates the reaction force (thrust) on the cathode as schematically shown in Figure 5.44. As earlier as in the 1930s, it was found experimentally that the plasma stream exerts a reaction force on the cathode, which is about 17 ± 3 dyn/A (or ∼0.2 mN/A) [186,187]. It is very important to note that since the reaction force is proportional to the arc current, the thrust can be controlled easily. Recent measurements confirmed those conclusions [188]. It should be pointed out that the force generated in the short vacuum arc exceeds that of the conventional cathodic arc [252].

In the late 1960s and early 1970s, several experimental groups investigated the use of vacuum arcs in electric thrusters for spacecraft propulsion (see review paper [189] and references therein). The main reason for considering vacuum arcs for electrical propulsion was the nature of the plasma plume expanding from the cathode, which exhibits high velocity and is highly directional. In general, a vacuum arc is established by using a coaxial geometry with a relatively small diameter cathode which is surrounded by an insulator and an anode. Vacuum arcs can be produced from any electrically conductive material. The suitability of various materials for vacuum arc propulsion was first studied by Dethlefsen [190]. In the case of a magnesium cathode, a plasma velocity of 50000 m/s was demonstrated, which is certainly very appealing to the propulsion community. Pulsed operation was demonstrated and the thrust efficiency was found to be dependent on the electrode material, the amplitude, and duration of the current pulse.

One reason for the material dependence is the fact that metals with low melting point and low thermal conductivity erode rather heavily, with a large part of erosion resulting in macrodroplets. These droplets tend to reduce the thrust efficiency because they are not accelerated to as high velocities as are the ions. While this reduces thrust efficiency, it has to be pointed out that the droplet emission does not depend on material properties alone and hence can be minimized through the action of thermal and mechanical inertia if a current pulse of sufficiently short duration on the order of 10 μs is used in a pulsed VAT. Apart from the high exit velocity and the fact that the cathode material can be chosen from a large variety of candidates, pulsed VATs were found to have other appealing characteristics, which include the fact that a solid fuel is used, the low-voltage operation, as the burning voltage of vacuum arcs does not exceed 100 V, and the simple control by adjusting arc current and repetition rate.

Interest in the vacuum arc source for propulsion applications was renewed [191] in 90s. The measured energy efficiency was about 80% which is comparable to that of xenon ion thruster. The development of the vacuum arc source for an ion thruster led to design of a thruster based on the pulsed vacuum arc without any ion extraction [192]. For later configuration, a specific impulse of about 1000 s was measured and 21% of efficiency was estimated. Further development of this technology is attributed primarily to Alameda Applied Science Corporation (AASC) [185].

The VAT is a simple, yet efficient electric propulsion device that combined with an inductive energy storage (IES) power processing unit (PPU) results in a low-mass (<300 g) system. The basic mechanism underlying the VAT is the production of micrometer-size fully ionized microplasmas (cathode spots) on the cathode surface, which expand into vacuum at high velocities, with the ions producing thrust. Every cathode spot carries a limited amount of current ($\sim 1 - 10$ A) and exists for a few nanoseconds, which makes the VAT very scalable with respect to pulsed operation and/or amount of plasma produced.

The need for a low-mass propulsion system motivated the development of the VAT system. A PPT that uses conductive cathode materials as propellant is combined with an energy storage PPU that takes 5−24 V from the bus and converts it into an adequate power pulse for the thruster. It is a system well suited to provide small impulse bits (≈ 1 μNs) at high specific impulse, I_{sp} of about 1000−3000 s. Applications include positioning and drag makeup for small, power, and mass limited satellites.

The performance of the VAT is determined by the propellant mass, the degree of ionization of the plasma, the angle of expansion, the average charge state, and the ion velocity. All these parameters have been measured repeatedly in the past and verified for numerous materials and operating conditions [166−169]. Typical values for the ion velocity vary between 10 and 30 km/s, the average arc/ion current ratio has been shown to be of the order 8%, and a cosine distribution has been found to emulate the plasma plume expansion very well.

In order to produce a low-mass system, the VAT was constructed using an IES circuit PPU and simple thruster head geometry. In the PPU, an inductor is charged through a semiconductor switch. When the switch is opened, a voltage peak $L \, dI/dt$ is produced, which breaks down the thin metal film coated anode cathode insulator surface at relatively low-voltage levels (≈ 200 V). The current that was flowing in the solid-state switch (for ≤ 1 ms) is fully switched to the vacuum arc load. Typical currents of approximately 100 A (for $\sim 100 - 500$ μs) are conducted with voltages of approximately 25−30 V. Consequently, most of the magnetic energy stored in the inductor is deposited into the plasma pulse. The efficiency of the PPU may thus be greater than 90%.

By varying the length of the trigger signal, the level of the current in the switch and thereby the energy stored in the inductor can also be adjusted. This in turn changes the amount of energy transferred to the arc and the impulse bit of the individual pulse. Obviously, the repetition rate of the individual pulse can be changed by varying the input signal as well.

The mass of the PPU is small (<300 g) resulting in a low-mass system. The plasma output is quasi-neutral; therefore, no additional neutralizer is needed. An electromagnetic interference (EMI) filter might be necessary due to the noisy characteristics of the discharge, high peak currents, and fast switching. A low-mass feed mechanism is available; therefore, even long missions can use this technology.

A picture of the VAT system with an equivalent circuit is shown in Figure 5.45. Additional information about the principle of the VAT can be found elsewhere [193−195].

While the system is simple and can be realized with an extremely low mass, the divergence of the plasma plume reduces the efficiency of the system due to the variations in the thrust vector and also increases the chances of contamination of essential parts of the spacecraft. Based on prior work by Gilmour and coworkers [196,197], who, by using a coaxial arc-diode configuration with an additional magnetic coil, achieved a total efficiency of 30% with a copper cathode and a magnetic field of 500 G and the fact that a magnetic field can be used to direct a vacuum arc plasma [194,198], an axial magnetic field was used to control the plasma flow. The magnetic field is produced by a coil wrapped around the anode of a coaxial thruster head. For the purpose of performing initial experiments, the currents in the coil and in the thruster are controlled independently, while a final design will utilize the magnetic field produced in the inductor of the IES PPU to control the plasma plume [199].

5.3.1.4 Microcathode arc thruster

Recently, a novel thruster design, the microcathode arc thruster (μ-CAT), was developed and investigated. This thruster improves on the vacuum arc discharge thruster by applying specially designed

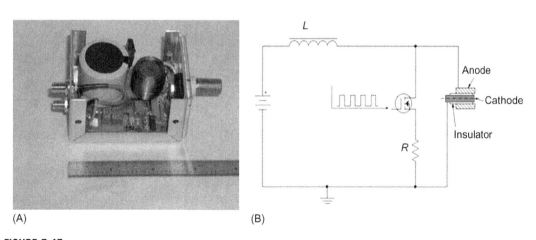

(A)   (B)

FIGURE 5.45

(A) VAT system including PPU (thruster head is on the right). (B) VAT equivalent circuit.

Source: *Reprinted with permission from Ref. [185]. Copyright (1996) by Institute of Physics.*

(A)

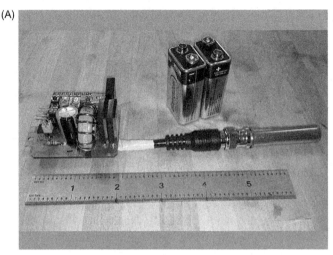

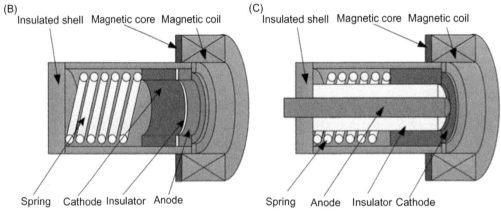

FIGURE 5.46

(A) μ-CAT with PPU. (B) Schematic design of the ring shape μ-CAT. (C) Schematic design of coaxial electrodes μ-CAT.

an external magnetic field. The unique magnetic field conditions achieved in a vacuum arc offer several potential advantages in these devices.

The μ-CAT is a simple electric propulsion device that combined with a magnetic coil and an IES PPU results in a low-mass (<100 g) system. A picture of the μ-CAT system and two types thruster are shown in Figure 5.46A−C. Figure 5.46B shows the schematic design of the ring electrodes μ-CAT (RE-μ-CAT), which consist of an annular titanium cathode and a same diameter annular copper anode with 1 mm width. The annular ceramic insulator tube having same inner and outer diameters and a width of about 1 mm was used as separator between the arc electrodes. Figure 5.46C shows the schematic design of the coaxial μ-CAT. Instead of the RE, this design employs cylindrically shaped cathode and anode.

Figure 5.47A shows a schematic of the thruster and the PPU system. The mass of the PPU is small (<100 g), resulting a low-mass system. PPU equipped with an IES system has been designed as shown in the Figure 5.46A and B. When the trigger pulse is applied to a semiconductor insulated gate bipolar transistor (IGBT) switch, the energy is accumulated in the inductor, while when trigger pulse ends, a surge voltage with the magnitude proportional to $L\,dI/dt$ is generated on the inductor and applied to the electrodes. This leads to a breakdown and initiation of arc discharge between the electrodes. A coil has been applied to the outside of the thruster to produce magnetic field as indicated in Figure 5.47B and C. The direction of the magnetic field could be simply reversed by reversing the coil current.

The μ-CAT operates by producing a fully ionized plasma at the inner surface of the electrode The plasma is formed in the cathode spots and expands into the vacuum zone under the applied magnetic field gradient. The plasma accelerates and expands into vacuum at a high velocity as the magnitude around tens thousands meter per second, which results in the impulse bit. The effect of the magnetic field on the thruster operation is clearly visible in Figure 5.47B. This figure shows a CCD camera observation of the RE μ-CAT firing in the vacuum chamber without and with an added magnetic field ($B = 0.3$ T). The yellow lines in the second image of Figure 5.47B indicate the simulation results of the magnetic field distribution.

The plasma formed by a vacuum arc is created on the cathode surface spots. Using optical methods it was observed that the spots consist of either a homogeneous bright region or consist of cells and fragments with a typical total size of about 10−100 μm [200−202]. The observation of cathode spots motion under magnetic field was studied in 1960s [203,204]. It is known that presence of a transverse magnetic field at the cathode surface produces cathode spot motion in the $-J \times B$ direction. Theoretical explanation of this effect was given by Beilis [200]. The observed vacuum arc cathode spot rotation has important implications for propulsion since the cathode spot rotation leads to a uniform cathode erosion, which is critical for assuring long thruster lifetime. The μ-CAT cathode spot rotation was measured by the four-probe assembly Langmuir probes [205]. Four single probes were located along the azimuth direction inside the thruster channel and the four probes ion current measured results were shown in Figure 5.48. The rotation speed was calculated using a quarter of circumference of the thruster inner surface divided by the delay time between each of two neighbor peaks. The average rotation speed is shown in Figure 5.49. It was found that the spot rotation speed increased by factor of 5 (from 20 to 100 m/s) as the magnet field strength increased from 0 T to 300 mT.

Magnetic field affects the total ion current (or jet current) collected in the thruster plume. The dependence of the ratio of jet current I_{jet} over arc current I_{arc} ($f = I_{jet}/I_{arc} \times 100\%$) on the magnetic field is shown in Figure 5.50. It was found that due to losses to walls, the total ion current was very low ($f = 0.06\%$) in the case without the magnetic field. Increase of the magnetic field resulted in significant increase in the total ion current (up to 50 times) and saturated at about $f \sim 3-3.5\%$. Note that f cannot exceed the ($f_{max} = I_{max}/I_{arc} \times 100\%$), where I_{max} is the total ion current generated by the cathodic spot. According to Kimblin [206], the f_{max} in the case of titanium is about 8%. Thus $f \sim 3-3.5\%$ observed in the experiment corresponds to the efficiency of cathodic jet transport through the thruster channel of about 37.5−43.8%.

Since total ion current increases with a magnetic field, one can expect increase in the impulse bit. Indeed recent measurements [207] indicated that the impulse bit increases with a magnetic field reaching about 1 μNs at 0.3 T as shown in Figure 5.51.

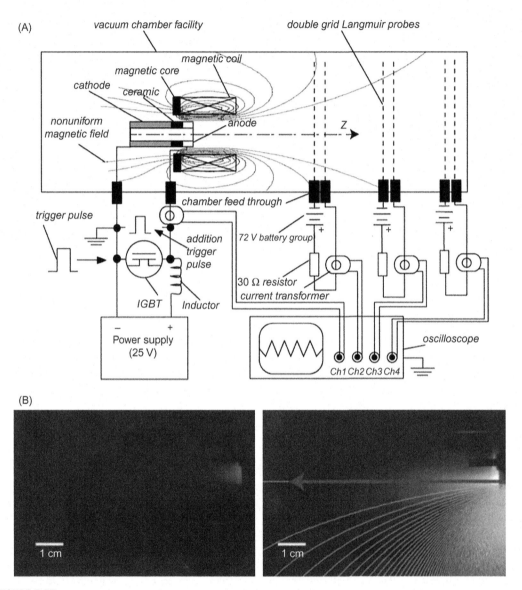

FIGURE 5.47

(A) Schematic of microcathode arc thruster experimental arrangement. (B) CCD camera observation of plasma plume with no magnetic field (left figure) and with magnetic field (right figure, $B = 300$ mT). (For interpretation of the references to color in this figure legend, the reader is referred to the web version of this book.)

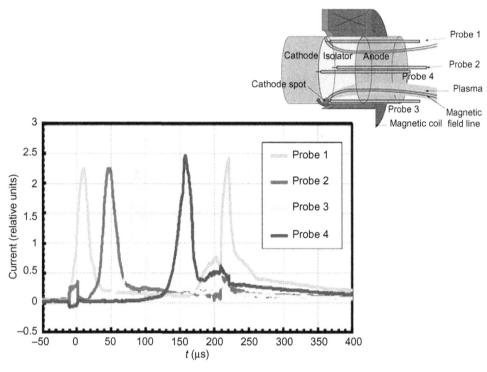

FIGURE 5.48

Example of four Langmuir probes method to measure the cathode spot rotation speed.

Source: *Reprinted with permission from Ref. [205]. Copyright (1996) by Institute of Physics.*

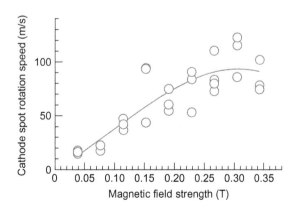

FIGURE 5.49

Dependence of cathode spot rotation speed on the magnetic field.

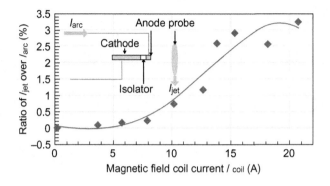

FIGURE 5.50

Ratio of I_{jet} (average over the pulse) over the total arc current as a function of a magnetic field.

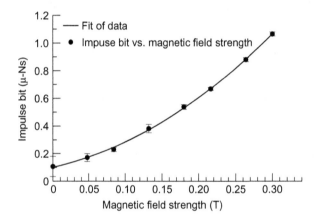

FIGURE 5.51

Impulse bit as a function of a magnetic field.

Figure 5.52 presents average ion velocity as a function of axial distance without and with magnetic field strengths of 0.17 and 0.3 T. It is seen that without magnetic field, the ions propagate at almost constant velocity as indicated in Figure 5.52 by the lower curve, while application of magnetic field leads to ion acceleration in the axial direction as demonstrated by two upper curves. The ion velocities saturate on the level of around $3-3.5 \times 10^4$ m/s [208].

Plasma flow is subject to the electromagnetic force in the divergent magnetic field region:

$$Mn\frac{dV}{dt} = j_\theta B_r$$

where M is the ion mass, n is the plasma density, j_θ is the azimuthal electron current, and B_r is the radial component of the magnetic field. Azimuthal electron current density can be estimated as

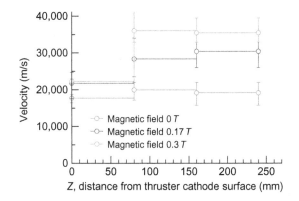

FIGURE 5.52

Ion velocity as a function of a magnetic field.

Source: Reprinted with permission from Ref. [208]. Copyright (2012) by American Institute of Physics.

$$j_\theta = \frac{\omega_e}{\nu_{ei}} j_r$$

where j_r is the electron radial current density. In the plasma jet outside of the interelecrode gap, the total current is zero and therefore the electron current is equal to the ion current. Thus

$$j_{er} = j_{ir} = enV_r$$

where V_r is the radial component of the ion current. Radial component of the plasma velocity is developed in course of plasma expansion [147] and could be about 0.3 of the axial velocity component in a high magnetic field. Based on the above scenario, one can estimate the ion velocity change in the axial direction as

$$\Delta V_z \approx \sqrt{\frac{2e^2 B_z B_r V_r \, \Delta z}{mM\nu_{ei}}}$$

Using the following typical parameters of the plasma jet: B (i.e., both B_r and B_z) ~ 0.1 T, $n \sim 10^{20}$ m^{-3} $V_r \sim 10^4$ m/s and $\Delta z \sim 0.01$ m, one can estimate that $\Delta V_z \sim 10^4$ m/s. This estimation is in good agreement with experimental measurements.

5.3.2 Microthrusters based on liquid propellants

Various micropropulsion concepts utilize liquid propellants since use of a liquid has some advantages over solid propellant especially in a miniature device.

5.3.2.1 Liquid PPT

In a microscale device with a high throughput, a solid propellant feeding mechanism may become impractical (unless propellant recession is organized in a proper manner, see previous section as an example). In that case, a liquid-based thruster can take advantage of fluid handling possibilities, such as capillary effect and adsorption. Several PPTs using liquid propellant were demonstrated over the last decade. Scharlemann and York [209] show that water is very effective propellant producing much higher specific impulse than Teflon. Microliquid PPT was demonstrated at Johns Hopkins University (Applied Physics Lab) [210]. JHU/APL thruster prototype weights about 13.5 g (without PPU) and also uses water as a propellant. This thruster accumulates water vapor in a small cavity as it diffuses through a membrane. An orifice at the opposite end of cavity leads to the main thruster chamber. Discharging a spark across two electrodes heats the vapor and injects it into the main chamber causing sustained discharge.

5.3.2.2 Field emission electric propulsion

Another type of thruster that may be relevant to nanopropulsion is field emission electric propulsion (FEEP). FEEP uses field emission ionization to extract and accelerate ions from the propellant. The propellant, which is typically a liquid metal, is stored in a reservoir. To initiate thruster operation, the propellant is first heated to its melting temperature in preparation for propellant extraction. A needle utilizes pressure gradients and propellant surface tension to extract the propellant from the reservoir and brings it in proximity of an extractor electrode. A large potential is applied between the needle and an extractor electrode. As the voltage increases, the propellant forms a cone (called a Taylor cone) that protrudes toward the extractor. The small radius of curvature at the protrusion tip increases the local electric field there, which causes further propellant displacement from the needle orifice. Once an electric field threshold of approximately 10^9 V/m is reached, atoms at the tip of the Taylor cone are ionized and accelerated through the gate electrode. Recently micro-FEEP thruster was demonstrated [211]. By utilizing electrostatic acceleration principles, FEEP has electrical efficiencies exceeding 90% [212,213]. However, the field emission process requires high voltages (10 kV range), leading to very high specific impulse up to 10,000 s and thus relatively high specific power values. In addition, the use of liquid metal propellant raises concerns regarding spacecraft contamination. While pure ion generation and acceleration are often sought, FEEP charge extraction instabilities can result in the formation of charged microscopic droplets. The amount of droplets formed is a strong function of emission current. A high droplet content decreases propellant utilization efficiency (called mass efficiency) and overall thruster specific impulse and electrical efficiency. In addition, a key limitation of FEEP has been the neutralization cathode. Typically, heated filament cathodes are employed for ion beam neutralization. In addition to their rather short lifetimes, filament cathodes consume an enormous amount of power.

5.3.2.3 Colloid thruster

Colloid thrusters are similar to FEEP featuring emitter tips and accelerating electrodes. However, unlike FEEP, colloid system does not accelerate individual ions but rather accelerate fine charged droplets. A key component of colloid thrusters is the phenomenon referred to as "electrospray." Among the different spraying modes that can be attained, the so-called cone-jet produces a stable spray of monodisperse droplets. Droplets with diameters ranging from hundreds of microns to a few nanometers may be generated by tuning the physical properties of the electrosprayed solution. A strong electric field applied between the sharp-edged exit of the capillary and an external electrode causes charge separation inside the liquid propellant, which is doped with an additive to increase its electric conductivity. Various instabilities causes jet breakup into small liquid droplets, and by the action of the electric field on the conductive liquid charged droplets are extracted from the capillary at high voltages, producing thrust.

Colloid thrusters are able to deliver a very stable, fine thrust (of the order of 1 μN per cone-jet) with high specific impulse, I_{sp} up to 1500 s. This type of electrostatic propulsion makes possible and enhances the use of microsatellites. Colloid thrusters also enable satellite flight formation and other missions requiring very precise positioning.

In principle, some combination of FEEP and colloid thruster is possible to provide flexibility in specific impulse and thrust. The idea of the hybrid approach is described in the next section.

5.3.2.4 FEEP operation in the colloid mode with low I$_{sp}$ range

To extend the range to lower I_{sp} and higher thrust while preserving high efficiency, the FEEP thruster can be purposely operated in a droplet extraction mode.

Droplet emission is a phenomenon inherent to the operation of liquid metal ion source [214]. Several mechanisms of droplet emission were identified. One possible mechanism is related to the fact that at higher ion currents, the liquid jet becomes unstable, the so-called Rayleigh instability [215]. Another possibility is excitation of capillary waves at the sides of the Taylor cone [216]. Both effects form droplets with size of about 1−100 nm. Typically in liquid metal ion sources, droplets are generated when Rayleigh instability is set that requires electric field higher than some critical one (or current higher than critical) [217,218].

In the indium FEEP, it was shown that the mass efficiency, which is the ratio of the ion mass fraction to the total flux, decreases from 99.5% at 10 μA down to 10% at 200 μA in the case of indium. This means that when current greater than 200 μA, about 90% of the propellant mass is emitted in the droplet form.

In addition, Crowley [219] showed that the temperature rise at the emitter tip enhances droplet emission. According to Crowley, the temperature rise depends upon the electrical resistivity to thermal conductivity ratio. The smaller ρ/k ratio corresponds to smaller temperature rise and therefore smaller droplet fraction. Increasing the propellant temperature would lead to higher ρ/k and therefore enhance the droplet emission. Again, in the indium FEEP, it was shown that the droplet mass fraction increases up to 90% when indium temperature increases from melting point (156.6°C) up to about 280°C.

The approach may be to use a combination of fine-tuning emission current and temperature control to operate the micron-scale FEEP in charged droplet mode. In this case, specific impulse produced by the thruster can be adjusted dependent on mission requirement from being very high of about 10,000 s in the ion extraction mode down to 1000 s in the droplet emission mode.

5.3.2.5 Other concepts

Other electrostatic and electromagnetic schemes can be considered for micropropulsion, such as Hall thruster and ion thruster.

A microion engine is under development at Jet Propulsion Laboratory. This is a MEMS-hybrid device in which the bulk of the engine is fabricated using conventional machining, while several components are machined using MEMS technology. The engine has a discharge chamber of 3 cm. However, further miniaturization is not feasible because thruster becomes extremely inefficient or cannot maintain a plasma within given power limit [220]. Small-scale Hall thruster with power level of 50−200W was also demonstrated recently by MIT [221] and Princeton University PPPL [222]. In this respect, Princeton cylindrical Hall thruster has promise due to decreased surface-to-volume ratio.

5.4 Plasma plumes from thrusters

It is accepted that plasma propulsion can offer significant advantages for many space missions; however, integration of electric thrusters on spacecraft can present significant challenges. In this respect, assessing the plasma thruster plume interactions with the spacecraft serves as a major factor in determining thruster location. Schematically, plume interaction is shown in Figure 5.53. Plume of the plasma thruster typically consists of energetic ions, neutral gas, low-energy ions, and electrons. In order to fully assess plasma plume contamination, one needs to take into account the interaction between each of the plume components and other spacecraft systems.

For instance, Hall thruster is an attractive form of electric propulsion and it is one the most efficient plasma thruster nowadays. However, one of the shortcomings of Hall thrusters is the large exhaust-beam divergence, which may cause electrostatic charging and communication interference of satellites. Thus, the structure of the plasma plume exhaust from the thruster is of

FIGURE 5.53

Schematic of the plume components.

great interest. In this section, we will discuss issues related to plasma plume and its interaction with spacecraft.

5.4.1 Description of the plume model

In general, vast majority of the plume models are based on hybrid approach, which includes particle simulation of the ions and neutrals flow and fluid description of electrons flow. Particle simulation method such as PIC was employed. Charge exchange (CEX) collisions between ions and neutrals are typically taken into account. Fast ions from the main beam undergo CEX collisions with neutrals, resulting in slow-moving ions and fast-moving neutrals. Presence CEX ions will generate secondary plasma which is responsible for large plume divergence and backflow.

Schematically, this process is shown in Figure 5.53. The CEX ions have much lower velocities; therefore, they are more influenced by the self-consistent electric field in the plume. These fields may cause them to interact with spacecraft surfaces.

The charged particles may impact solar arrays or interfere with transmission from the satellite. A critical engineering issue in the employment of various devices is potentially hazardous interaction of the plume with spacecraft system. Thus, it is important to understand their dynamics in the plume. Energy distribution of the CEX is important for understanding possible effect on spacecraft such as sputtering of spacecraft components. In order to address this issue, a 2D simulation tool has been developed. The plume model that will be described below consists of two main components: a Lagrangian algorithm for determining the expansion of the main beam and a PIC solver for calculation of the dynamics of the secondary CEX plasma. This approach is similar to one described in Ref. [223].

A 2D, axisymmetric, and steady-state flow was assumed. We will use Hall thruster as an example of the plasma thruster for plume analysis. Inlet conditions for the plume will be generated from the plasma flow at the exit of Hall thruster. These conditions include the plasma density and velocity distribution in radial direction at the thruster exit. The main ion beam exhaust from the thruster is collisionless, singly ionized, quasi-neutral expanding under the influence of the electric field. Electron flow will be analyzed assuming ambipolar, isothermal, collisionless, and unmagnetized [224]. The contribution of cathode mass in the total mass flow rate was assumed to be small and therefore will not be taken into account.

5.4.1.1 Ion beam generated from the thruster

Lagrangian approach will be employed here for ion simulations. It should be pointed out that the Lagrangian technique reduces the numerical noise associated with PIC method. Moreover, PIC technique is not accurate in the case of the problem with large-scale variation such as from millimeter to meter, which is the case of the thruster plume.

Ion momentum equation can be expressed as follows:

$$m_i \frac{dV_i}{dt} = eE \tag{5.113}$$

where V_i is the ion velocity and E is the electric field. Mass elements or macroparticles are ejected from the thruster exit and are tracked using trajectory analysis. The location of each macroparticle is determined by the following system of equations:

$$r = r_0 + V_i t + \frac{eE}{m_i} t^2$$

$$V(t) = V_0 + \frac{eE}{m_i} t \tag{5.114}$$

Each time the ion density is updated using calculated $V(t)$ with the conservation of mass:

$$\nabla \cdot (nV) = 0 \tag{5.115}$$

Density is updated on each simulation time step as well as electric field. The iteration process ensures that ion trajectories are consistent with local electric field.

Typically the calculations of the ion density in the plume relay on the condition at the exit plane from the thruster.

5.4.1.2 Electron flow

Electron density distribution is given by a balance of electrostatic force and electron pressure while the magnetic field effects can be neglected. Recall that in the near-plume region of the electromagnetic thruster (such as Hall thruster), a magnetic field may affect the electron motion. However, it was found that this effect is small in a typical case since magnetic field decreases strongly outside of the Hall thruster channel [152]. Effect of the magnetic field on the plume in the near-plume field will be described in Section 5.4.4.1. Mathematically, the balance for electrons under considered conditions can be written as follows:

$$n_e eE = - \frac{dP}{dr} \tag{5.116}$$

where P is the electron pressure. Equation (5.116) comes from the general electron momentum conservation taking into account the currentless and unmagnetized electron flow. For densities and temperatures of the plasmas, which are typical for the plumes of Hall thrusters, the electron−ion collision term is negligible. The ratio of the collision frequency to the plasma frequency is much less than one. Further assuming isothermal conditions yields the Boltzmann relation:

$$n_e = n_{ref} \exp(e\varphi/kT_e) \tag{5.117}$$

where n_{ref} is the reference electron density defined where potential is zero.

5.4.1.3 Neutral density

Typical scale of the plume of interest for spacecraft propulsion application is in the order of several meters which is about the size of the solar panels of a communication satellite. Neutral density decreases due to expansion into vacuum. As a result, in the most part of the plume, the flow is collisionless. The most common approach to describe the neutral density in the plasma plume is a Monte Carlo simulation such as Direct Simulation Monte Carlo (DSMC).

In some case, a simple analytical model for the neutral expansion can be employed as well [223]. The profile of neutrals in the thruster plume is computed using the emission by the solid angle $\Omega(r,z)$ [223]:

$$n(r, z) = \frac{2\Gamma_0}{m_a U_a \pi (r_2^2 - r_1^2)} \Omega(r, z) \tag{5.118}$$

where Γ_o is the mass flow rate of neutrals, U_a is the neutral velocity, and Ω is the solid angle [223].

The ion–neutral interactions include momentum and CEX collisions. The most important collision for plume contaminations is the CEX. The fast ions from the main beam undergo CEX collisions with neutral particles, resulting in slow-moving ions and fast-moving neutrals:

$$Xe_{fast}^+ + Xe_{slow} \rightarrow Xe_{slow}^+ + Xe_{fast} \tag{5.119}$$

CEX collision rate is calculated as follows:

$$\frac{dn}{dt}_{(CEX)} = n_e V \sigma_{CEX} n_a \tag{5.120}$$

A value for σ_{CEX} of 5.5×10^{-19} m^2 has been assumed following the previous comparison with experimental data. The data for collision cross section has been analyzed by Pullins et al. [225].

5.4.1.4 Potential distribution

The PIC algorithm solves the 2D Poisson equation and iterates until the steady-state solution is reached. The sum of the primary ion density (n_{mean}^i) and the CEX ion density is used in the Boltzmann relation:

$$\varphi(r, z) = T_e \ln \left(\frac{n_{mean}^i + n_{CEX}^i}{n_{ref}} \right) \tag{5.121}$$

The primary ion density is calculated using the Lagrangian method.

The overall approach to calculate the primary ion density and CEX ion density is shown in Figure 5.54. Most simulations of the plasma plume employ about 100,000–500,000 particles in 300×300 PIC cells. These simulations required about 50–100 CPU time on single workstation.

5.4.2 Example of plasma plume simulation: Hall thruster plume

In this section, we present example of calculation of the plasma plume expanding from the Hall thruster called SPT-100 and compare these calculations with available experimental data for ion current density angular distribution at 50 cm. SPT-100 (i.e., 100 mm diameter thruster) is the Hall thruster developed by Fakel which has the longest flight history. The nominal operational conditions for this thruster are summarized in Table 5.1.

Calculated ion density distribution in the plume is shown in Figure 5.55 and neutral density distribution is shown in Figure 5.56. The domain that considered has the size of 0.7×0.7 m^2.

A comparison of the calculated and measured ion current density distributions are shown in Figure 5.57. The radius at which data were taken is shown in Figure 5.55. One can see that the primary ion beam analysis predicts the narrow plume of about 30°, while CEX ions are

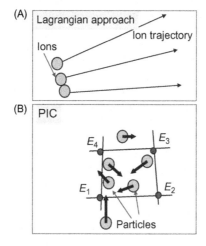

FIGURE 5.54

Approach of calculation of the primary ion density using the Lagrangian technique (A) and the CEX ion density using the PIC technique (B).

Table 5.1 Operational Conditions and Parameters of the Plasma in SPT-100	
Discharge current	5 A
Discharge voltage	300 V
Xenon mass flow rate	4.5 mg/s
Plasma density	$2 \times 10^{17} \, \text{m}^{-3}$
Neutral xenon density	$3.3 \times 10^{18} \, \text{m}^{-3}$
Axial ion velocity	$1.9 \times 10^{4} \, \text{m/s}$
Electron temperature	3 eV

spread over large angles. Due to CEX process, slow ions are found at 80°. In general, the calculated results are in good agreement with experimental data especially for a primary ion beam. One can see that ion current density drops at about 40−50°. This is consistent with previous simulations of the Hall thruster plume [223]. It should be pointed out that more detailed neutral particle models such as DSMC can lead to better reproduction of the experimental data [226].

It can be seen from Figure 5.57 that ion current drops at about 40−50°. This drop is due to the fact that most CEX ions are generated near the exit plane in a relatively dense region. As a result of potential distribution, these ions are driven toward the satellite (backflow). Thus ion density has second peak at large angles. Recall that previous analysis show that assuming higher electron temperature (>10 eV) eliminate this effect.

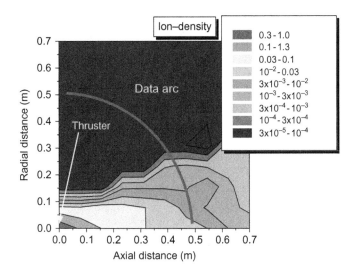

FIGURE 5.55

Normalized ion density distribution in the plume. The thruster exit plane and the data arc at 0.5 m are shown.

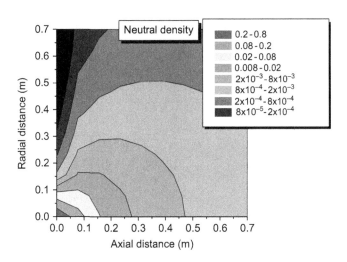

FIGURE 5.56

Normalized neutral density distribution in the plume.

5.4.3 Plasma plume ejected from microlaser plasma thruster

The μ-LPT developed by Phipps, Luke, and coworkers [228−230] with power from 1 to 10 W uses the high-brightness diode laser irradiating various absorbing material and substrate combinations (e.g., black ink on paper, black PVC on Kapton). Laser coupling coefficients on the order of 60 μN/W and specific impulses on the order of 500 s with a 1 W laser are achieved. One of the

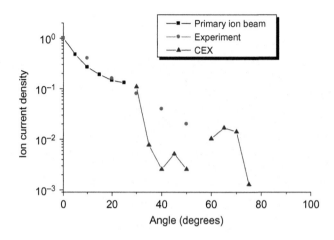

FIGURE 5.57

Normalized ion current density distribution in the plume.

Source: Experimental data are taken from Ref. [227].

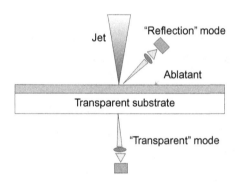

FIGURE 5.58

Schematics of the μ-LPT.

major advantages of the μ-LPT is its large dynamic range of impulse bit that can be varied from 0.4 to 16 μN-s by simply increasing the laser pulse duration. In addition, selection of absorber and substrate materials allows the specific impulse and the laser characteristics to be tailored for specific mission requirements.

As illustrated in Figure 5.58, the μ-LPT can be operated in two different modes. In reflection mode (R), the laser is incident on the target and the ablated material "reflects" from the surface. This mode has the potential problem of leading to deposition of plume effluent on the laser optics. In transmission mode (T), the laser passes through a transparent substrate film from the back. The substrate is coated on the other side with an absorbing material that ablates material away from the laser. T-mode circumvents the problem of optics contamination found with the R-mode. However, the dynamic range of impulse bit available in T-mode is more restrictive. On the other hand, coating of laser optics by plume deposition is one of the major lifetime limitations of the μ-LPT in R-mode. Such coating reduces the lifetime of the thruster. Thus T-mode operation has advantage from the practical point of view.

The principle of operation of this device is shown schematically in Figure 5.58. A lens focuses the laser diode output onto a 25 μm diameter spot on the transparent side of a fuel tape. The beam heats an absorbing coating to high temperature, producing a miniature ablation jet. The material that is ablated is usually PVC (carbon-doped PVC, ranging from 2−5% C). Typical parameters of operation are power of 2−14 W, pulse duration of 3−10 ms. Typically Q^* (energy of laser light required to ablate 1 kg of target material) is about 2×10^7 J/kg and the momentum coupling coefficient is about $C_w = 60-100$ μN/W. These parameters will be used for evaluation of the boundary conditions for the plume simulations.

5.4.3.1 Laser generated plasma expansion

The laser generated plasma plume has been described by Itina et al. [231] and by Franklin and Thareja [232]. In each case, Monte Carlo methods were employed. The focus of these studies was on use of laser ablation for thin film deposition.

In the example that will be described in the next section, a hybrid fluid-particle approach is employed for the plasma plume simulations. The heavy-particle products of the laser ablation of PVC (neutral atoms of C and H, and ions C+ and H+) are modeled as particles. Particle collisions are computed using the DSMC method [233]. Both momentum-exchange and CEX collisions are taken into account. Momentum-exchange cross sections follow the model of Dalgarno et al. [234] and the collision dynamics follows the normal DSMC procedures. The implementation of this algorithm for the unsteady plasma plume is described in Ref. [235]. The CEX processes employ the cross sections proposed by Sakabe and Izawa [236].

Acceleration of the charged particles in self-consistent electric fields is simulated using the PIC method [237]. The plasma potential, φ, is obtained from the Boltzmann relation assuming that the plasma is quasi-neutral. Recall that the ion density is known from the PIC simulations.

By further assuming the electrons are adiabatic, the electron number density, n_e, is then used in the Boltzmann relation to obtain the plasma potential:

$$\varphi - \varphi^* = T^* \frac{\gamma}{\gamma - 1} \left[\left(\frac{n_e}{n*} \right)^{\gamma-1} - 1 \right] \tag{5.122}$$

where φ^*, T^*, and n^* are reference values and $\gamma = 5/3$. In the case of the μ-LPT, the reference point for the Boltzmann relation is taken as the target surface. It is assumed that the potential here is constant.

The 2D axisymmetric simulation uses a single grid for both the collision and ion motion calculations. Since charge neutrality is assumed, the PIC cells are not required to be of the order of the Debye length. Instead they are chosen to be small enough to resolve in a reasonable way the gradients in the potential. At the same time, the cells satisfy the DSMC requirement that their size be less than a mean free path. The background pressure of the order of $10^{-5}-10^{-2}$ Torr, typical for Laboratory conditions, is simulated. The background pressure gas is assumed to be fully composed of hydrogen atoms at a temperature of 300 K.

5.4.3.2 Example of calculation of expanding plume from microlaser plasma thruster

For simplicity, we consider here a two-component plasma consisting of carbon and hydrogen. The chlorine component of PVC is omitted due to its relatively low abundance. Two examples of a 2.5–8 W diode laser beam focused on a 25 μm radius spot were considered. The pulse duration is 3–10 ms, and the experimentally measured thrust-to-power (coupling coefficient) of about 70–200 μN/W is used in order to estimate the recoil pressure.

The particles injected into the DSMC–PIC simulations are sampled from the equilibrium velocity distribution [178]. The grid employed in the plume computation contains 300 by 300 nonuniform rectangular cells. The flow domain extends to about 3 cm in both the axial and radial directions from the center of the ablation spot. A constant time step of 2×10^{-10} s is employed that is smaller than the smallest collision and plasma timescales (the inverses of the maximum collision and plasma frequencies, respectively). A steady state is reached after 40,000 iterations and final results are obtained by averaging over a further 50,000 steps. A maximum of more than 1 million particles is employed in the simulation.

Contours of plasma densities are shown in Figure 5.59. One can see that background gas pressure strongly affects the plume expansion due to CEX and momentum transfer elastic collisions.

The distribution of plume deposition on a witness plate was obtained experimentally. The schematic of the experiment is shown in Figure 5.60. The axes in this figure are labeled in degrees, with 0° corresponding to the direction normal to the tape surface.

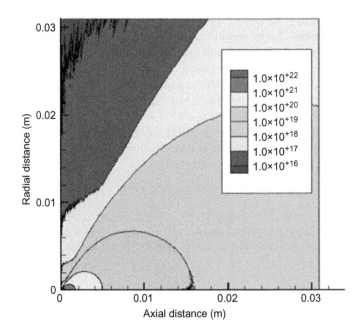

FIGURE 5.59

Plasma density (in m^{-3}) distribution in the plume exhaust of the μ-PPT.

Source: *Reprinted with permission from Ref. [178]. Copyright (2004) by American Institute of Physics.*

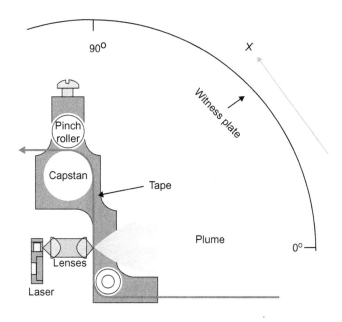

FIGURE 5.60

Schematics of witness plate experiment.

Source: *Reprinted with permission from Ref. [178]. Copyright (2004) by American Institute of Physics.*

The distance from the laser focus to the cylinder on which the witness plate is rolled is 6 cm. The cylinder is made of metal and mounted to the vacuum chamber so that it is always in the same location. The X-axis in the images is the distance around the circumference of the cylinder. The Y-axis is the distance along the length, and the degree marks are not equally spaced. The deposit is attached to the metal cylinder as shown in Figure 5.60.

The normalized flux profiles are shown in Figures 5.61 and 5.62 for all cases (low and high power and low and high pressure, and vacuum). It is assumed that the black deposition material on the witness plate is carbon (this assumption was verified by chemical element analysis of the plume signatures using SEM) and so the simulation results in Figures 5.61 and 5.62 contain the total fluxes of carbon ions and atoms. One can see differences in the two deposition signatures due to the background pressure effect. High background pressure leads to narrow plume, while low-pressure case corresponds to large plume expansion. In turn, the strongest effect of the background pressure can be seen in the higher pressure case, while in the low-pressure case calculations in the vacuum case produces similar result as the one for low pressure. One can see that the simulations generally predict a narrowing (focusing) of the plume. These results are comparable to previous experimental and theoretical study in the similar background pressure range [238,239]. In this relatively low-pressure regime ($<10^{-2}$ Torr), the effect of ablated atom collisions with background gas (hydrogen in our case) leads to redistribution of the ablated atom velocity toward the axial direction. These predictions are confirmed by present experiments as shown in Figures 5.61 and 5.62 showing the plume narrowing.

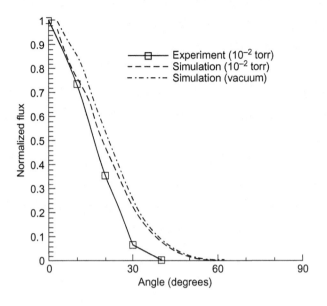

FIGURE 5.61

Normalized flux distribution. High-pressure (10^{-2} Torr) case, distribution along the x-direction.

Source: *Reprinted with permission from Ref. [178]. Copyright (2004) by American Institute of Physics.*

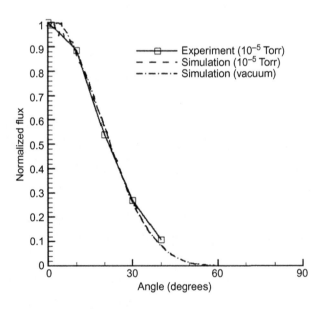

FIGURE 5.62

Normalized flux distribution. Low-pressure case (10^{-5} Torr), distribution along the x-direction.

Source: *Reprinted with permission from Ref. [178]. Copyright (2004) by American Institute of Physics.*

The above example of the code validation illustrate that the hybrid PIC–DSMC approach is the appropriate tool for plasma plume simulations.

5.4.4 Magnetic field effects on the plasma plume

In this section, we describe effects related to magnetic field that is present in the near-field plume region of plasma thrusters. For instance, in a Hall thruster, the magnetic field leaks outside the acceleration region.

In the case of electromagnetic PPT, self-magnetic field is formed and it is essential for plasma acceleration. Thus it is important to describe some peculiarities induced by the magnetic field in the near-field region of plasma thrusters.

The effect of the magnetic field on the near-field plume for Hall thrusters was modeled elsewhere [152] under steady-state conditions. It was found that the magnitude of the magnetic field near the thruster exit has an important effect on the plasma potential distribution in the plume. In the PPT, on the other hand, the self-magnetic field drives the plasma acceleration and thus one needs to include the electromagnetic effects on the near-field plume of unsteady plasma flow.

5.4.4.1 Effect of the magnetic field on the plasma plume

In this section, we describe effect of the magnetic field on the plasma potential distribution in the plume of the Hall thruster. A quasi-1D model of the three-fluid (ions, electron, and neutrals) plasma flow is considered. The model is described elsewhere [152]. The effect of the magnetic field appears in the equation for potential that can be obtained by solving the electron momentum equation for the potential:

$$\frac{\partial \varphi}{\partial z} = \frac{kT_e}{e} \frac{d\ln(n)}{dz} + BV\beta_e \tag{5.123}$$

where B is the magnetic field, V is the flow velocity, and β_e is the Hall parameter. The solution procedure and the approach is described in Ref. [152]. In this section, the main results of this model will be highlighted.

The axial potential distribution is shown in Figure 5.63 with the electron temperature as a parameter. One can see that the potential sharply increases initially and then saturates. For comparison, the experimental data [240] on the lines of 0° and 45° off the thruster centerline are plotted. It can be seen that, in general, the predicted results are in agreement with experiment. For instance, in the experiment, plasma potential saturation with axial distance was observed in the case of 0° as is predicted by the model.

The plasma potential distribution with magnetic field as a parameter is shown in Figure 5.64. In the case of a zero magnetic field, the plasma potential is negative and decreases with axial distance according to the Boltzmann relation. The plasma potential becomes positive when the magnetic field is finite. One can see that, in general, the plasma potential increases with increasing magnetic field.

The potential increase across the magnetic field can be understood in the following manner. The cases considered correspond to a plasma jet entering a transverse magnetic field with a high supersonic directed velocity in the condition that the magnetic field is relatively weak. This means that only electrons are magnetized, while ions are not magnetized on the spatial scale of interest.

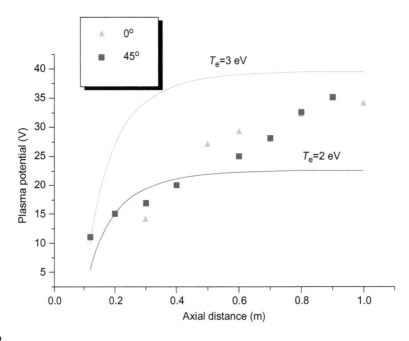

FIGURE 5.63

Potential distribution along centerline. Distance is measured from the Hall thruster exit plane.

Source: *Reprinted with permission from Ref. [152]. Copyright (1999) by American Institute of Physics.*

Since plasma flow across magnetic field is ambipolar, an electric field is formed that leads to the potential increase across the magnetic field.

5.4.4.2 Near-field plasma plume of PPT

The model is based on a hybrid approach involving a DSMC description of neutrals, a PIC approach for ions, and a fluid description of the electrons. In these methods, the potential distribution is usually calculated by reducing the electron momentum equation to the Boltzmann relation in the absence of a magnetic field. When the magnetic field was presented in the plasma plume domain, the electron momentum equation including the magnetic field effect should be used instead of Boltzmann relation.

5.4.4.3 Boundary conditions for the plume simulation

During the discharge, the plasma density near the propellant face is large (on the order of $10^{23}-10^{24}$ m^{-3} [25,241]) and therefore a fluid approach can be used. The plasma layer model includes the following features (similar to the model of an electrothermal PPT [25,241]): Joule heating of the plasma, heat transfer to the Teflon surface, and Teflon ablation. Mechanisms of energy transfer from the plasma column to the Teflon wall include heat transfer by particle convection and by radiation. The Teflon ablation computation is based on a kinetic ablation model described in Chapter 1.

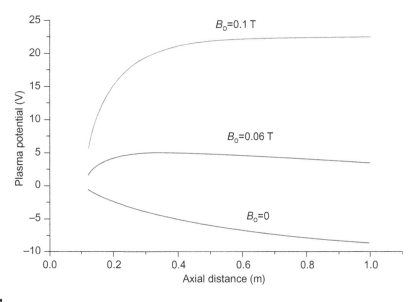

FIGURE 5.64

Potential distribution along centerline with a magnetic field as a parameter. Distance is measured from the Hall thruster exit plane.

Source: *Reprinted with permission from Ref. [152]. Copyright (1999) by American Institute of Physics.*

5.4.4.4 Plasma plume electrodynamics

As it was mentioned above, the hybrid fluid-particle approach was used for the plasma plume description in which the neutrals and ions are modeled as particles, while electrons are treated as a fluid. Elastic (momentum transfer) and nonelastic (CEX) collisions were taking into account. The particle trajectory due to collisions is calculated using the DSMC method [233]. Momentum-exchange elastic cross sections use the model of Dalgarno et al. [242], while CEX cross sections was calculated according to Sakabe and Izawa [236]. Acceleration of the charged particles is computed using the PIC method [237]. A single grid employing nonuniform, rectangular cells is used for both the DSMC and PIC steps. Since the flow is assumed to be quasi-neutral, there is no requirement to resolve the Debye length. Hence, the cells are scaled by the local mean free path. A single time step given by the reciprocal of the maximum plasma frequency is used throughout.

In the presence of a strong magnetic field, the electron density distribution deviates from that of the Boltzmann distribution [84]. In the case of a magnetic field, the electron momentum equation reads (neglecting electron inertia):

$$0 = -eN_e(E + V_e \times B) - \nabla P_e - N_e \nu_{ei} m_e(V_e - V_i) \tag{5.124}$$

We have assumed quasi-neutrality therefore $N_e = N_i = N$. Using a definition $j = eN((V_e - V_i)$, Eq. (5.125) is usually referred to as the generalized Ohm's law. The electric and magnetic field distributions in the plume were calculated from the set of Maxwell's equations (see Chapter 2). We further assume that the magnetic field has only an azimuthal component and also neglect the displacement current. The combination of Maxwell's equations and electron momentum conservation (5.125) gives the following equation for the magnetic field:

$$\partial B/\partial t = 1/(\sigma\mu)\nabla^2 B - \nabla \times (j \times B/(eN)) + \nabla \times (V \times B) \tag{5.125}$$

where $\sigma = e^2 N_e/(\nu_{ei} m_e)$ is the plasma conductivity which has only a weak dependence on the plasma density (in the Coulomb logarithm). Using this fact, it was assumed that plasma conductivity depends only on the electron temperature in the plasma plume. A similar approach regarding the assumption about conductivity was also employed in the plasma flow simulation in the PPT [243]. The electron temperature in the plume was assumed to be constant. The plasma conductivity is also constant (mainly due to the fact that electron temperature is assumed to be constant in the near plume [226]) and that the density gradient does not affect the magnetic field diffusion. The last assumption comes from the fact that the main density gradient is developed in the direction of magnetic field transport (as clear from the results below) and therefore does not affect magnetic field transport [32].

A scaling analysis shows that in the case of the near plume of the μ-PPT with a characteristic scale length L of about 1 mm, the magnetic Reynolds number $Re_m = \mu\sigma LV \ll 1$ (where V is the characteristic velocity $\sim 10^4$ m/s as shown below) and therefore the last term in Eq. (5.126) can be neglected. In addition, depending on the plasma density, the Hall effect (second term on the right-hand side of Eq. (5.125)) may be important for the magnetic field evolution. One of the first calculations of the plasma flow with Hall effect was performed by Brushlinski and Morozov (see Ref. [244] and references therein), who considered isothermal flow. The plasma densities become high at the cathode and lower at the anode. The Hall effect has a particularly noticeable influence on the magnetic field distribution. The field near the anode increases and near the cathode decreases. As a result, the current is deflected to the side and grazes the anode. Our estimations of the Hall parameter show that $\omega\tau \ll 1$ if the plasma density near the Teflon surface $N > 10^{23}$ m^{-3}. This is usually the case in the μ-PPT (PPT with the scale length of about few mm and low power), so the Hall effect is expected to be small and will not be considered in this analysis. Therefore Eq. (5.126) is reduced to the simple magnetic transport equation:

$$Re_m \partial B/\partial t = \nabla^2 B \tag{5.126}$$

Having the magnetic field distribution, one can calculate the current density distribution from Ampere's law:

$$\mu j = \nabla \times B \tag{5.127}$$

The magnetic field and current distributions calculated from this model are used in PIC for ion motion calculations. The ion velocity distribution depends upon the magnetic field distribution, and the ion motion is calculated using following momentum equation:

$$m \, dV/dt = Z_i e(E + V \times B) + \nu_{ei} m_e (V_e - V_i) \tag{5.128}$$

The electric field in this equation was determined using the electron momentum equation and therefore the last equation reduces to the following simplified form:

$$m \, dV/dt = j \times B/N \tag{5.129}$$

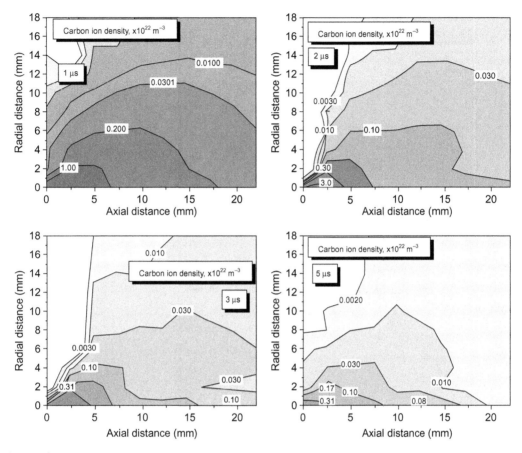

FIGURE 5.65

Evolution of the carbon ion density during the pulse.

5.4.4.5 Example of plasma plume simulations

The results are presented for a 3.6 mm (0.141″) diameter axisymmetric μ-PPT, which has a 0.9 mm diameter central electrode, 3.1 mm propellant diameter, and 0.24 mm anode wall [174]. Results presented below correspond to energy of about 2.25 J and $C = 0.5$ μF.

Figure 5.65 shows evolution of the carbon ion (C +) component of the plasma plume during the main part of the pulse. One can see that a high-density region is developed at a few millimeters from the thruster exit plane.

This dense region exists during the entire pulse as shown in Figure 5.65, but the plasma density decreases from about 2×10^{22} m^{-3} at the beginning of the pulse down to 0.3×10^{22} m^{-3} at 5 μs. At the beginning (first 2 μs), the C + density mainly develops a gradient in the radial direction that is a result of high directed velocity in the axial direction. Later, during the discharge pulse, the axial density gradient becomes comparable to the radial one.

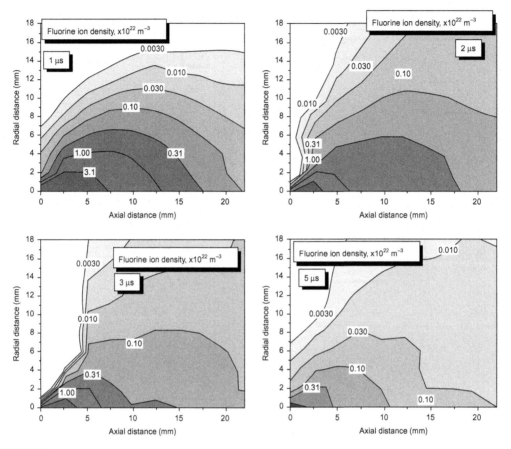

FIGURE 5.66

Evolution of the fluorine ion density during the discharge pulse.

Source: *Reprinted with permission from Ref. [245]. Copyright (2004) by AIAA.*

The fluorine ions (F+), due to their larger mass, have different dynamics as shown in Figure 5.66. They have smaller acceleration in the axial direction even at the beginning of the pulse and therefore both axial and radial density gradients are developed. The F+ density in the plume and in the high plasma density region is larger than that of C+, because originally Teflon has composition C_2F_4 with F density twice that of C. Additionally, F ions experience less acceleration in the plume because of their higher mass that also contributes to their relative density increase.

The μ-PPT is essentially an electromagnetic accelerator as shown in the velocity phase plots (Figures 5.67 and 5.68). The phase plot of the carbon ions at 5 μs is centered at 10 km/s in the axial direction (note that the sound speed is $\sim$5 km/s for a 4 eV plasma). Ions also experience radial expansion due to the magnetic field structure and expansion. The radial velocity in the negative direction is related to the focus formation along the axis, as shown in Figures 5.67. The fluorine

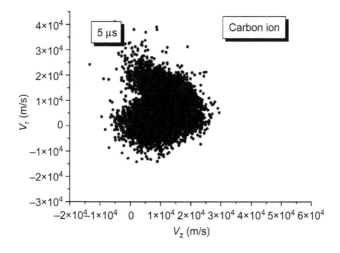

FIGURE 5.67

Velocity phase plot (radial-axial velocity components) for carbon ions.

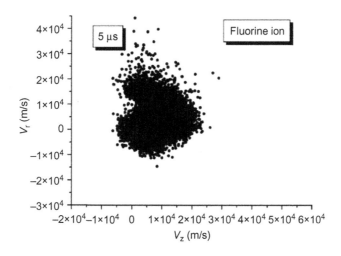

FIGURE 5.68

Velocity phase plot (radial-axial velocity components) for fluorine ions.

Source: Reprinted with permission from Ref. [245]. Copyright (2004) by AIAA.

ions have generally smaller axial and radial velocities comparing to carbon ions due to their higher mass.

During the entire pulse, there is a population of ions having a negative axial velocity with magnitude up to about 10 km/s (see Figures 5.67 and 5.68). This population creates the backflow contamination that occurs mainly onto the thruster itself. The carbon ions have a larger negative velocity due to their higher mobility that results in their domination in the backflux. This backflux may be mainly responsible for charring phenomena observed in the μ-PPT [173].

5.4.4.6 Comparison with experiment

In this section, the comparison of measured and predicted electron and neutral density distributions in the near-field plume for one μ-PPT design is discussed.

An experimental basis for comparison is provided using results from a two-color interferometer [245]. Electron and neutral density measurements are taken on a 6.35 mm (1/4") diameter μ-PPT at AFRL. The interferometer uses 488 and 1152 nm wavelengths and a quadrature heterodyne technique described by Spanjers et al. [246]. The two-color technique uses the difference in phase shifts from the two wavelengths to calculate electron and neutral densities. This technique tends to ensure that both lasers are sampling the same portions of the plume [247]. Older data with less focusing is shown with typical uncertainty in Antonsen et al. [248] along with a more thorough treatment of the two-color diagnostic. The data in Figure 5.69 has a maximum uncertainty of $\pm 1.15 \times 10^{16}$ cm^{-3} for electron density and $\pm 1.76 \times 10^{16}$ cm^{-3} for neutral density at 2 μs into the discharge. For the data shown here, the beam center is located 3 mm from fuel face on the thruster centerline with a beam diameter of 6 mm. It should be noted that the discharge in the μ-PPT may be azimuthally nonuniform (nonaxisymmetric) [249]. Therefore a comparison between an axisymmetric model and experiment in this case may be questionable. However, the mentioned nonuniformity decreases with discharge energy[173] and experimental data are reproducible during many pulses. This suggests that any effect of nonuniformity may be small and our comparison is justified.

Figure 5.69 shows the experimental data coplotted with model predictions. Plasma density peaks at about 2×10^{22} m^{-3} and decreases by several orders of magnitude toward the pulse end. The neutral density is significantly higher and peaks at about 14×10^{22} m^{-3}. The experimental data were taken at discharge energy of 6.0 J from a 0.417 μF capacitor. One can see that plume model predicts very well magnitude and temporal variation of both electron and neutral densities.

Homework problems

Section 1

1. **Calculate the velocity (normalized by local sound speed) at the outer edge of the Knudsen layer as a function of plasma density (10^{22}–10^{24} m^{-3}).** Use iteration to perform analysis. Use the following conditions at the surface: $n_0 = 10^{23}$ m^{-3}; $T_0 = 1000$ K and plasma temperature is 0.5 eV. Compare results with and without effect of magnetic field. Current is about 10^3 A and characteristics size of the electrode is about 1 cm.
2. **Compute the Alfven speed** in a magnetic field of about 0.1 T and plasma density of about 10^{20} m^{-3}.
3. **Thickness of the ionization layer in ablative pulsed plasma thrusters.** Demonstrate that the thickness of the ionization layer decreases with the magnetic field strength using the analysis of plasma acceleration in pulsed plasma accelerators.
4. **Compute the ablation rate of the Teflon** in the case of surface temperature of about 900 K (use system of equation from Chapter 1, Section 5). Teflon surface is in contact with plasma having density of about 10^{22} m^{-3} and temperature of about 3 eV.

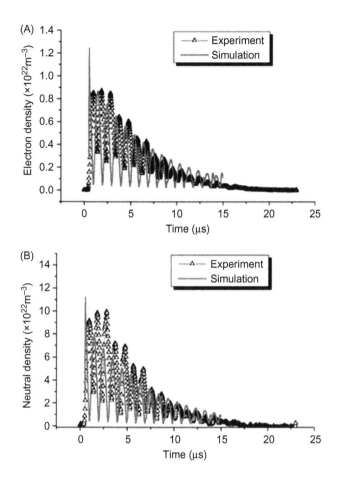

FIGURE 5.69

Comparison of predicted and measured electron and neutral densities time variation at 3 mm from the propellant face at the axis in the case of the 6.35 mm diameter μ-PPT firing at 6 J, $\theta = 0.2$: (A) plasma density and (B) neutral density.

Source: *Reprinted with permission from Ref. [245]. Copyright (2004) by AIAA.*

Section 2

1. Estimate the frequency of near wall conductivity across the channel having width of about 2 cm. Consider xenon plasma having electron temperature of about 30 eV. (**use material from Section 5.2.2**)

2. Plot the dependence of the current density distribution on potential drop across the sheath. Consider the near wall conductivity. Comment on effect of the sheath. (**use material from Section 5.2.2**)

3. Calculate the deviation from the equipotential condition along the magnetic line in the plasma having electron temperature of about 10 eV. Consider the plasma density decrease along the magnetic field by factor of 10.

4. Compute the dependence of the sheath potential drop near the dielectric wall in the Hall thruster channel on the SEE coefficient (0–0.9). Consider Xenon plasma flow with electron temperature of about 20 eV.

5. Assume that SEE coefficient has linear dependence on the electron energy (5–20 eV). Plot the dependence of the sheath potential drop on the electron energy. Consider krypton as a propellant.

6. **Anodic plasma analysis.** Calculate the near anode potential drop assuming that the electron density near the anode is about 10^{17} m^{-3}, electron temperature is about 7 eV. Discharge current is 3 A and the anode collection area is about 6 cm^2.

7. **Consider the sheath thickness** near the channel wall of the thruster with anode layer. Electron temperature can be assumed to be about 100 eV. Consider the discharge voltage of about 3 kV, bismuth as a propellant and the channel wall having the cathode potential. Plot the dependence of the maximum sheath thickness on the electron density (10^{16}–10^{18} m^{-3}).

Section 4

1. **Compute the potential drop in the near field plume of the Hall thruster.** Consider a plasma flow across the 1 cm layer of a magnetic field of about 0.005 T, Hall parameter of about 10 and density decrease across the layer by factor of 2. Electron temperature is about 5 eV and ion velocity is about 14 km/s.

2. **A Hall thruster** 10 cm in diameter produces a Xe ion beam at 300 eV. Assume that initially ions have only axial component of velocity. Compute the effective divergence of the beam (angle) if diameter of the beam doubles in a distance of 50 cm and radial component of ion velocity.

3. **Estimate the rate of particle production due to charge exchange** in the near field plume of Hall thruster having outer diameter of channel of about 10 cm and the inner diameter of about 6 cm. Consider the mass flow rate of about 5 mg/s and the ionization degree of about 10 %. Assume that the ion velocity is about 16 km/s and the neutral particles velocity is about 1 km/s. Plume expansion can be neglected.

4. **Compute the sputtering rate of the Kapton plate** placed normal to the plasma plume. Consider conditions of the problem #2. Calculate the sputtering at following radial locations: 0, 10, 15 and 20 cm. The molecular weight of Kapton is 382 g/mol and that its mass density is 1.42×10^3 kg/m^3. Use the sputtering yield of Kapton as a function of ion energy and angle.

References

[1] M. Keidar, A.D. Gallimore, Y. Raitses, J.P. Bouef (Eds.), Special issue on plasma propulsion, IEEE Trans. Plasma Sci. 36 (5) (2008) 1962–1966.

[2] R.G. Jahn, E.Y. Choueiri, Electric propulsion, Encyclopedia of Physical Science and Technology, third ed., vol. 5, pp. 125−141. The Academic Press, San Diego, 2001.

[3] I.I. Beilis, State of the theory of vacuum arcs, IEEE Trans. Plasma Sci. 29 (2001) 657.

[4] M. Keidar, I.D. Boyd, I.I. Beilis, Electrical discharge in the Teflon cavity of a co-axial pulsed plasma thruster, IEEE Trans. Plasma Sci. 27 (2001) 376.

[5] L. Muller, Modelling of an ablation controlled arc, J. Phys. D Appl. Phys. 26 (1993) 1253−1259.

[6] E. Domejean, P. Chevrier, C. Fievet, P. Petit, Arc−wall interaction modelling in a low-voltage circuit breaker, J. Phys. D Appl. Phys. 30 (1997) 2132−2142.

[7] S.V. Kukhlevsky, J. Kaiser, O. Samek, M. Liska, J. Erostyak, Stark spectroscopy measurements of electron density of ablative discharges in Teflon−$(CF_2)_n$ capillaries, J. Phys. D Appl. Phys. 33 (2000) 1090−1092.

[8] D. Hong, R. Dussart, C. Cachoncinlle, W. Rosenfeld, S. Gotze, J. Pons, et al., Study of a fast ablative capillary discharge dedicated to soft X-ray production, Rev. Sci. Instrum. 71 (2000) 15−19.

[9] I.I. Beilis, Analysis of the cathode spots in a vacuum arc, Sov. Phys. Tech. Phys. 19 (1974) 251−256.

[10] X. Zhou, J. Heberlein, Analysis of the arc-cathode interaction of free-burning arcs, Plasma Sources Sci. Technol. 3 (1994) 564.

[11] B. Rethfeld, J. Wendelstorf, T. Klein, G. Simon, A self-consistent model for the cathode fall region of an electric arc, J. Phys. D Appl. Phys. 29 (1996) 121.

[12] R.J. Vondra, K.I. Thomassen, Flight qualified pulsed plasma thruster for satellite control, J. Spacecraft Rockets 11 (9) (1974) 613−617.

[13] M. Keidar, I.D. Boyd, I.I. Beilis, On the model of teè on ablation in an ablation-controlled discharge, J. Phys. D Appl. Phys. 34 (2001) 1675.

[14] M. Keidar, I.D. Boyd, F.S. Gulczinski III, E. Antonsen, G.G. Spanjers, in: 27th International Electrical Propulsion Conference, Pasadena, CA, 2001 (IEPC-01-155).

[15] G.G. Spanjers, D.R. Bromaghim, C.J. Lake, M. Dulligan, D. White, J.H. Schilling, et al., in: 28th AIAA Joint Propulsion Conference, Indianapolis, IN, 2002 (Paper AIAA-2002-3974).

[16] S.I. Anisimov, Vaporization of metal absorbing laser radiation, Sov. Phys. JETP 27 (1968) 182.

[17] I.I. Beilis, On the theory of the erosion processes in the cathode region of an arc discharge, Sov. Phys. Dokl. 27 (1982) 150.

[18] C.J. Knight, Theoretical modeling of rapid surface vaporization with back pressure, AIAA J. 17 (1979) 519.

[19] M. Keidar, J. Fan, I.D. Boyd, I.I. Beilis, Vaporization of heated materials into discharge plasmas, J. Appl. Phys. 89 (2001) 3095.

[20] I. Mikellides, Theoretical modeling and optimization of ablation-fed pulsed plasma thrusters, Ph.D Thesis, Ohio State University, Columbus, 2000.

[21] P.J. Turchi, I.G. Mikellides, P.G. Mikellides, H. Kamhawi, in: M.M. Micci, A.D. Ketsdever (Eds.), Micropropulsion for Small Spacecraft, vol. 187, AIAA, 2000Progress in Astronautics and Aeronautics

[22] H. Alfven, On the Origin of the Solar System, Oxford University Press, Oxford, 1964.

[23] R.L. Burton, et al., Pulsed plasma thruster performance for microspacecraft propulsionProgress in Astronautics and Aeronautics in: M. Michael, Micci, A.D. Ketsdever (Eds.), Micropropulsion for Small Spacecraft, vol. 187, Reston VA AIAA, 2000.

[24] L. Danielson, Experiment on the interaction between a plasma and a neutral gas, Phys. Fluids 13 (1970) 2288.

[25] M. Keidar, I.D. Boyd, I.I. Beilis, Model of an electrothermal pulsed plasma thruster, *J. Propul. Power* 19 (3) (2003) 424−430.

[26] D.E. Potter, Numerical studies of the plasma focus, Phys. Fluids 14 (1971) 1911.

[27] A. Fruchtman, N.J. Fisch, Y. Raitses, Control of the electric-field profile in the Hall thruster, Phys. Plasmas 8 (2001) 1048.

[28] E. Ahedo, P. Martinez-Cerezo, M. Martinez-Sanchez, One-dimensional model of the plasma flow in a Hall thruster, Phys. Plasmas 8 (2001) 3058.

[29] A.I. Zemskov, V.V. Prut, V.A. Khrabrov, Pulsed discharge in dielectric chamber, Sov. Phys. Tech. Phys. 17 (1972) 285–289.

[30] G.I. Kozlov, V.A. Kuznetsov, V.A. Masyukov, Radiative losses by argon plasma and the emissive model of a continuous optical discharge, Sov. Phys. JETP 39 (1974) 463–468.

[31] Y.P. Raizer, Gas Discharge Physics, Nauka, Moscow, 1987 (in Russian).

[32] A.S. Kingsep, Yu.V. Mokhov, K.V. Chukbar, Sov. J. Plasma Phys. 10 (4) (1984) 495.

[33] T.E. Markusic, K.A. Polzin, E.Y. Choueiri, M. Keidar, I.D. Boyd, N. Lepsetz, Ablative Z-pinch pulsed plasma thruster, J. Propul. Power 21 (2) (2005) 392–400.

[34] IAEA Aladin database <http://www-amdis.iaea.org/aladdin.html>.

[35] R.J. Vondra, (Paper 76-998) The MIT Lincoln Laboratory Pulsed Plasma Thruster, AIAA, Reston VA, 1976

[36] P.J. Turchi, Directions for improving PPT performance, in: Proceedings of the 25th International Electric Propulsion Conference, vol. 1, Worthington, OH, 1998, pp. 251–258.

[37] E. Antonsen, R.L. Burton, F. Rysanek, Energy Measurements in a Co-Axial Electromagnetic Pulsed Plasma Thruster, Reston VA (Paper AIAA-1999-2292).

[38] F. Gulczinski III, M. Dulligan, J. Lakes, G. Spanjers, Micropropulsion research at AFRL, Reston VA (Paper AIAA-2000-3255).

[39] R.L. Burton, S.S. Bushman, Probe measurements in a co-axial gasdynamic PPT, in: 35th Joint Propulsion Conference, AIAA, Los Angeles, CA, 1999, pp. 99–2288.

[40] F. Rysanek, R.L. Burton, in: 36th Joint Propulsion Conference, AIAA, Huntsville AL, 2000 (Paper 2000-3429).

[41] M. Keidar, I.I. Beilis, Non-equilibrium thermal boundary layer in a capillary discharge with an ablative wall, Phys. Plasmas 13 (2006) 114503.

[42] S.S. Bushman, Investigations of a coaxial pulsed plasma thruster, Master's Thesis, University of Illinois Urbana-Champaign, 1999.

[43] M. Keidar, I.D. Boyd, I.I. Beilis, On the model of Teflon ablation in an ablation-controlled discharge, J. Phys. D Appl. Phys. 34 (June) (2001) 1675–1677.

[44] I.I. Beilis, Parameters of the kinetic layer of arc discharge cathode region, IEEE Trans. Plasma Sci. PS-13 (1985) 288–290.

[45] C.B. Ruchti, L. Niemeyer, Ablation controlled arc, IEEE Trans. Plasma Sci. 14 (1986) 423–434.

[46] P. Kovatya, J.J. Lowke, Theoretical predictions of ablation-stabilised arcs confined in cylindrical tubes, J. Phys. D Appl. Phys. 17 (1984) 1197–1212.

[47] P. Kovatya, Thermodynamic and transport properties of ablated vapors of PTFE, alumina, perspex and PVC in the temperature range 5000–30000 K, IEEE Trans. Plasma Sci. 12 (1984) 38–42.

[48] C.S. Schmahl, P.J. Turchi, Development of equation-of-state and transport properties for molecular plasmas in pulsed plasma thrusters. Part I: A two-temperature equation of state for Teflon, Proc. Inter. Electr. Propul. Conf (1997) 781–788.

[49] R.L. Burton, B.K. Hilko, F.D. Witherspoon, G. Jaafari, Energy-mass coupling in high-pressure liquid-injected arcs, IEEE Trans. Plasma Sci. 19 (1991) 340–349.

[50] G.G. Spanjers, J.S. Lotspeich, K.A. McFall, R.A. Spores, J. Propul. Power 14 (4) (1998) 554–559.

[51] A. Shapiro, The Dynamics and Thermodynamics of Compressible Fluid Flow, Wiley, New York, NY.

[52] V. Kim, The Main Physical Features of SPT Thruster and Classification (IEPC-2011-007).

[53] V. Kim, Main physical features and processes determining the performance of stationary plasma thrusters, J. Prop. Power 14 (5) (1998) 736–743.

[54] E.E. Yushmanov, Radial distribution of the potential in cylindrical trap with magnetron ion injection, in: M.A. Leontovich (Ed.), Plasma Physics and Problem of Controlled Fusion, vol. IV, USSR Academy of Science, Moscow, 1958, pp. 235–237 (in Russian).

[55] A.V. Zharinov, Electric double layer in strong magnetic field, Kurchatov Institute Report, 1961 (in Russian).

[56] G.R. Seikel, E. Reshotko, Hall current ion accelerator, Bull. Am. Phys. Soc. Sr. II 7 (1962) 414.

[57] F. Salz, R.G. Meyerand, E.C. Lary, Ion acceleration in a gyro-dominated neutral plasma-theory, Bull. Am. Phys. Soc. Sr. II 7 (1962) 441.

[58] G.S. Janes, R.S. Lowder, Anomalous electron diffusion and ion acceleration in a low-density plasma, Phys. Fluids 9 (6) (1966) 1115−1123.

[59] A.V. Zharinov, Yu.S. Popov, Acceleration of plasma by a closed Hall current, Sov. Phys. Tech. Phys. 12 (2) (1967) 208−211.

[60] A.I. Morozov, Effect of near-wall conductivity in magnetized plasma, J. Appl. Math Tech. Phys. 3 (1968) 19−22.

[61] V.V. Zhurin, H.R. Kaufman, R.S. Robinson, Plasma Sources Sci. Technol. 8 (1999) 1.

[62] A.I. Morozov, Focusing of the cold quasineutral beams in electromagnetic fields, Doklady Akademii Nauk 163 (36) (1965) 1363 (in Russian).

[63] E. Choueiri, Fundamental difference between the two Hall thruster variants, Phys. Plasmas 8 (11) (2001) 5025−5033.

[64] M.A. Vlasov, A.V. Zharinov, Yu.A. Kovalenko, On the theory of discharge in crossed fields, Sov. Phys. Tech. Phys. 46 (12) (2001) 1522−1529.

[65] A.Yu. Kovalenko, Yu.A. Kovalenko, Modeling of a discharge in crossed fields with account of collisions and ionization, Sov. Phys. Tech. Phys. 48 (11) (2003) 1413−1418.

[66] A. Yu., Yu. Kovalenko, Kovalenko, M. Keidar, Conditions for formation of a kinetic anode layer in crossed $E \times B$ fields, Appl. Phys. Lett. 89 (2006) 051503.

[67] I.I. Beilis, A. Fruchtman, Y. Maron, A mechanism for ion acceleration near the anode of a magnetically insulated ion diode, IEEE Trans. Plasma Sci. 26 (3) (1998) 995−999.

[68] Y. Raitses, J. Ashkenazy, G. Appelbaum, M. Guelman, 25th International Conference on Electric Propulsion, Cleveland, OH, The Electric Rocket Propulsion Society, Worthington, OH, 1997 (IEPC 97-056).

[69] Y. Raitses, L. Dorf, A. Litvak, N. Fisch, Plume reduction in segmented electrode Hall thruster, J. Appl. Phys. 88 (3) (2000) 1263−1270.

[70] Y. Raitses, D. Staack, A. Smirnov, N.J. Fisch, Space charge saturated sheath regime and electron temperature saturation in Hall thrusters, Phys. Plasmas 12 (7) (2005) 073507.

[71] J.M. Fife, Hybrid PIC modeling and electrostatic probe survey of Hall thrusters, Ph.D. Thesis, MIT, Cambridge MA, 1998.

[72] J.P. Bouef, L. Garrigues, Low-frequency oscillations in a stationary plasma thrusters, J. Appl. Phys. 84 (7) (1998) 3541−3550.

[73] I.D. Boyd, L. Garrigues, J. Koo, M. Keidar, Progress in development of a combined device/plume model for Hall thrusters, in: 36th AIAA Joint Propulsion Conference, Huntsville, AL, 2000, American Institute of Aeronautics and Astronautics, Washington, DC, 2000 (AIAA-2000-3520).

[74] A. Fruchtman, N.J. Fisch, Modeling of a Hall thruster, in: 34th AIAA Joint Propulsion Conference, Cleveland, OH, 1998, American Institute of Aeronautics and Astronautics, Washington, DC, 1998 (AIAA-1998-3500).

[75] E. Ahedo, P. Martinez, M. Martines-Sanches, One dimensional model of the plasma flow in a Hall thruster, Phys. Plasmas 8 (6) (2001) 3058−3067.

[76] K. Makowsky, Z. Peradzynski, N. Gascon, M. Dudeck, in: 35th AIAA Joint Propulsion Conference, Los Angeles, CA, 1999, American Institute of Aeronautics and Astronautics, Washington, DC, 1999 (AIAA-99-2295).

[77] S. Locke, U. Schumlak, J.M. Fife, in: 37th AIAA Joint Propulsion Conference, Salt Lake City, UT, 2001, American Institute of Aeronautics and Astronautics, Washington, DC, 2001 (AIAA-01-3327).

[78] M. Keidar, I.D. Boyd, I.I. Beilis, Plasma flow and plasma—wall transition in Hall thruster channel, Phys. Plasmas 8 (12) (2001) 5315—5322.

[79] M. Keidar, I.D. Boyd, I.I. Beilis, Modeling of a high-power thruster with anode layer, Phys. Plasmas 11 (4) (2004) 1715—1722.

[80] M. Keidar, A.D. Gallimore, Y. Raitses, I.D. Boyd, On the potential distribution in Hall thrusters, Appl. Phys Lett. 85 (13) (2004) 2481—2483.

[81] Y. Raitses, M. Keidar, D. Staack, N.J. Fisch, Effects of segmented electrode in Hall current plasma thrusters, J. Appl. Phys. 92 (9) (2002) 4611—4906.

[82] M. Keidar, I.D. Boyd, On the mirror effect in Hall thrusters, Appl. Phys. Lett. 87 (Sept) (2005) 121501.

[83] Y. Raitses, D. Staack, M. Keidar, N.J. Fisch, Electron—wall interaction in Hall thrusters, Phys. Plasmas 12 (2005) 057104.

[84] I.I. Beilis, M. Keidar, S. Goldsmith, Plasma—wall transition: the influence of the electron to ion current ratio on the magnetic presheath structure, Phys. Plasmas 4 (10) (1997) 3461—3468.

[85] O. Bohm, in: A. Guthrue, R.K. Wakerling (Eds.), The Characteristics of Electrical Discharges in Magnetic Fields, McGraw Hill Book Co., New York, NY, 1949.

[86] E.Y. Choueiri, Plasma oscillations in Hall thrusters, Phys. Plasmas 8 (4) (2001) 1426—1441.

[87] A.A. Litvak, N.J. Fisch, Resistive instabilities in Hall current plasma discharge, Phys. Plasmas 8 (2) (2001) 648—651.

[88] E. Chesta, N.B. Meezan, M.A. Cappelli, Stability of a magnetized Hall plasma discharge, J. Appl. Phys. 89 (6) (2001) 3099—3107.

[89] A.A. Litvak, Y. Raitses, N.J. Fisch, Experimental studies of high-frequency azimuthal waves in Hall thrusters, Phys. Plasmas 11 (4) (2004) 1701—1705.

[90] A.A. Litvak, N.J. Fisch, Rayleigh instability in Hall thruster, Phys. Plasmas 11 (4) (2004) 1379—1383.

[91] A. Lazurenko, V. Vial, M. Prioul, A. Bouchoule, Experimental investigation of high-frequency drifting perturbations in Hall thrusters, Phys. Plasmas 12 (1) (2005) 013501.

[92] J.C. Adam, A. Heron, G. Laval, Study of stationary plasma thrusters using two-dimensional fully kinetic simulations, Phys. Plasmas 11 (1) (2004) 295—305.

[93] C.A. Thomas, M.A. Cappelli, Gradient transport processes in $E \times B$ plasmas, in: 41th AIAA Joint Propulsion Conference, Tucson, AZ, 2005, American Institute of Aeronautics and Astronautics, Washington, DC, 2005 (AIAA paper 2005-4063).

[94] M.W. Alcock, B.E. Keen, Experimental observation of the drift dissipative instability in an afterglow plasma, Phys. Rev. A 3 (3) (1971) 1087—1096.

[95] S. Chable, F. Rogier, Numerical investigation and modeling of stationary plasma thruster low frequency oscillations, Phys. Plasmas 12 (2005) 033504.

[96] G.D. Hobbs, J.A. Wesson, Heat flow through a Langmuir sheath in the presence of electron emission, Plasma Phys. 9 (1967) 85—87.

[97] V.I. Baranov, Yu.S. Nazarenko, V.A. Petrosov, A.I. Vasin, Yu.M. Yashnov, Energy balance and role of walls in ACDE, in: 25th International Conference on Electric Propulsion, Cleveland, OH, The Electric Rocket Propulsion Society, Worthington, OH, 1997 (IEPC-97-060).

[98] A.I. Morozov, A.P. Shubin, Electron kinetics in the near-wall conductivity regime, Sov. J. Plasma Phys. 10 (1984) 1262—1269.

[99] A.I. Morozov, V.V. Savel'ev, Theory of the near-wall conductivity, Plasma Phys. Rep. 27 (7) (2001) 570—575.

[100] A.A. Ivanov, A.A. Ivanov Jr, M. Bacal, Effect of plasma—wall recombination on the conductivity in Hall thrusters, Plasma Phys. Control. Fusion 44 (2002) 1463—1470.

[101] I. Levchenko, M. Keidar, Visualization of ion flux neutralization effect on electrical field and atom density distribution in Hall thruster channel, IEEE Trans. Plasma Science 33 (2) (2005) 526−527.

[102] S. Barral, K. Makowski, Z. Peradzynski, and M. Dudeck, Is nearwall conductivity a misnomer_B plasmas, in: Proc. 39th AIAA Joint Propulsion Conference, Huntsville, AL, Reston VA, 2003, AIAA 2003-4855.

[103] M. Keidar, I.I. Beilis, Electron transport phenomena in plasma devices with $E \times B$ drift (invited review), IEEE Trans. Plasma Sci. 34 (3) (2006) 804−814.

[104] V.V. Egorov, V. Kim, A.A. Semenov, I.I. Shkarban, Near-wall process and its influence on operation of accelerators with closed electron drift, Ion Injectors and Plasma Accelerators, Energoatomizdat, Moscow, 1990 (in Russian).

[105] L.A. Schwager, Effects of secondary and thermionic electron emission on the collector and source sheaths of a finite ion temperature plasma using kinetic theory and numerical simulation, Phys. Fluids B5 (2) (1993) 631−645.

[106] A.I. Morozov, Effect of near-wall conductivity in a strongly magnetized plasma, Sov. Phys. Dokl. 10 (1966) 775.

[107] A.I. Morozov, V.V. Savel'ev, in: 24th International Conference on Electric Propulsion, Moscow, Russia, The Electric Rocket Propulsion Society, Worthington, OH, 1995 (IEPC-95-161).

[108] N.B. Meezan, W.A. Hargus, M.A. Cappelli, Anomalous electron mobility in a coaxial Hall discharge plasma, Phys. Rev. E63 (2001) 026410.

[109] J.M. Fife, S. Locke, AIAA Aerospace Sciences Meeting and Exhibit, 2001, Reno, NV, American Institute of Aeronautics and Astronautics, Washington, DC, 2001 (AIAA-2001-1137).

[110] A.I. Morozov, Effect of near-wall conductivity in a strongly magnetized plasma, Prikl. Mekh. Tekh. Fiz. 3 (1968) 19.

[111] L.B. King, A.D. Gallimore, C.M. Marrese, Transport property measurements in the Plume of an SPT-100 Hall thruster, J. Prop. Power 14 (1998) 327.

[112] J.P. Bugeat, C. Koppel, Development of a second generation of SPT, in: 24th International Conference Electric Propulsion, Moscow, Russia, The Electric Rocket Propulsion Society, Worthington, OH, 1995 (IEPC-95-35).

[113] S. Bechu, G. Perot, N. Gascon, P. Lasgorceix, A. Hauser, M. Dudeck, in: 35th AIAA Joint Propulsion Conference, Los Angeles, CA, 1999, American Institute of Aeronautics and Astronautics, Washington, DC, 1999 (AIAA-99-2567).

[114] A.I. Morozov, V.V. Savel'ev, One-Dimensional hybrid model of a stationary plasma thruster, Plasma Phys. Rep. 26 (2000) 219.

[115] S. Kim, J.E. Foster, A.D. Gallimore, in: 32th AIAA Joint Propulsion Conference, Orlando, FL, American Institute of Aeronautics and Astronautics, Washington, DC, 1996 (AIAA-96-2972).

[116] M. Keidar, I. Beilis, R.L. Boxman, S. Goldsmith, 2-D expansion of the low-density interelectrode vacuum arc plasma jet in an axial magnetic field, J. Phys. D Appl. Phys. 29 (1996) 1973.

[117] A.M. Bishaev, V. Kim, Sov. Phys. Tech. Phys. 23 (9) (1978) 1055.

[118] J.M. Haas, A.D. Gallimore, An investigation of internal ion number density and electron temperature profiles in a laboratory-model hall thruster, in: 36th Joint Propulsion Conference (JPC) Huntsville, AL, 2000, American Institute of Aeronautics and Astronautics, Washington, DC, 2000 (AIAA-00-3422).

[119] R.C. Weast (Ed.), Handbook of Chemistry and Physics, The Chemical Rubber Co., Boca Raton FL 1965, p. E-138.

[120] J.M. Fife, Hybrid PIC modeling and electrostatic probe survey of Hall thrusters, Ph.D. Thesis, MIT, Cambridge, MA, 1998.

[121] A. Fruchtman, N. J. Fisch, Modeling of a Hall thruster, in: 34th AIAA Joint Propulsion Conference, Cleveland, OH, 1998, American Institute of Aeronautics and Astronautics, Washington, DC, 1998 (AIAA-1998-3500).

[122] K. Makowsky, Z. Peradzynski, N. Gascon, M. Dudeck, in: 35th AIAA Joint Propulsion Conference, Los Angeles, CA, 1999, American Institute of Aeronautics and Astronautics, Washington, DC, 1999 (AIAA-99-2295).

[123] J. Ashkenazy, Y. Raitses, G. Appelbaum, Parametric studies of the Hall current plasma thruster, Phys. Plasmas 5 (1998) 2055.

[124] S. Barral, K. Makowski, Z. Peradzyński, N. Gaskon, M. Dudeck, Wall material eects in stationary plasma thrusters. II. Near-wall and in-wall conductivity, Phys. Plasmas 10 (2003) 4137−4152.

[125] M. Keidar, I. Beilis, R.L. Boxman, S. Goldsmith, 2-D expansion of the low-density interelectrode vacuum arc plasma jet in an axial magnetic field, J. Phys. D Appl. Phys. 29 (1996) 1973−1983.

[126] E. Ahedo, J.M. Gallardo, M. Martinez-Sanchez, Effects of the radial plasma−wall interaction on the Hall thruster discharge, Phys. Plasmas 10 (2003) 3397.

[127] M. Keidar, I. Boyd, I.I. Beilis, Plasma flow and plasma−wall transition in Hall thruster channel, Phys. Plasmas 8 (2001) 5315.

[128] N. Meezan, M. Cappelli, Kinetic study of wall collisions in a coaxial Hall discharge, Phys. Rev. E 66 (2002) 036401.

[129] O. Batishchev, M. Martinez-Sanchez, Charged particles transport in the Hall effect thruster, in: Proc. 28th International Electric Propulsion Conference, 2003, Toulouse, France, Electric Rocket Propulsion Society, Cleveland, OH 2003 (IEPC paper 2003-188).

[130] D.Y. Sydorenko, A.I. Smolyakov, Simulation of secondary electron emission effects in a plasma slab in crossed electric and magnetic fields, in: APS DPP 46th Annual Meeting, Savannah, GA, 2004 (NM2B.008).

[131] D. Sydorenko, A. Smolyakov, I. Kaganovich, Y. Raitses, Kinetic simulation of secondary electron emission effects in Hall thrusters, Phys. Plasmas 13 (2006) 014501.

[132] E. Ahedo, F. Parra, Partial trapping of secondary-electron emission in a Hall thruster, Phys. Plasmas 12 (2005) 073503.

[133] Y. Raitses, D. Staack, A. Smirnov, N.J. Fisch, Space charge saturated sheath regime and temperature saturation in Hall thrusters, Phys. Plasmas 12 (2005) 073507.

[134] J.M. Haas, A.D. Gallimore, IEEE Trans. Plasma Sci. 30 (2002) 687.

[135] J.M. Haas, A.D. Gallimore, Phys. Plasmas 8 (2001) 652.

[136] T.B. Smith, D.A. Herman, A.D. Gallimore, R.P. Drake, Deconvolution of axial velocity distributions from Hall thruster LIF spectra, in: 27th International Electric Propulsion Conference, Pasadena, CA, 2001 (IEPC-01-019).

[137] L. Dorf, Y. Raitses, N.J. Fisch, V. Semenov, Effect of anode dielectric coating on Hall thruster operation, Appl. Phys. Lett. 84 (2004) 1070.

[138] I. Melikov, Soviet Phys. Tech. Phys. 22 (1977) 452.

[139] I.V. Melikov, Soviet Phys. Tech. Phys. 19 (1974) 35.

[140] M. Keidar, I.D. Boyd, I.I. Beilis, Analyses of the anode region of a Hall thruster channel, in: Proc. 38th Joint Propulsion Conference and Exhibit, 2002, Indianapolis, IN, American institute of Aeronautics and Astronautics, Reston, VA, 2002 (AIAA-2002-4107).

[141] J. Koo, M. Keidar, I.D. Boyd, Boundary conditions for a 2-D hybrid stationary plasma thruster model, in: Proc. 40th Joint Propulsion Conference and Exhibit, 2004 Fort Lauderdale, FL. American institute of Aeronautics and Astronautics, Reston, VA, 2002 (AIAA-2004-3781).

[142] E. Ahedo, J. Rus, J. Appl. Phys. 98 (2005) 043306.

[143] L. Dorf, Y. Raitses, N.J. Fisch, J. Appl. Phys. 97 (2005) 103309.

[144] A. Anders, S. Anders, Plasma Sources Sci. Technol. 4 (1995) 571.

[145] I. Katz, J.R. Anderson, J.E. Polk, J.R. Brophy, J. Prop. Power 19 (2003) 595.

[146] M. Keidar, I.I. Beilis, IEEE Trans. Plasma Sci. 33 (2005) 1481.

[147] M.A. Lieberman, A.J. Lichtenberg, Principles of Plasma Discharges and Materials Processing, Wiley, New York, NY, 1994.

[148] L. Dorf, Y. Raitses, N.J. Fisch, Rev. Sci. Instrum. 75 (2004) 1255.

[149] I.I. Beilis, A. Fruchtman, I. Maron, IEEE Trans. Plasma Sci. 26 (1998) 995.

[150] D.T. Jacobson, R.S. Jankovsky, V.K. Rawlin, D.H. Manzella, High voltage TAL performance, in: 37th AIAA Joint Propulsion Conference, Salt Lake City, UT, 2001, American Institute of Aeronautics and Astronautics, Washington, DC, 2001 (AIAA-2001-3777).

[151] S. Tverdokhlebov, A. Semenkin, J. Polk, Bismuth propellant option lor very high power tal thruster, in: 40th AIAA Aerospace Sciences Meeting, Reno, NV, 2002, American Institute of Aeronautics and Astronautics, Washington, DC, 2002 (AIAA Paper 2002-0348).

[152] M. Keidar, I.D. Boyd, Effect of a magnetic field on the plasma plume from Hall Thruster, J. Appl. Phys. 86 (1999) 4786−4791.

[153] A.E. Soloduchin, A.V. Semenkin, Study of discharge channel erosio in multi mode anode layer thruster in: 28th International Conference on Electric Propulsion, Toulouse, France, The Electric Rocket Propulsion Society, Worthington, OH, 2003 (IEPC-2003-0204).

[154] I.I. Beilis, M. Keidar, Phys. Plasmas 5 (1998) 1545.

[155] V.S. Erofeev, L.V. Leskov in: Physics and Application of Plasma Accelerators, A.I. Morozov, (Ed.), Minsk, 1974 (in Russian).

[156] C.D. Child, Phys. Rev. 32 (1911) 492.

[157] I. Langmuir, Phys. Rev. Ser. II 2 (1913) 450.

[158] M. Keidar, O.R. Monteiro, A. Anders, I.D. Boyd., App. Phys. Lett. 81 (**7**) (2002) 1183.

[159] Y. Raitses, D. Staack, L. Dorf, N.J. Fisch, Experimental study of acceleration region in A 2 Kw Hall thruster, in: 39th AIAA Joint Propulsion Conference, Huntsville, AL, 2003, American Institute of Aeronautics and Astronautics, Washington, DC, 2003 (AIAA paper-2003-5153).

[160] L. Brieda, M. Keidar, Multiscale modeling of Hall thrusters, in: 32nd International Electric Propulsion Conference in Wiesbaden, Germany, 2011 (Paper IEPC-2011-101).

[161] L. Brieda, S. Pai, M. Keidar, Kinetic analysis of electron transport in a cylindrical Hall Thruster, IEEE Trans. Plasma Sci. 39 (11) (2011) 2946−2947.

[162] R.A. Spores, R.B. Cohen, M. Birkan, The USAF electric propulsion program, in: Proc. 25th International Electric Propulsion Conference, vol. 1, Worthington, OH, 1997, p. 1.

[163] M. Birkan, Formation flying and micro-propulsion workshop, Lancaster, CA, 1998.

[164] P.J. Turchi, Directions for improving PPT performance, Proceeding of the 25th International Electric Propulsion Conference, vol. 1, 1988, pp. 251−258. The Electric Rocket Propulsion Society, Worthington, OH, 1997.

[165] C.A. Scharlemann, R. Corey, I.G. Mikellides, P.J. Turchi, P.G. Mikellides, Pulsed plasma thruster variations for improved mission capabilities, in: 36th Joint Propulsion Conference (JPC) Huntsville, AL, 2000, American Institute of Aeronautics and Astronautics, Washington, DC (AIAA Paper-00-3260).

[166] E.Y. Choueiri, System optimization of ablative pulsed plasma thruster for stationkeeping, J. Spacecraft Rockets 33 (1) (1996) 96−100.

[167] M. Keidar, I.D. Boyd, Device and plume model for an electrothermal pulsed plasma thruster, in: 36th Joint Propulsion Conference (JPC) Huntsville, AL, 2000, American Institute of Aeronautics and Astronautics, Washington, DC (AIAA Paper 2000-3430).

[168] J.W. Dunning, S. Benson, S. Oleson, NASA's electric propulsion program, in: 27th International Electric Propulsion Conference, Pasadena, CA, 2001 (IEPC-01-002).

[169] C. Zakrzwsky, S. Benson, P. Sanneman, A. Hoskins, On-orbit testing of the EO-1 pulsed plasma thruster, in: 38th Joint Propulsion Conference, 2002, Indianapolis IN (AIAA-2002-3973).

[170] G.G. Spanjers, D. White, J. Schilling, S. Bushman, J. Lake, M. Dulligan, AFRL MicroPPT development for the TechSat21 flight, in: 27th International Electric Propulsion Conference, Pasadena, CA, 2001 (IEPC paper 2001-166).

[171] G.G. Spanjers, D.R. Bromaghim, J. Lake, M. Dulligan, D. White, J.H. Schilling, et al., AFRL microPPT development for small spacecraft propulsion, in: 38th AIAA Joint Propulsion Conference, Indianapolis, IN, 2002 (Paper AIAA-2002-3974).

[172] F.S. Gulczinski, M.J. Dulligan, J.P. Lake, G.G. Spanjer, Micropropulsion research at AFRL, in: 36th AIAA Joint Propulsion Conference, Huntsville, AL, 2000 (Paper AIAA-2000-3255).

[173] M. Keidar, I.D. Boyd, E.L. Antonsen, F.S. Gulczinski III, G.G. Spanjers, Propellant charring in pulsed plasma thrusters, J. Propulsion Power, 20 (6) (2004) 978.

[174] M. Keidar, I.D. Boyd, E.L. Antonsen, R.L. Burton, G.G. Spanjers, Optimization issues for a micro-pulsed plasma thruster, J. Propulsion Power 22 (1) (2006) 48−55.

[175] C. Phipps, J. Luke, Diode laser-driven microthrusters: a new departure for micropropulsion, AIAA J. 40 (2002) 310−317.

[176] D.A. Gonzalez, R.P. Baker, Microchip laser propulsion for small satellites, in: 37th AIAA/ASME/SAE/ASEE Joint Propulsion Conference, 2001 (AIAA Paper 2001-3789).

[177] I.D. Boyd, M. Keidar, Simulation of the plasma plume generated by a microlaser ablation thruster, in: C.R. Phipps (Ed.), High-Power Laser Ablation IV, vol. 4760, Proc. SPIE, 2002, p. 852.

[178] M. Keidar, I.D. Boyd, J. Luke, C. Phipps, Plasma generation and plume expansion for a transmission-mode micro-laser plasma thruster, J. Appl. Phys. 96 (1) (2004) 49−56.

[179] J. E. Polk, M. Sekerak, J. K. Ziemer, J. Schein, N. Qi, R. Binder et al., A theoretical analysis of vacuum arc thruster performance, in: 27th International Electric Propulsion Conference, Pasadena, CA, 2001 (Paper IEPC-2001-211).

[180] N. Qi, S. Gensler, R.R. Prasad, M. Krishnan, A. Visir, I.G. Brown, A pulsed vacuum arc ion thruster for distributed small satellite systems, in: 34th AIAA Joint Propulsion Conference, Cleveland OH, 1998 (AIAA-98-3663).

[181] N. Qi, J. Schein, R. Binder, M. Krishnan, A. Anders, J. Polk, Compact vacuum arc micro-thruster for small satellite systems, in: 37th Joint Propulsion Conference, Salt Lake City, UT, 2001 (AIAA Paper 2001-3793).

[182] J. Schein, N. Qi, R. Binder, M. Krishnan, J.K. Ziemer, J.E. Polk, A. Anders, Low mass vacuum arc thruster system for station keeping missions, in: 27th International Electric Propulsion Conference, Pasadena, CA, 2001 (IEPC-01-228).

[183] M. Au, J. Schein, A. Gerhan, K. Wilson, B. Tang, M. Krishnan, Magnetically enhanced vacuum arc thruster (MVAT), AIAA-2004-3618, in: 40th AIAA/ASME/SAE/ASEE Joint Propulsion Conference and Exhibit, Fort Lauderdale, Florida, July 11−14, 2004.

[184] M. Keidar, J. Schein, Modeling of a magnetically enhanced vacuum arc thruster, in: 40th AIAA Joint Propulsion Conference, Fort Lauderdale, FL, 2004 (Paper AIAA − 04-4117).

[185] M. Keidar, J. Schein, K. Wilson, A. Gerhan, M. Au, B. Tang, et al., Magnetically enhanced vacuum arc thruster, Plasma Source Sci. Technol. 14 (2005) 661−669.

[186] R. Tanberg, On the cathode of an arc drawn in vacuum, Phys. Rev. 35 (1930) 1080.

[187] R. Robertson, The force on the cathode of a copper arc, Phys. Rev. 53 (1938) 578.

[188] H. Marks, I.I. Beilis, R.L. Boxman, Measurement of the vacuum arc plasma force, IEEE Trans. Plasma Sci. 37 (7) (2009) 1332−1337, Part 2.

[189] A.M. Dorodnov, Pulsed metallic plasma generators Sov. Phys. Tech. Phys. 23 (9) (1978) 1058−1064.

[190] R. Dethlefsen, Performance measurements of a pulsed vacuum arc thruster, AIAA J. 6 (1968) 1197−1199.

[191] N. Qi, S. Gensler, R.R. Prasad, M. Krishnan, A. Visir, I.G. Brown, A pulsed vacuum arc ion thruster for distributed small satellite systems, in: 34th Joint Propulsion Conference, 1998 (AIAA-Paper-98-3663).

[192] M. Keidar, I. Beilis, R. Boxman, S. Gold-smith, Voltage of the vacuum Arc with a ring anode in an axial magnetic field, IEEE Trans. Plasma Sci. 25 (4) (1997) 580−585.

[193] J. Schein, A. Anders, R. Binder, M. Krishnan, J.E. Polk, N. Qi, J. Ziemer, Inductive energy storage driven vacuum arc thruster, Rev. Sci. Instrum. 72 (2002) 925−927.

[194] J. Schein, M. Krishnan, J. Ziemer, J. Polk, Adding a "Throttle" to a clustered vacuum arc thruster, in: Nanotech Conference, 2002 (AIAA paper 2002-5716).

[195] J. Schein, A. Gerhan, F. Rysanek, M. Krishnan, Vacuum arc thruster for cubesat propulsion, in: 28th International Electric Propulsion Conference, Toulouse, France, 2003 (IEPC-0276, 28th IEPC).

[196] A.S. Gilmour, Concerning the feasibility of a vacuum arc thruster, in: A.I.A.A. Fifth Electric Propulsion Conference, San Diego, CA, 1966 (AIAA Paper 66-202).

[197] A.S. Gilmour, R.J. Clark, H. Veron, Pulsed vacuum-arc microthrusters, in: AIAA Electric Propulsion and Plasmadynamics Conference, Colorado Springs, CO, 1967 (AIAA Paper 67-737).

[198] A. Anders, S. Anders, I.G. Brown, Transport of vacuum arc plasmas through magnetic macroparticle filters, Plasma Sources Sci. Technol. 4 (1995) 1−12.

[199] M. Au, J. Schein, A. Gerhan, K. Wilson, B. Tang, M. Krishnan, Magnetically enhanced vacuum arc thruster (MVAT), in: 40th AIAA Joint Propulsion Conference, Fort Lauderdale, FL, 2004 (AIAA Paper 2004-3618).

[200] I.I. Beilis., Vacuum arc cathode spot grouping and motion in magnetic fields, IEEE Trans. Plasma Sci. 30 (6) (2002) 2124−2132.

[201] B.E. Djakov, R. Holmes, J. Phys. D Appl. Phys. 7 (1974) 569.

[202] I.I. Beilis, B.E. Djakov, B. Juttner, Pursch, Structure and dynamics of high-current arc cathode spots in vacuum, J. Phys. D: Appl. Phys. 30 (1) (1997) 119−130.

[203] D. Murphree, Phys. Fluids 13 (1747) (1970) 1970.

[204] G. Ecker, K.G. Muller, J. Appl. Phys. 29 (1958) 1606.

[205] T. Zhuang, A. Shashurin, M. Keidar, I.I. Beilis, Circular periodic motion of plasma produced by a small-scale vacuum arc, Plasma Sources Sci. Technol. 20 (2011) 015009.

[206] C.W.J. Kimblin, Appl. Phys. 44 (1973) 3074.

[207] T. Zhuang, et al., J. Prop. Power (2012) (in review).

[208] T. Zhuang, A. Shashurin, I.I. Beilis, M. Keidar, Ion velocities in a micro-cathode arc thruster, Phys. Plasmas 19 (2012) 063501.

[209] C.A. Scharemann, T.M. York, Pulsed plasma thruster using water propellant, in: 39th AIAA/ASME/SAE/ASEE Joint Propulsion Conference, Huntsville AL, 2003 (Paper AIAA-2003-5022).

[210] D. Simon, H.B. Land III, Micro pulsed plasma thruster technology development, in: 40th AIAA/ASME/SAE/ASEE Joint Propulsion Conference, Fort Lauderdale, 2004 (Paper AIAA-2004-3622).

[211] Tajmar, MEMS Indoum F.E.E.P. thruster: manufacturing study and first prototype results, in: 40th AIAA/ASME/SAE/ASEE Joint Propulsion Conference, Fort Lauderdale, 2004 (AIAA-2004).

[212] M. Tajmar, K. Marhold, S. Kropatsch, Three-dimensional In-FEEP plasmadiagnostics, in: Proc. International Electric Propulsion Conference, Toulouse, France, 2003 (IEPC-2003-161).

[213] E. Chesta, D. Nicolini, D. Robertson, G. Saccoccia, Experimental studies related to field emission thruster operation: emission impact on solar cell performances and neutralization electron backstreaming phenomena, in: Proc. International Electric Propulsion Conference, Toulouse, France, 2003 (IEPC-2003-102).

[214] M. Tajmar, A. Genovese, Appl. Phys. A76 (2003) 1003.

[215] L. Rayleigh, Math. Soc. 10 (1878) 4.

[216] M. Faraday, Phil. Trans. Roy. Soc. 121 (1831) 299.

[217] V.V. Vladimirov, V.E. Badan, V.N. Gorshkov, Surf. Sci. 266 (1992) 185.

[218] G.L.R. Mair, J. Phys. D Appl. Phys. 29 (1996) 2186.

[219] J.M. Crowley, J. Appl. Phys. 48 (1) (1977) 145.

[220] J. Mueller, et al., An overview of MEMS-base micropropulsion development at JPL, in: Third International Symposium on Small Satellites for Earth Observation, IAA, Berlin, Germany, 2001.

[221] V. Khayms, M. Martinez-Sanchez, Design of a miniaturized hall thruster for microsatellites, in: 32nd AIAA/ASME/SAE/ASEE Joint Propulsion Conference, Buena Vista FL, 1996 (AIAA Paper 96-3291).

[222] Y. Raitses, N.J. Fisch, Phys. Plasmas 8 (2001) 2579.

[223] I.G. Mikellides, et al., J. Spacecraft Rockets 39 (6) (2002) 894.

[224] D.B. VanGilder, I.D. Boyd, M. Keidar, Particle simulation of a Hall thruster plume, J. Spacecraft Rockets 37 (1) (2000) 129−136.

[225] S. Pullins, R.A. Dressler, Y. Chiu, D.J. Levandier, Ion dynamics in hall effect and ion thrusters: Xe Xe symmetric charge transfer, in: 36th AIAA Joint Propulsion Conference, Huntsville, AL. 2000 (AIAA Paper 2000-0603).

[226] I.D. Boyd, Review of hall thruster plume modeling, J. Spacecraft Rockets 38 (2001) 381−387.

[227] L.B. King, A.D. Gallimore, 1996 (AIAA Paper 96-2712).

[228] C.R. Phipps, J.R. Luke, G.G. McDuff, A diode-laser-driven microthruster, in: 27th International Conference on Electric Propulsion, Pasadena, CA, 2001, The Electric Rocket Propulsion Society, Worthington, OH, 2001 (Paper IEPC-01-220).

[229] C.R. Phipps, J.R. Luke, G.G. McDuff, A Laser- Ablation-based micro-rocket, in: Proc. 33rd AIAA Plasmadynamics and Lasers Conference, Maui 2002, American Institute of Aeronautics and Astronautics, Washington, DC, 2002 (Paper AIAA 2002-2152).

[230] C.R. Phipps, J.R. Luke, G.G. McDuff, T. Lippert, Appl. Phys. A77 (2003) 193.

[231] T.E. Itina, W. Marine, M. Autric, J. Appl. Phys. 82 (**7**) (1997) 3536.

[232] S.R. Franklin, R.K. Thareja, Appl. Surf. Sci. 177 (2001) 15.

[233] G.A. Bird, Molecular Gas Dynamics and the Direct Simulation of Gas Flows, Clarendon Press, Oxford, 1994.

[234] A. Dalgarno, M.R.C. McDowell, A. Williams, Proc. Royal Soc. Lond. 250 (1958) 411.

[235] I.D. Boyd, M. Keidar, W. McKeon, J. Spacecraft Rockets 37 (2000) 399.

[236] S. Sakabe, Y. Izawa, Simple formula for the cross sections of resonant charge transfer between atoms and their ions at low impact velocity, Phys. Rev. A General Phys. 45 (3) (1992) 2086−2089.

[237] C.K. Birdsall, A.B. Langdon, Plasma Physics via Computer Simulation, Adam Hilger Press, Bristol, UK, 1991.

[238] D.J. Lichtenwalner, O. Auciello, R. Dat, A.I. Kingon, J. Appl. Phys. 74 (1993) 7497.

[239] J.C.S. Kools, J. Appl. Phys. 74 (1993) 6401.

[240] D. Kusamoto, K. Mikami, K. Komurasaki, A.D. Gallimore, Trans. Japan Soc. Aeron. Space Sci. 40 (1998) 238.

[241] M. Keidar, I.D. Boyd, I.I. Beilis, Electrical discharge in the Teflon cavity of a co-axial pulsed plasma thruster, IEEE Trans. Plasma Sci. 28 (2000) 376−385.

[242] A. Dalgarno, M.R.C. McDowell, A. Williams, The modalities of ions in unlike gases, Proc. Royal Soc. Lond. 250 (April) (1958) 411−425.

[243] G.A. Popov, M.N. Kazeev, V.F. Kozlov, Two dimensional numerical simulation of co-axial APPT, in: 27th International Electric Propulsion Conference, Pasadena, CA, 2001 (IEPC paper 2001-159).

[244] K.V. Brushlinskii, A.I. Morozov, Calculation of two-dimensional plasma flows in channels, in: M.A. Leontovich (Ed.), Rev. Plasma Physics, vol. 8, Consultants Bureau, New York, NY, 1980.

[245] M. Keidar, I.D. Boyd, E.L. Antonsen, G.G. Spanjers, Electromagnetic effects in the near field plume exhaust of a micro-pulsed plasma thruster, J. Propul. Power 20 (6) (2004) 961−969.

[246] G.G. Spanjers, K.A. McFall, F. Gulczinski III, R.A. Spores, in: Investigation of Propellant Inefficiencies in a Pulsed Plasma Thruster, 1996 (AIAA Paper 96-2723).

[247] E.L. Antonsen, Herriott cell interferometry for pulsed plasma density measurements, MS Thesis, University of Illinois at Urbana-Champaign, 2001.

[248] E. Antonsen, R. Burton, G.G. Spanjers, High resolution laser diagnostics in millimeter-scale micro pulsed plasma thrusters, in: 27th International Electric Propulsion Conference, Pasadena, CA, 2001 (IEPC paper 2001-157).

[249] M. Keidar, I.D. Boyd, E.L. Antonsen, G.G. Spanjers, Progress in development of modeling capabilities for a micro-pulsed plasma thruster, in: 39th AIAA Joint Propulsion Conference, Huntsville, AL, 2003 (Paper AIAA-03-5166).

[250] M. Keidar, I.D. Boyd, I.I. Beilis, Ionization and ablation phenomena in an ablative plasma accelerator, Journal of Applied Physics 96 (10) (2004) 5420−5428.

[251] M. Keidar, Anodic plasma in Hall thrusters, J. Appl. Phys. 103 (2008) 053309.

[252] I.I. Beilis, Modeling of a microscale short vacuum Arc for a space propulsion thruster, IEEE Trans. Plasma Sci. 36 (5) (2008) 2163−2166, Part 1.

Plasma Nanoscience and Nanotechnology

6.1 Plasmas for nanotechnology

6.1.1 Definitions

Nanoscience is defined as a research field that encompasses objects having dimensions smaller than 100 nanometers. Nanoscience addresses basic organization principles of the nanoscopic objects and describes their unique properties. Objects at this scale exhibit very different properties and physics than that of the bulk objects of the same material. At this scale, quantum mechanic effects become very important. *Nanotechnology* deals with synthesis of nanoscopic objects and devices as well as various applications. Nanoscopic objects could be formed from the precursors in various states (atoms, molecules, clusters, exited states, etc.). *Plasma nanoscience and nanotechnology* deal with synthesis of nanoparticles from the ionized gas or plasma state.

In general, *nanoscience* and *nanotechnology* study nanoscopic objects used across many scientific fields, such as chemistry, biology, physics, materials science, and engineering. By encompassing nanoscale science, engineering, and technology, nanotechnology involves imaging, measuring, modeling, and manipulating matter at this length scale.

Historically, many ideas and concepts behind *nanoscience* and *nanotechnology* as they are known today started with a talk entitled "There's Plenty of Room at the Bottom" by physicist Richard Feynman at an American Physical Society meeting at the California Institute of Technology on December 29, 1959, long before the term nanotechnology was used [1]. In that, now famous talk, Feynman described a process in which scientists would be able to manipulate and control individual atoms and molecules.

This section is devoted to description of the plasma-based nanoscience and nanotechnology, which is emerging as one of the promising field.

6.1.2 Plasma-based synthesis of nanoparticles

Deterministic synthesis of nanoparticles and nanodevices is the most pressing demand of today's nanotechnology. At the elementary level, this means high fidelity control over the precursor density and energy distribution as well as a high degree of control over precursor position. In this respect, presence of the charged particles allows particle manipulation using electric and magnetic fields.

As a result, plasma-based nanoparticle synthesis and fabrication could offer a better degree of controllability in the size, shape, and pattern uniformity as compared to neutral-based process such as chemical vapor deposition (CVD) [2]. An observation made in numerical simulations demonstrates this statement (Figure 6.1). One can see that the nanotips grown on plasma-exposed surfaces (*plasma-enhanced CVD*) are much taller and sharper than those grown by the CVD process under the same deposition conditions.

Plasma-based techniques were demonstrated to be effective in synthesis of various nanomaterials such as carbon nanotubes (CNTs), nanofibers, graphene, graphene nanoribbons, graphene nanoflakes, nanodiamond and related carbon-based nanostructures; metal, silicon, and other inorganic nanoparticles and nanostructures; soft organic nanomaterials; nanobiomaterials; biological objects and nanoscale plasma etching [3]. To this end, various types of plasmas and plasma reactor systems are utilized in nanotechnology, including low-temperature nonequilibrium plasmas at low and high pressures, thermal plasmas, high-pressure microplasmas, plasmas in liquids and plasma—liquid interactions, high-energy-density plasmas, and ionized physical vapor deposition to name just a few.

6.1.3 Synthesis of carbon nanoparticles

Carbon is one of the few elements known from antiquity and the one that is mostly used nowadays. There are several allotropes of carbon in the world, which can be categorized by dimensions, such as diamond and graphite in three dimensions, graphene in two dimensions, CNTs in one dimension, and fullerene in zero dimension. Carbon with sp^3 hybridization will form a tetrahedral lattice, thus giving rise to diamond. Carbon with sp^2 hybridization may form graphite, graphene, CNT, or fullerene, depending on the conditions of their formation. Different structures and hybridizations of carbon atoms can determine the unique properties of each carbon allotropes.

Among possible forms of carbon, CNT and graphene attracted significant interest nowadays. CNT was first discovered in carbon deposits by the arc-discharge method by Iijima in 1991 [4], and the advance of graphene appears when Novoselov et al. [5] were able to extract it from bulk graphite by micromechanical cleavage or the Scotch tape approach in 2004. CNTs have unique structures with cylindrical walls of carbon. According to the number of wall layers, CNT can be categorized as *single-walled carbon nanotubes* (SWCNTs) and *multiwalled carbon nanotubes* (MWCNTs), while graphene is one-atom-thick hexagonal-lattice planar carbon layer.

Most SWCNTs have a diameter of around few nanometers, with a tube length that could be millions of times longer. SWCNT can be formed by rolling up a one-layer graphene into a cylinder.

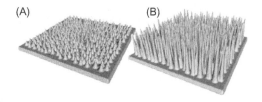

(A)   (B)

FIGURE 6.1

Developed carbon nanotip patterns (A) grown by CVD and (B) grown by plasma-enhanced CVD in a plasma with density of about 3.0×10^{18} m^{-3}.

The way the graphene sheet is wrapped is represented by a pair of indices (n,m) [6]. The integers of n and m denote the number of unit vectors along two directions in the honeycomb crystal lattice of graphene shown in Figure 6.2. The (n,m) nanotube naming scheme is a vector (C_h) in an infinite graphene sheet that describes how to "roll up" the graphene sheet to make the nanotube as shown in Figure 6.2. An SWCNT can be imagined as graphene sheet rolled at a certain *chiral angle* with respect to a plane perpendicular to the tube's long axis. Consequently, SWCNT can be defined by its diameter and chiral angle. The chiral angle can range from $0°$ to $30°$. The SWCNT with index $m = 0$ are named *zigzag nanotubes*, and the SWCNT with index $n = m$ are called *armchair nanotubes*. The electrical property of SWCNT is determined by the indices of n and m. If the difference of indices n and m is integral multiple of three, the SWCNTs have metallic properties. The others are semiconducting materials. It will be shown below that various synthesis techniques produce both metallic and semiconductor nanotubes.

The unique structures of SWCNT and graphene lead to excellent mechanical, electrical, and thermal properties. SWCNT and graphene appear to be one of the strongest and stiffest materials in terms of Young's modulus and tensile strength. This strength results from the covalent sp^2 bonds formed between the individual carbon atoms. The Young's modulus of SWCNT and graphene can reach over 1 TPa, which is tens of times higher than that of aluminum [7]. The recent measurements have shown that graphene has a breaking strength which is 200 times greater than steel [8]. Metallic SWCNT can carry an electric current density of 4×10^9 A/cm^2 in theory, which is more than 1000 times greater than those of metals such as copper [9]. Intrinsic graphene is a semimetal or zero-gap semiconductor. Experimental results from transport measurements show that graphene has remarkable electron mobility at room temperature, with reported values in excess of

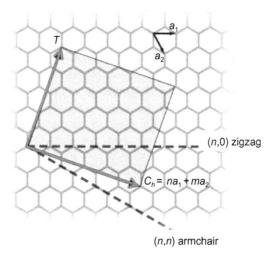

FIGURE 6.2

The (n,m) nanotube naming scheme can be thought of as a vector (C_h) in an infinite graphene sheet that describes how to "roll up" the graphene sheet to make the nanotube. T denotes the tube axis, and a_1 and a_2 are the unit vectors of graphene in real space.

15,000 cm^2/V/s. The corresponding resistivity of the graphene sheet would be 10^{-6} Ω cm, which is less than the resistivity of silver [10]. SWCNTs have the thermal conductivity of 300 W/(mK) in axial direction. The measurement by a noncontact optical technique demonstrated that the near-room temperature thermal conductivity of graphene was measured to be between $(4.84 \pm 0.44) \times 10^3$ and $(5.30 \pm 0.48) \times 10^3$ W/(mK) [11], which is significantly higher than that of SWNT.

6.1.3.1 Carbon nanotubes

CNTs are tubular carbon-based structures that are produced from graphitic carbon. Since their discovery, an interest in CNTs has been stimulated by their unique mechanical, thermal, and electrical properties, and various potential applications that exploit these properties, such as field-emission displays [12,13], nanoelectronics [14], hydrogen storage [15], and chemical gas sensors [16]. SWNTs have the greatest stiffness, both in tension and bending [17]. The combination of stiffness and toughness makes single-wall nanotubes (SWNTs) the strongest known fibers. Thus one of the important applications of SWNTs is creation of new materials. Continuum mechanics calculations have shown that SWNTs are among the strongest tubular tensile members available [18]. Several applications that purport to exploit the remarkable mechanical, electrical, and thermal properties of SWNT have been investigated. While the preponderance of applications involving nanotubes are in fields that can be broadly categorized as "life sciences," real challenges associate with making practical, useful materials of large amounts with superior and unusual mechanical, electrical, and thermal properties.

Several techniques have been developed for CNT synthesis such as arc-discharge, chemical vapor deposition, and laser ablation [19,20]. A progress in arc-discharge SWNT synthesis was motivated by Journet et al. [21] who showed that the SWNTs can be efficiently produced by this technique. The use of anodic arc for CNT synthesis is based on ablation of the anode material and deposition of the ablated material on the cathode. Two different textures and morphologies can be observed in the cathode deposit: the gray outer shell and dark-soft inner core deposit. MWNTs as well as graphitic particles are found typically in the inner core [22]. SWNTs produced by the anodic arc discharge are found in a "collaret" around the cathode deposit, cloth-like soot suspended in the chamber walls, and the weblike structure suspended between cathode and walls [22]. MWNTs and SWNTs produced in arc discharge are dependent on gas background and arc conditions [23,24]. In the He$-$Ar mixture, it was found that the argon mole fraction affects the SWNT diameter [25], while SWNT diameter was found to be fairly independent of pressure in the pure helium environment [26]. In addition to SWNT diameter, two other parameters are important for SWNT applications, namely chirality and aspect ratio. The chiral angle (as defined at Figure 6.2) determines whether SWNT has metallic or semiconductor electrical conductivity [27]. Some challenges regarding control of the SWNTs chirality and radius were reported [26,28,29]. The issues related to large-scale and high-purity synthesis of SWNT by arc discharge are very important objectives nowadays [30$-$34].

Among several methods for preparing CNTs, arc discharge is the most practical one for scientific and technological purposes due to the number of advantages in comparison with other techniques. Firstly, arc-discharge method yields highly graphitized tubes with very small defects, because the manufacturing process occurs at a very high temperature (which is about 1200$-$1500 K, see next two sections) and arc-grown SWNTs demonstrate the highest time of emission capability degradation than those produced by other techniques [35]. Secondly, nanotubes produced in arc usually demonstrate a high flexibility, thus eventually providing higher strength characteristics [36].

The lack of control of the SWNT growth in arc is the main disadvantage of the arc-discharge technique for nanotube. The controllability and flexibility of the arc-plasma-based process may be significantly improved by the use of a magnetic field, which strongly influences the plasma parameters [37]. It was shown that the high-purity multiwall nanotubes (MWNTs) can be grown in the magnetically enhanced arc discharge [38]. It was also demonstrated that the use of the magnetic-field-enhanced arc discharge is very promising for the production of the long SWNTs [39].

SWNTs produced in the arc has aspect ratios typically in the range of 100−1000 [24,39], while there is a tremendous interest in production of ultralong SWNTs (with aspect ratio greater than 10^5) which will enable new types of Micro-elecro-mechanical and nano-electromechanical (MEMS/NEMS) systems, such as microelectric motors, and can act as a nanoconducting cable [40]. In addition, it was demonstrated that the thermal conductivity of an individual SWNTs increases with length [41], thus making ultralong SWNTs an ideal structure for thermal control.

6.1.3.2 Graphene

Graphene is a one-atom-thick planar sheet of sp^2-bonded carbon atoms that are densely packed in a honeycomb crystal lattice [42]. This new material, which combines aspects of semiconductors and metals, could be a leading candidate to replace silicon in applications ranging from high-speed computer chips to biochemical sensors. Large-area graphene films are of enormous interest for electronic and optical applications, namely, their potential was recently demonstrated for field effect transistors (FETs) [43−48] (and in particular for transistors operating at GHz frequencies [49]) and for conductive films on transparent plastic electrodes required for development of concept of flexible and stretchable electronics [50−53].

Single-layer graphene was first synthesized using regular Scotch tape by mechanical exfoliation of layers from bulk graphite [42,54,55], and its creators were awarded by Nobel Prize in Physics for 2012 "for groundbreaking experiments regarding the two-dimensional (2D) material graphene" [56]. Since the method utilized originally by Geim and Novoselov is extremely expensive and characterized by low output, a very active search for more efficient ways of graphene synthesis was facilitated. Among other methods created in following years, one can mention epitaxial growth on SiC, CVD, and colloidal suspensions [45,57−62]. Generally, these methods allow synthesizing two types of the graphene, namely large-area pristine graphene films (i.e., graphene films on Si wafers) and micron-sized graphene platelets (bulk graphene).

The first type of graphene, namely large-area uniform graphene films, is synthesized using epitaxial growth and CVD, and being applied in ultrahigh-speed, low-power graphene-channel FETs, and transparent electrodes [45,63−66]. The graphene films are being first synthesized on the hot metal substrates and then being transferred to desired substrate, e.g., Si wafer or transparent electrode. Currently the main challenges in this field are creation of high-quality continuous and uniform graphene films in CVD process (greater than 8 inches in diameter), and reduction of damages to the graphene film at its transfer [45,63]. Currently there are two widely recognized and well-studied mechanisms of CVD synthesis of graphene. First mechanism is dominant when materials characterized by high solubility of carbon are utilized as a growth substrate, e.g., Ni [52,66]. In this case, the carbon atoms dissolve in the substrate material first and then precipitate on the substrate surface during the substrate's cooling. These precipitated atoms form graphene film on the substrate surface. Such grown graphene films are usually limited to the grain size of substrate and contains several layers. Second mechanism of graphene growth is employed when materials characterized by low solubility of carbon are utilized for the growth substrate, e.g., Cu [45]. As proposed by Ruoff et al. [45], the graphene growth is surface catalyzed in this case. It was shown that synthesis on Cu substrates is basically self-limited on production of single-layer graphene film of extraordinary quality. Various conditions of the growth substrate were shown to have significant impact on the properties of the final graphene film. Indeed, the crystal structure of the substrate plays an important role in quality of graphene films grown on that substrate and Cu(111) was indicated to yield best quality uniform, monolayer graphene growth [64]. It was shown that degree of substrate polishing changes the homogeneity and electronic transport properties of the grown graphene film and recommendation to utilize the electropolished metal surfaces was made [65].

The second type of graphene, the bulk graphene (graphene platelets, flakes), is usually characterized by several layers (with thickness of about 2—20 nm) graphene pieces having characteristic sizes of about microns. The main application of bulk graphene is electrochemical energy storage devices including ultracapacitors and fuel cells. Currently, the main method for synthesis of bulk graphene is chemical exfoliation, where chemicals are utilized to separate the graphene sheets from the piece of graphite. Current production capability are quite limited and estimated to be around several tens of tons of material annually worldwide [67].

Application of plasmas is a well-known tool to improve properties of the CVD grown films and it makes up the vast category of various plasma-enhanced and plasma-assisted CVD techniques [67,68]. The benefits of the plasma-enhanced CVD techniques are resulted from significant enhancement of the reactivity of species involved in synthesis by means of plasmas. In particular, this approach allows significantly reducing the synthesis temperature (usually to about 300°C), improving adhesion of the films to the substrate and providing higher deposition rates [69,70]. Potential utilization of plasmas in graphene synthesis was recently demonstrated by showing that temperature of synthesis can be reduced by several hundred degrees centigrade from about 1000°C at conventional CVD to about 500—650°C in plasma-enhanced CVD [71,69].

Recently, a new method of graphene synthesis in magnetically controlled anodic arc discharge was discovered [70]. Method utilizes pure carbon vapor ablation from the solid carbon electrode by means of arc discharge in the atmosphere of helium and its following delivery to the heated growth substrate. Preliminary studies demonstrated that a few-layer graphene of superior quality can be synthesized with high efficiency in this plasma-assisted process.

6.1.4 Controlled synthesis of carbon nanostructures in arc plasmas: theoretical premise

Several models of a cathodic carbon arc were developed in past dealing with electrode phenomena [72—74] or interelectrode plasma [75]. A 1D model (in axial direction) of the SWNT formation was developed and the SWNT growth rate in an anodic arc discharge was calculated [76]. According to the existed model predictions, the nanotube formation occurs in the region of relatively small plasma temperature (1300—1800 K [77]) where carbon reacts to form large molecules and clusters. No detailed model for relationship between the discharge parameters and SWNT formation was developed as mentioned in a review paper [78]. A model of SWNT interaction with discharge plasma and SWNT formation in the cathode region (collaret) was developed [79]. It was shown that under certain conditions, SWNT can be deposited on the cathode surface. This process depends on SWNT charging in the arc plasma. In turn, the charging phenomena depend on the electron temperature.

Gamaly and Ebbesen [80] proposed that the bimodal carbon velocity distribution (ions with drift velocity and isotropic neutrals) determines the nanotube creation process near the cathode. They suggested that isotropic distribution leads to fullerene formation, while directed flux results in nanotubes. Iijima et al. [81] proposed an open-ended growth model. In this model, carbon atoms and small carbon clusters add on to the reactive dangling bonds at the edges of the open-ended nanotubes. Other researchers argue that CNTs are elongated by electrostatic forces along the electric field near the cathode [82,83]. However, it seems like a high-resolution transmission electron microscopy (TEM) analysis does not support this hypothesis. Alternatively, a two-step growth model has been proposed [84]. According to this model, different carbon structures are formed first. Then, during the cooling process, the graphitization occurs from the surface toward the interior of the assemblies. Several workers developed a growth model of SWNT explaining the root growth of nanotube bundles emerging from catalyst particles [85]. These models include a catalyst phase diagram of carbon metal. The investigation of mechanism for the catalytic synthesis methods of CNTs in arc plasma is still subject to ongoing research. The vapor−liquid−solid (VLS) mechanism was first proposed by Wagner and Ellis and can be utilized to demonstrate the growth model of SWCNT [86]. According to the VLS framework, Ding et al. [87] simulated the nucleation processes of SWCNT associated with catalyst particles by molecular dynamics method, presenting the dependent relationship between diameters of SWCNT and catalyst particles theoretically. Based on the analysis of diffusion model of carbon atom and calculation by Monte Carlo technique, Keidar et al. [88] also demonstrated that the SWCNT diameter is determined by the size of the molten catalyst core in arc discharge. Chiang and Sankaran [89] reported the very important experimental results suggesting that the variation of the element composition of Ni_xFe_{1-x} catalyst particles strongly affects the SWCNT chirality. The link between the composition-dependent catalyst structure and the chirality of SWCNT would improve the *in situ* controllability of SWCNT synthesis. The results indicate the important role of the catalyst particles in the SWCNT synthesis.

6.1.4.1 SWNT interaction with arc plasma

In this section, a simple model of SWNT interactions with arc plasma and predictions based on this model will be described.

Typically in the interelectrode gap of the arc-discharge plasma, temperature varies from about 5000 K in the center of the channel down to 1000 K at the periphery [90]. Probability of atomic collisions and therefore nanotube seed formation is higher in the center of the interelectrode gap, i.e., in the region with highest carbon atoms and ions density. Recall that SWNT seeds formed in the plasma are subject to interaction with plasma particles that include charge, momentum, and energy transfer. As a result of these interactions, high heat fluxes may lead to overheating and preventing formation of the stable nanostructures. Thus, the nanotube formation occurs in the region of relatively small plasma temperature (1300−1800 K) where carbon reacts to form large molecules and clusters as will be shown in the following. Although the mechanism of the formation and growth of SWNTs in an arc discharge was studied for a decade [30], location of the region in arc discharge in which SWNT synthesis occurs and the temperature range favorable for SWNT growth remains unclear. According to previous work [30,45], the nanotube formation occurs on the periphery of an arc column at a moderate temperature range of 1200−1800 K. Other studies suggested that it is the cathode sheath adjacent to hot arc column (~5000 K) is the arc region where the nanotube growth occurs [78,91−93]. Recall that in the cathode sheath region, the temperature might be well above the reported critical temperatures of thermal stability of the nanotubes. In this respect, a question about possible CNT growth in cathode sheath region remains open. Moreover, there are no consistent data on the thermal stability of SWNT in the arc discharge, including the temperature ranges of SWNT synthesis and destruction. Thermal stability of SWNTs produced in helium arc was studied [94]. Using a furnace, temperature conditions closely resembling the natural conditions of SWNT growth in an arc were created. Based on these experimental data, it can be concluded that SWNT produced by an anodic arc discharge and collected in the web area outside the arc plasma most likely originated from the arc-discharge peripheral region, i.e., plasma−gas interface.

Let us calculate the residence time of SWNT cluster in the growth region. Carbon clusters diffuse with diffusion coefficient D_{SWNT} from the region of origin without any chemical reactions [95]. In the diffusion approximation, one can determine the diffusion coefficient of carbon clusters and SWNT as follows [95]

$$D_{SWNT} = \frac{4}{3} \frac{\sqrt{2}}{\pi^{3/2}} (1/m_{SWNT} + 1/m_{He})^{1/2} \frac{(kT)^{3/2}}{p(d_{SWNT} + d_{He})^2} \tag{6.1}$$

where m_{SWNT} and m_{He} are SWNT and He molecular masses, respectively, T is the plasma temperature, p is the pressure, and d_{SWNT} and d_{He} are effective diameters of the SWNT seed and He molecule, respectively. In this formulation, the radial diffusion velocity of SWNT seed can be estimated as $V_{SWNT} \sim D_{SWNT}/R_a$, where R_a is the anode radius, which is the characteristic dimension of the plasma region. In addition, an initial SWNT velocity can be estimated from experimental measurements and it is about 0.01−0.5 m/s.

The charge transfer from the plasma to SWNT is due to electron and ion fluxes to SWNT seed and due to thermoionic emission from the SWNT:

$$\frac{dQ_{SWNT}}{dt} = I_i - I_e + I_{em} \tag{6.2}$$

where Q_{SWNT} is SWNT charge, I_i, I_e, I_{em} are ion, electron, and thermoionic current, respectively. The electron current is given by $I_e = Sj_e$, where S is SWNT surface area and j_e is the electron current density. The electron current density absorbed by SWNT depends upon SWNT potential with respect to the surrounding plasma: $j_e = j_{eo} \exp(-e\varphi_{SWNT}/kT)$ if $\varphi_{SWNT} < 0$ and $j_e = j_{eo} \exp(1 + e\varphi_{SWNT}/kT)$ in the opposite case, where j_{eo} is the electron thermal current density, φ_{SWNT} is the SWNT potential with respect to the plasma. The ion current density at the SWNT surface is given by $j_i = j_{io}(1 + \alpha)$ if $\alpha \geq 0$ and $j_i = j_{io}$ if $-1 < \alpha < 0$ and $j_i = 0$ if $\alpha < -1$, where $\alpha = -2e\varphi_{SWNT}/m_i V_i^2$ and $j_{io} = en_e V_i$ is the ion current density in the plasma and V_i is the ion velocity.

The current of thermoionic emission is given by Richardson−Duschman equation:

$$I_{em} = SAT_s^2 \exp\left(-\frac{e[\Phi + \varphi_{SWNT}]}{kT_s}\right) \tag{6.3}$$

where Φ is the work function and T_s is the SWNT temperature.

Following Ref. [96], the electric capacitance of the cylindrical particle is calculated as $C = 4\pi\varepsilon_o(L/\ln(2L/a))$, where L is SWNT length and a is the SWNT radius ($a = 1.4$ nm, [92]). The capacitance does not depend on the inner radius of SWNT since it is calculated between the SWNT and the surrounding plasma. In addition, we take into account that SWNT is a conductor (having either metallic or semiconductor properties), thus SWNT charge is equal: $Q_{SWNT} = C\varphi_{SWNT}$.

SWNT growth rate is determined by carbon atoms and ions precipitation to the nanotube surface and chemical reactions, which depends on the density of carbon species in the vicinity of SWNT and the electron temperature. It is assumed that influx of carbon ions and atoms to SWNT causes an increase in SWNT length. We further assume that growing SWNT has C–C spacing of about 1.4 A° [92]. Flux of the carbon atoms to the SWNT surface can be calculated as follows:

$$\Gamma = 0.25 n_C \sqrt{\frac{8kT}{\pi m}} \tag{6.4}$$

SWNT interaction with plasma in the interelectrode region leads to momentum transfer, which can be accounted as follows:

$$M_{SWNT} \frac{dV_{SWNT}}{dt} = F_D + Q_{SWNT}E \tag{6.5}$$

where F_D is the drag force, $F_D = \rho d_{SWNT}(V_i - V_{SWNT})^2$, ρ is the plasma density, and E is the electric field in the region of SWNT formation. Equation (6.5) is supplemented by equation for SWNT trajectory: $(dr/dt) = V_{SWNT}$. Initial velocity of SWNT, $V_{SWNT}(r = 0)$ is calculated from Eq. (6.1). System of equations (6.2–6.5) was solved to calculate SWNT trajectory in arc-discharge plasma and SWNT growth. These results are shown in Figure 6.4.

As it is mentioned above, one possibility to affect SWNT growth in the SWNT formation region is to apply an electric field. Due to the fact that SWNT accumulates charge in course of interaction

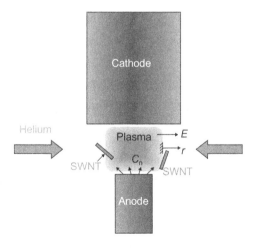

FIGURE 6.3

Schematic of the arc-discharge, plasma interactions with SWNT, and simulation geometry.

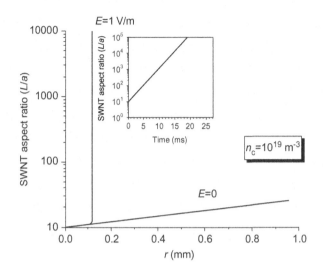

FIGURE 6.4

SWNT aspect ratio as a function of distance with electric field as a parameter. Insert shows SWNT aspect ration as a function of residence time.

with plasma, it is expected that this electric field may affect SWNT motion. In fact, simulations show that the electric field in the SWNT formation region has strong effect on SWNT motion. SWNT relative length (aspect ratio) is shown in Figure 6.4 with electric field in the SWNT growth region as a parameter. It can be seen that a relatively small electric field significantly affects SWNT growth and leads to a large SWNT aspect ratio in comparison to zero electric field case. This is due to trapping of SWNT in region of the preferable growth. In the enlarged section of Figure 6.4, one can see time evolution of SWNT charge and SWNT aspect ratio.

6.2 Magnetically enhanced synthesis of nanostructures in plasmas
6.2.1 Arc-discharge plasma system for synthesis of SWNT

A typical arc-discharge system consists of anode–cathode assembly installed in a stainless steel flanged chamber capped at both ends as shown in Figure 6.5. The arc discharge is sustained with a constant power supply, using a feedback connected to the linear drive of the anode and the power supply generating the arc. Linear drive allows to keep constant interlectrode gap of about 1 mm during the arcing time. The anode hole is packed with various metal catalysts. Quanta sizing and microscope examinations of arc-discharge products for equal arc runtime has revealed that the catalyst combination yielding the largest amount of nanotubes was Y−Ni in a 1−4 ratio [91]. The nanotube samples are typically produced at constant helium pressures ranging from 500 to 700 Torr and arc current ranging between 70 and 80 A. Magnetic field can be used to enhance arc-discharge technique. Schematically application of a nonuniform magnetic field is shown in Figure 6.6.

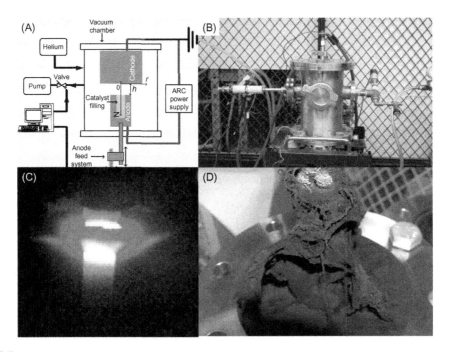

FIGURE 6.5

(A) Schematics of the arc discharge, (B) photo of the experimental setup, (C) typical photograph of the arc, and (D) image of the collected nanotube web.

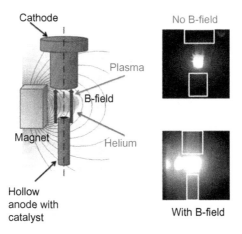

FIGURE 6.6

Schematics of magnetically enhaced arc discharge.

Photos show that magnetic field modifies arc discharge. Magnetic field leads to formation of a plasma jet that will be describe below.

6.2.2 Synthesis of SWCNTs in a magnetic field

Permanent magnets $(5 \times 5 \times 2.5)$ cm^3 with different strengths were used to create and vary the strength of a nearly axial magnetic field in the discharge gap of about 0.5 cm as shown schematically in Figure 6.6. Magnetic field varies in the range of 0.2–2 kG. Samples with SWCNTs collected from locations were subsequently analyzed in the solid state with SEM and Raman spectroscopy, and postprocessing into aqueous dispersion by ultraviolet–visible–near infrared (UV–Vis–NIR) and PL spectroscopy techniques.

Collected samples were analyzed under SEM and the length distribution of individual SWNT was measured from SEM images. SEM data indicate that the presence of a magnetic field makes a significant difference in the average length of the SWNTs produced by the arc discharge as shown in Figure 6.7. Typical SEM image used for length measurements is shown in the inset in Figure 6.7. It should be pointed out that in many cases, it was confirmed that we measured length of a single nanotube, while in other cases we refered to an image of SWNT bundle. Left inset shows a typical SEM image used for length measurements; right inset shows the TEM image of the SWNT and bundle of SWNT. TEM image also indicates that SWNTs in a bundle have the same length. Of the entire measurements of SWNT length taken without the magnetic field, 50% were under 600 nm and 90% were under 1300 nm; while 50% of the measurements taken from samples with the magnetic field were under 1100 nm and 90% were under 2600 nm. Thus, the presence of the magnetic field seemingly doubles the length of SWNTs produced [39].

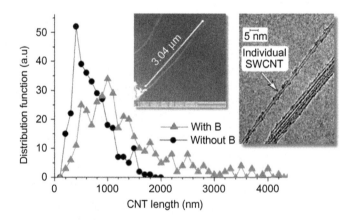

FIGURE 6.7

Comparison of the SWCNTs lengths distribution with and without magnetic field applied to the arc discharge. Inset shows a typical SEM image used for length measurements. Left inset shows a typical SEM image used for length measurements; right inset shows the TEM image of the single SWNT and a bundle of SWNTs.

A mathematical model describing the very complicated problem of the SWNT formation in arc plasma was developed [39,78,79,97,98]. It has been shown that the electrical charges influence the growth of nanostructures [2,99]. In the following, we outline the principle points of these models to explain the SWCNT length increase in the strong magnetic-field-enhanced arc plasmas. The detailed description of the SWCNT growth will be presented in the next section.

A nanotube immersed in the plasma accumulates an electric charge on its surface and eventually encloses by the sheath which thickness can be estimated as $\lambda = \gamma \sqrt{\varepsilon_0 T_e / e n_p}$, where ε_0 is the dielectric constant, T_e is the electron temperature, n_p is the plasma density, e is the electron charge, and γ is a constant in the range of $1-5$ [100]. In the sheath, there is an uncompensated electric charge which induced an electric field between the plasma bulk and the SWNT surface. In the calculations considered, it was assumed a maximum SWNT length of 5 μm that is in accordance with experimental data (Figure 6.7). Thus, for the typical plasma density ($10^{17}-10^{18}$ m^{-3}) and electron temperature (up to 1 eV) in the arc plasma, the sheath thickness is in the range of $15-25$ μm that significantly exceeds the SWNT length. While magnetic field can affect the sheath width [101], it does not affect the current collection and thus SWCNT growth. During the process of SWNT formation, the orientation of the nanotube is chaotic and changing in time, so the magnetic field cannot significantly decrease the electron current to the SWNT surface. As a result, the influence of the magnetic field on the sheath is moderate in this case.

In the sheath around SWNT, the ion motion is determined by the electrical field between SWNT and plasma bulk. The electric field is described by the Poisson equation for the electric potential $\Delta \varphi = \rho_e / \varepsilon_0$, where ρ_e is the density of electrical charge in the sheath. As a boundary condition for the Poisson equation, an equipotentiality of the entire SWNT surface was assumed, i.e., $\varphi(x,r,\alpha)|_{(x,R,\alpha)} = \Psi_{SWNT}$, where R is the SWNT radius. Ions enter the sheath with the Bohm velocity $\upsilon_B = \sqrt{T_e / m_i}$, where m_i is the ion mass. An ion trajectory in the sheath can be obtained by integrating a motion equation. More details on the electric field and ion motion influence on nanostructures can be found elsewhere [102,103].

Here, the following scenario of the SWNT growth in a plasma was implemented. It was assumed that the SWNT grow on the partially molten metal catalyst particle supplied to the plasma from ablated electrode. In plasma, the metal catalyst particle is a subject to the additional heating and ablation, which reduce the catalyst size, and then condition and molten the external layer creating a liquid shell. The carbon atom flux gets to the catalyst surface, diffuse through it, and eventually incorporate into the SWNT structure. An ion flux supplies carbon atoms to the SWNT and catalyst. Upon recombination, carbon adatoms migrate about the SWNT surface, eventually reach the molten catalyst shell or reevaporate to the plasma bulk (Figure 6.8). Today, the two main growth scenarios are mostly accepted: the VLS [104] and solid−liquid−solid (SLS) [105]. Both scenarios involve the carbon atom diffusion in the metal catalyst particle, and thus the process of the carbon supply to the external catalyst surface is a decisive factor that determines the SWNT growth kinetics. To calculate the carbon supply to the catalyst surface, we implemented a diffusion model which was used for simulation of the diffusion-driven growth of carbon nanostructures on surface [106]. For the ion motion calculations, a Monte Carlo technique to obtain an ion flux distribution over the SWNT−catalyst surface was used. The diffusion model is used to calculate adatom migration about SWNT surface and the carbon atoms diffusion in the molten catalyst [107].

Obtained ion flux distributions over the nanotube were used for simulation of the SWNT growth rates η (μm $\times$ s^{-1}). The results of the calculations are shown in Figure 6.9, with the plasma density as a parameter. We should point out that the SWNT growth rate strongly decreases with the SWNT length

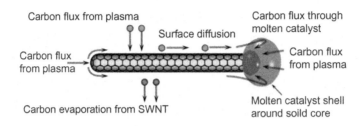

FIGURE 6.8

Growth of SWNT on molten metal catalyst particle in plasma. Carbon flux from plasma is nonuniformly distributed about SWNT and catalyst particle surface. Carbon adatoms diffuse to the catalyst end, and then incorporate into SWNT structure through molten catalyst shell.

Source: *Reprinted with permission from Ref. [39]. Copyright (2008) by American Institute of Physics.*

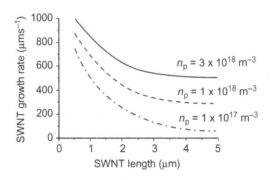

FIGURE 6.9

Dependence of SWNT growth rate η on SWNT length with plasma density as a parameter. SWNT diameter is 2 nm, catalyst particle diameter is 10 nm. The graph shows strong decrease of SWNT growth rate with the SWNT length.

Source: *Reprinted with permission from Ref. [39]. Copyright (2008) by American Institute of Physics.*

and increases with the plasma density. Let us try to interpret the results shown in Figure 6.9. Note that formation of new layers is not considered here; thus, the growth rate of the nanotube depends only on the total influx of the carbon atoms to the surface of catalyst particle, which in turn depends on the total carbon influx to the SWNT and catalyst surfaces, as well as on the influx distribution over the SWNT and catalyst. When an SWNT is short, its growth rate is determined by the total carbon influx and the adaom migration kinetics. An essential part of the carbon atoms gets into the catalyst shell and participate in the SWNT growth. With the SWNT length increasing, the carbon flux to catalyst decreases due to increased carbon loss by evaporation, thus causing the decrease in the SWNT growth rate.

6.2.3 Effect of magnetic field on SWNT chirality

Since the discovery of SWCNTs [4], significant efforts have been directed toward attempts to synthesize SWCNTs of controlled chiral angle as discussed in Section 6.1. In particular, interest in chirality control is driven by the strict requirement to have a narrow distribution of SWCNT diameters, or a small number of chiralities, for enabling nanoelectronic applications [108−110]. Recent works indicate that one of the key parameters for SWCNT chirality control is the initial characteristics of catalyst particle [89,111]. For CD techniques, Li et al. [111] demonstrated that changing the size of their Co-MCM-41 catalyst particle (by altering the synthesis temperature through the range of 550−950 °C) allows for control of the produced SWCNT diameters over the range from 0.6 to 2 nm. Chiang and Sankaran [89] reported that varying the composition of Ni_xFe_{1-x} catalyst particle strongly affects distribution of produced SWCNT chiralities, namely a decrease of x leads to narrower distribution of produced chiralities and a decrease in the mean SWCNT diameter. However, fine control of the chirality distribution through manipulation of the catalyst has proven to be highly demanding, and so alternative techniques for shaping the distribution during the production process are desirable.

Below tuning of the distribution of produced SWCNTs for the anodic arc production method using an applied magnetic field is described. As it was mentioned above, SWCNTs synthesized in anodic arc have properties superior to those produced by CVD, including on the typical measures for quality of nanotubes (smaller defects, higher flexibility and strength) as well as a significantly higher production rate and should thus be more advantageous for practical applications [78,112−114].

Recently, different methods for control of anodic arc synthesis have been reported. It was shown that the anode composition and structure [115], background gas composition and pressure [25,116], and electric field [117] affect the production yield, diameter range, and aspect ratio of the synthesized SWCNTs. Particularly significant progress in control of arc synthesis was demonstrated using the application of external magnetic fields to the arc [39,118]; this magnetically enhanced discharge was demonstrated to be able to control the aspect ratio of SWCNTs [39]. Nanotubes synthesized in magnetically enhanced arc were two times longer than those produced without magnetic field. By changing the strength of the applied static magnetic field, the diameter distribution of the arc product can be controlled as shown schematically in Figure 6.6. The SWCNT samples synthesized at different magnetic fields were analyzed using scanning electron microscopy (SEM), photoluminescence (PL), UV−Vis−NIR absorbance and Raman spectroscopy.

Produced materials were analyzed both as -produced and as an aqueous dispersion. Figure 6.10A shows Raman spectra and SEM images of the as-produced samples (not purified) obtained with and without magnetic field. The SEM images show that both samples are enriched with SWCNTs ropes. Raman spectra of the samples with ($B = 1.2$ kG) and without magnetic field had similar shapes. Detailed comparison of spectra, however, shows that a slight 1D peak was observed at about $1330\ cm^{-1}$ in the nonzero B sample, while no such peak was observed for the $B = 0$ sample [119,120]. Separately, the 1G line is observed to be located at different wave number shifts in the two samples, with the primary peak at $1580\ cm^{-1}$ and a shoulder at $1557\ cm^{-1}$ for the $B = 0$ sample and slightly upshifted peaks at 1585 and $1562\ cm^{-1}$ for the nonzero B samples. The presence of 1G indicates that the excitation at 514 nm is predominantly in resonance with the E_{33} semiconducting transitions and not metallic nanotubes [121]. 2D peaks were observed around $2665\ cm^{-1}$ for both samples. Full range UV−Vis−NIR spectra are shown in Figure 6.10B.

Detailed comparison of SWCNTs synthesized with and without magnetic field was carried out using UV−Vis−NIR and NIR fluorescence spectrometry. The evolution of UV−Vis−NIR and PL spectra of the purified samples produced at different magnetic field strengths from 0 to 2 kG is shown in Figure 6.11A and B.

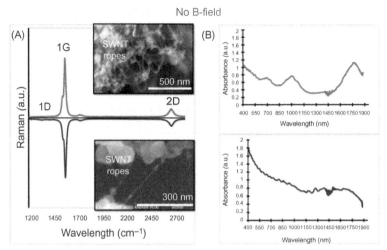

FIGURE 6.10

(A) Raman spectra of as-produced samples without/with magnetic field together with SEM images of SWCNT ropes. (B) Full range UV–Vis–NIR absorbance spectra of the purified samples produced without/with magnetic field. Although the Raman spectra are relatively unaffected, the presence of the field dramatically alters the distribution of chiralities observed via their optical electronic peak positions in the UV–Vis–NIR spectra.

Source: *Reprinted with permission from Ref. [181]. Copyright (2010) American Chemical Society.*

The UV–Vis–NIR spectrum of the $B = 0$ sample shows spectra typical for arc-produced SWCNTs with peaks observed in the optical absorption bands corresponding to metallic (in vicinity of 650 nm) and semiconducting (around 900 nm) SWCNTs respectively. As is typical for many synthesis methods, the $B = 0$ sample was enriched with semiconducting tubes around the roughly 2:1 ratio expected from the combinatorial probabilities when wrapping the graphene sheet. The apparent purity by the Haddon method, revised denominator $= 0.141$ [122], was $\approx 72\%$ for this sample, indicating that the dispersed SWCNTs are well purified by the dispersion and centrifugation process steps. It should be noted that the spectrofluorometer used in this study is unable to detect SWCNTs produced by the arc without a magnetic field due to their relatively large diameter, ≈ 1.5 nm typical for arc method production [116], which fluoresce from their S_{11} transitions at wavelengths beyond the long wavelength range of the InGaAs array detector (≈ 1600 nm).

Now let us consider the evolution of spectra with increase of magnetic field. Firstly, the UV–Vis–NIR spectrum of the nonzero B sample demonstrates overall decrease of peak intensities corresponding to decrease of SWCNT production yield of both metallic and semiconducting nanotubes. Secondly, both UV–Vis–NIR and PL spectra indicate that increase of B-field leads to production of greater variety of semiconducting SWCNT diameters with an overall shift to smaller diameters. This is evidenced by the appearance of new peaks in the nonzero B samples with peak positions around 800 nm on UV–Vis–NIR spectra and new chiralities observed on PL spectra.

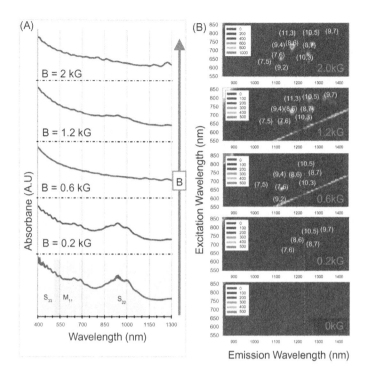

FIGURE 6.11

The evolution of UV−Vis−NIR (A) and PL spectra (B) of the purified samples produced at different magnetic field strengths, (0−2) kG, is shown. With increasing magnetic field strength, the diameter distribution is increasing skewed toward smaller diameter nanotubes that are visible both in the shifting of peak positions (absorbance) and in the observation of fluorescence.

Source: *Reprinted with permission from Ref. [181]. Copyright (2010) American Chemical Society.*

To better characterize the produced materials, an additional processing to separate enriched semiconducting and metallic fractions [123] was performed. Below results obtained using separation of semiconducting and metallic SWCNTs are described.

Three layers formed in the test tube after electronic type separation are schematically shown in Figure 6.12C and D by green (bottom layer containing mostly semiconducting SWCNTs), blue (upper layer—metallic SWCNTs), and red (medium—mixture) colors.

UV−Vis−NIR spectra of three layers are also presented in Figure 6.12C and D (green curve from semiconducting SWCNTs layer, blue—metallic SWCNTs layer, and red—from mixture layer). It is seen in Figure 6.12C ($B = 0$ sample) that well pronounced peaks in semiconducting and metal SWCNT bands were observed in corresponding layers of the test tube. In contrast, the UV−Vis−NIR spectrum of the nonzero B sample showed reduced peak features in the layer where the typical metallic arc was separated and was similar to that from graphenic-like structures [125]. This indicates that the population of typical arc diameter metallic SWCNTs was significantly reduced by the application of the magnetic field. The spectrum from semiconducting layer had

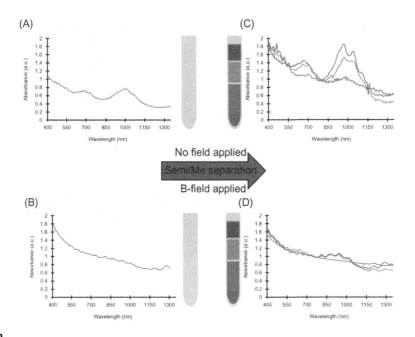

FIGURE 6.12

UV—Vis—NIR spectra of purified nonseparated samples without (A) and with (B) magnetic field. Semiconducting/metal separated samples without (C) and with (D) magnetic field. The sample without an applied magnetic field separates in a manner typical for electric arc synthesized nanotubes as previously reported in the literature [25,124]. The sample synthesized in the magnetic field separates differently due to the altered distribution of diameters; this effect is driven by both the intrinsic change in buoyancy with diameter and the altered interactions with cosurfactants by the diameter change. (For interpretation of the references to color in this figure legend, the reader is referred to the web version of this book.)

Source: *Reprinted with permission from Ref. [181]. Copyright (2010) American Chemical Society.*

greater variety of peaks in comparison with the $B = 0$ sample, which correspond to production of semiconducting SWCNTs with smaller diameter.

Thus both UV—visible—NIR and PL indicate that magnetically enhanced anodic arc yields broader spectrum of diameters of synthesized SWCNTs and smaller diameters compared with that without magnetic field. Such behavior is closely related to the change of catalyst particle motion in the presence of magnetic field. One possible pathway that can explain effect observed is the effect of magnetic field on catalysis particle nucleation [126]. The mechanism leading to catalyst nanoparticle diameter decrease is related to acceleration of the nickel-contained (magnetic) particles by the magnetic force toward the magnet when the temperature drops below the Curie point. In the center of the arc, the temperature is about 3000 K, while in the catalyst particle growth region the temperature is below 1800 K and can reach the Curie point toward the outer region.

Upon reaching the temperature below the Curie point, catalyst particles are accelerated toward the magnet thus reducing the residence (i.e., growth) time. As a result, catalyst particle diameter is

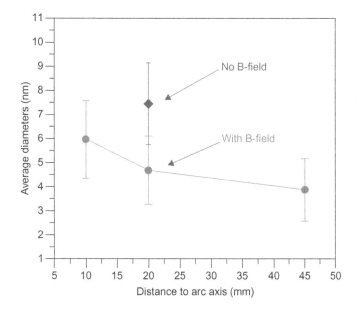

FIGURE 6.13

The relationship between average diameters of catalyst particles and the sample location with standard deviation as error bar. The red dots stand for the synthesis condition with a magnetic field, and the blue one shows the average diameter without a magnetic field. (For interpretation of the references to color in this figure legend, the reader is referred to the web version of this book.)

expected to be smaller in the case of a magnetic field as compared to the case without a magnetic field as shown in Figure 6.13.

6.2.4 Synthesis of graphene in arc plasmas

With an external magnetic field applied to the discharge, the plasma temperature and density significantly increase. The plasma density strongly increases due to the effect of the magnetic-field-related focusing of the plasma jet. Indeed, magnetic confinement restricts the plasma boundaries and prevents the plasma from expansion. Another reason is the magnetization of plasma electrons which leads to more effective ionization of the neutral gas atoms by electron impact. The plasma temperature, in turn, increases in the magnetic field due to the stronger electric field in the magnetized plasma, in contrast to the nonmagnetic conditions [39,127]. Schematically, the magnetically controlled process is shown in Figure 6.14. The carbon samples were collected from the discharge enhancing/separating magnet unit (DESMU) side and top surfaces, and from the chamber walls.

In the growth zone, the ambient temperature is much higher than the Curie point of the catalyst nanoparticles which therefore remain hot and nonmagnetic. This is why the growth conditions are determined by the high catalyst temperature and also a strong incoming flux of carbon material. Outside of the optimum growth zone, the plasma temperature and hence the catalyst temperature decrease sharply. Further away, the temperature decreases below the Curie point, the catalyst

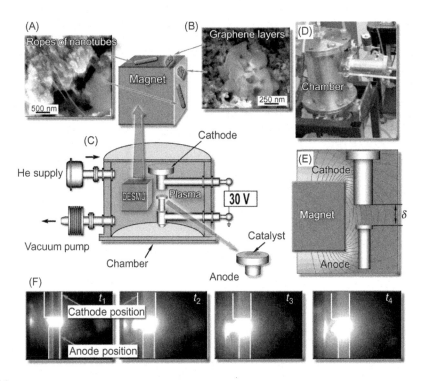

FIGURE 6.14

Experimental setup, photo of the plasma reactor and discharge, and SEM micrographs of representative graphene flakes. (A, B) Representative SEM images of the carbon deposit collected from different collection areas. Ropes of CNTs found on the top and side surfaces of the DESMU, in the areas close to the discharge; graphene layers found on the top and side surfaces of the DESMU, in the areas remote from the discharge. An effective separation of the two different carbon nanostructures was ensured. (C) Schematic of the experimental setup. (D) Photograph of the experimental setup. (E) Schematic of the mutual position of the cube-shaped magnet, anode and cathode, and the computed 2D map of the magnetic field (field strength of 1.2 kG in the discharge gap was optimized for the highest yield of both graphene particles and CNTs). (F) Consecutive photographs of the discharge development in the nonuniform magnetic field.

Source: Reprinted with permission from Ref. [70]. Copyright (2010) by Royal Society of Chemical.

particles become ferromagnetic, respond to the magnetic field, and the separation process starts. Thus, the boundary between the growth and the magnetic separation zones is determined by the catalyst alloy and the plasma parameters. Indeed, in the high-density plasma, the catalyst is hot and nonmagnetic; both graphene particles and CNT are developing in the optimum growth zone with no magnetic separation [71,128].

In the separation zone, the plasma density and the temperature are low, and the catalyst is cold. Hence, while the growth is disabled, the magnetic separation starts. To this end, the optimized composition of the two transition metals, yttrium (which is paramagnetic) and nickel (ferromagnetic with the Curie temperature of about 350°C), was used. Nickel exhibits very high carbon solubility but does not form carbon-containing compounds without oxygen, thus ensuring an efficient carbon supply to the nanostructures [129]. On the other hand, yttrium easily forms carbides and as such enables a very quick nucleation of the carbon nanostructures. Note that the melting points for both these metals are very close, so the catalyst alloy nanoparticles have a stable aggregate structure. In this way, the Y−Ni catalyst alloy was customized to exhibit the excellent nucleation/growth support ability when hot (in the optimum growth zone) and the ferromagnetic response when cooled down below 350°C (in the magnetic separation zone). Experiments have proven the effectiveness of this catalyst alloy for the large-scale carbon nanostructure production [39].

Figure 6.15 shows the images of the representative structures produced. The nanostructured carbon (nanotubes and graphene flakes) could be found on the magnet surfaces only, whereas lacey carbon was found only on the chamber walls. The carbon samples were collected and then analyzed with the SEM, TEM, atomic force microscopy (AFM), and micro Raman techniques. In Figure 6.15, we show representative SEM and TEM images of carbon samples collected from various parts of the setup. Figure 6.15A−C shows the low-, medium-, and high-magnification SEM images, respectively, of the samples containing graphene layers, collected from the top and side surfaces of the magnet (see Figure 6.15).

The estimated size of the graphene flakes is approximately 500−2500 nm, with up to 10 graphitic layers. Some graphene flakes show explicit crystallographic faceting (e.g., clearly visible hexagon sections in Figure 6.15C). It is also seen that the graphene flakes are surrounded and partially covered by loose carbon. Figure 6.15D and E shows TEM images of folded graphene layers in the carbon samples collected from the top and side surfaces of the magnet, respectively. It is seen that these fragments contain a few flake-like graphene layers, up to 3. In Figure 6.15F, it is shown the TEM image of the sample containing CNTs, collected from the side surfaces (remote from the discharge) of the DESMU. It should also be pointed out that a typical catalyst size found by the TEM was approximately 2−10 nm. The SEM analysis of the deposits found on the magnet allows a rough estimate of the production rate to be about 1 cm^2 of graphene per hour of operation.

The results that characterize the samples collected at the top surface of the DESMU by the AFM, Raman, and selected area electron diffraction (SAED) techniques are shown in Figure 6.16. The AFM clearly revealed the presence of flake-like structures with the surface size of around 1 μm and a height variation of 1−5 nm (Figure 6.16A and B). The Raman characterization of the specimens collected from the side surfaces of the magnet showed the occurrence of a weak D-peak at around 1325 cm^{-1}, which is related to the amount of defects in sp^2 bonds (Figure 6.16C) [130]. The SAED TEM pattern from a similar specimen collected from the top surface of the magnet is shown in Figure 6.16E. It reveals the pattern expected for a hexagonal close-packed crystal with the incident beam close to (0001) plane.

It should be pointed out that the magnetic field strongly enhances the arc discharge. Indeed, with the DESMU installed, the plasma arc (normally confined between the cathode and the anode) is stretched toward the magnet as shown in Figure 6.17. In the video snapshots shown in Figures 6.17, it can be noted that the presence of the magnet results in deviation of arc plasma in the direction of $J \times B$ force. It was also observed that the geometry of arc plasma column did not change by removing the nickel catalyst from the anode. This means that the influence of magnetic field on nickel catalyst particles motion does not affect overall geometry of plasma column. It is

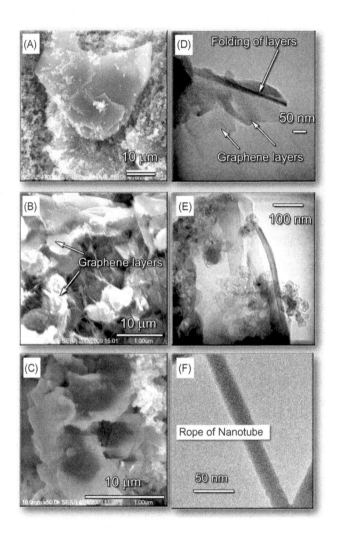

FIGURE 6.15

Representative SEM and TEM images of various carbon deposits collected in different collection areas. (A–C) Low-, medium-, and high-magnification SEM images of the samples containing graphene layers, collected from the top and side surfaces of the magnet. (D, E) TEM image of folded graphene layers in the carbon sample collected from the top and side surfaces of the magnet, respectively. (F) TEM image of the sample containing CNT bundles, collected from the side surfaces (remote from the discharge) of the magnet.

Source: *Reprinted with permission from Ref. [70]. Copyright (2010) by Royal Society of Chemical.*

possible to control distribution of magnetic field by changing the position of permanent magnet, and consequentially the growth region of carbon nanostructures can be easily manipulated according to the $J \times B$ direction. SWCNT and graphene flakes are collected in the different areas. The sample collected from the surface of Mo sheet where the arc plasmas jet was directed contains

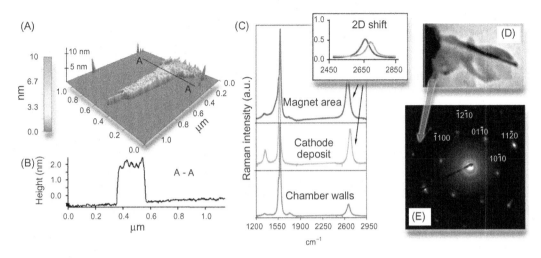

FIGURE 6.16

Microanalysis of the samples shown in Figures 6.1 and 6.2 (A and B). 3D reconstruction and profile of the specimens collected at the top side of the magnet. The presence of flake-like structures with the surface size of around 1 μm^2 and a height variation of 1–2 nm, as well as the occurrence of "bumps/wrinkles" with the height variation of about ∼0.5 nm are clearly revealed. (C) Raman spectra of the samples collected from the side surfaces of the magnet, cathode, and chamber walls. (D) Fragment of TEM photo of the folded graphene layers. (E) SAED pattern generated by the specimen collected from the top surface of the magnet.

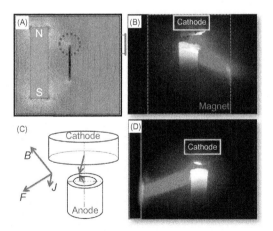

FIGURE 6.17

Distribution of magnetic field simulated by FEMM 4.2 (A), simultaneous photographs of arc plasmas jets from the front (B) and right (D) viewports, and schematic diagram (C) of electrodes position and direction magnetic field for the case when the interelectrode gap is positioned about 75 mm above the bottom of magnet.

high-quality and large-scale graphene. Experimental observations suggest that the graphene is growing by a surface precipitation mechanism [131].

6.2.5 Current state of the art of plasma-based synthesis of carbon nanostructures

In this section, we describe most pressing issues and current state of the art associated with synthesis of carbon nanoparticles in plasma-based synthesis technique.

6.2.5.1 Large-scale production

Large-scale and high-purity synthesis of SWNT by arc discharge stills remain very important objectives of the nanotechnology research [132–138]. Indeed, the majority of the surface-based methods, such as micromechanical exfoliation [139], epitaxial growth on electrically insulating surfaces [140] and graphene formation by thermal decomposition [141], or thermal annealing of silicon carbide [142] have not reached the expected yields [143]. Some promising results of graphene production in arc discharge [144] and separation of graphene and SWNTs were published recently, pushing further state of the art [128,144].

6.2.5.2 Control of synthesis

For a long time, arc-discharge technique was based on a trial-and-error approach and this is why ability to control and tailor the synthesis process is one of the most highly topical and pressing issues. To large extend, problem with control of the synthesis arises from the complicated nature of arc-discharge process preventing for the fixing of the elementary process of catalyst formation, carbon precipitation, and nanoparticle nucleation in space and time domain. Although the mechanism of the formation and growth of SWNTs in an arc discharge was studied for a decade, the region in arc discharge in which SWNT synthesis occurs and the temperature range favorable for SWNT growth remains unclear. According to some authors [78], the nanotube formation occurs on the periphery of an arc column at a moderate temperature range of 1200–1800 K, while other studies suggested that it is the cathode sheath adjacent to hot arc column (~5000 K) where the nanotube growth occurs [80,145,146]. Recall that in the cathode sheath region, the temperature might be well above the reported critical temperatures of thermal stability of the nanotubes. Thermal stability of SWNTs produced in helium arc was studied [94]. Using a furnace, temperature conditions (for SWNT sample) closely resembling the natural conditions of SWNT growth in the arc plasma were created. The maximum temperature determined from electrical resistance measurements combined with SWNT dynamics analysis was used for predicting SWNT synthesis region. It was concluded that SWNTs produced by an anodic arc discharge and collected in the web area outside the arc plasma are originated from the arc-discharge peripheral region, i.e., plasma−gas interface.

6.2.5.3 Outlook

In a quest for optimization of the synthesis technique and control of the SWNT diameter and chirality, the detailed comparison of SWNTs synthesized with and without magnetic field was carried out using UV−Vis−NIR and NIR fluorescence spectrometry as shown in Section 6.2.3. It is accepted that SWNTs are created by rolling up a hexagonal lattice of carbon (graphite). Rolling the lattice at different angles creates a visible twist, chirality, or spiral in the SWNT's molecular structure, though the overall shape remains cylindrical. The SWNT's chirality, along with its diameter, determines its electrical properties with the chiral numbers uniquely defining the SWNT diameter [147]. The armchair structure has metallic characteristics. Both zigzag and chiral structures produce band gaps, making these nanotubes semiconductors and, thus, dependent on chirality SWNT can have metallic or semiconductor conductivity. UV–Vis–NIR diagnostics demonstrated that application of the magnetic field strongly changes the outcome product with the diameter range broadens toward the smaller diameter. The data given in Table 6.1 suggest that the length, diameter, and thus chirality of arc-produced SWNTs can be controlled by external magnetic field applied to the discharge [148]. Magnetic field of relatively small magnitude of several kG was found to result in dramatically increased production of smaller diameter (about 1 nm) SWNTs and broaden of spectrum of diameters/chiralities of synthesized SWNTs.

In spite of a decade-long intense research, some basic understanding of the arc-discharge technique is still lacking and, as such, warrants detailed basic studies. Recent research advance demonstrates that CNT parameters can be controlled by a magnetic field. The summary of these results is given in Table 6.1. It is clear that SWNT parameters are coupled with properties of catalyst nanoparticle. This leads to the conclusion that the control of the arc-discharge synthesis is directly related to the fundamentals of the catalyst formation and interaction of catalyst with the active carbon species. Most critical areas where research is needed fall within the broad program of basic understanding of the arc-discharge technique by utilizing most advances experimental techniques and simulations.

Several experimental techniques under development can be utilized to probe the plasma and nanostructures in an arc. One of the possible techniques is the Langmuir probe [149]. The applicability of Langmuir probe technique for highly collisional plasma of atmospheric anodic arc producing SWNTs remains subject of active ongoing investigation. A limitation of the application of Langmuir probes in the conditions of nanostructure producing arc is caused by the very fast contamination of the probe with the synthesized nanoproducts [150]. In this respect, fast-moving probes providing exposure times to the plasma environment in the millisecond range was shown to be robust technique for plasma diagnostics [150]. Recent application of laser-induced fluorescence (LIF) and laser-induced incandescence (LII) for conditions of nanotube synthesis using laser ablation and Rayleigh microwave scattering for small-scale atmospheric plasmas opens up wide spectra of new prospects for *in situ* diagnostics of arc SWNT synthesis [151].

Table 6.1 Parameters of GWNT and Graphene with and without a Magnetic Field

	Without Magnetic Field	With Magnetic Field
SWCNT length	Average: 0.4 µm	Average: 1 µm
	Maximum: 2 µm	Maximum: 6 µm
SWCNT diameter	∼1.5 nm	∼1 nm
SWCNT bundle diameter	14 nm	6.5 nm
Catalyst particle diameter	7.5 nm	4.7 nm

6.3 Nanoparticle synthesis in electrical arcs: modeling and diagnostics

6.3.1 Arc-discharge plasma

It was already mentioned above that the arc-discharge synthesis of carbon nanoparticles is a very promising plasma-based technique [117]. On the other hand, the controllability and flexibility of the arc-plasma-based process may be significantly improved by the use of a magnetic field (Section 6.2), which strongly influences the plasma parameters [127]. In fact, it was shown that the high-purity MWNTs can be grown in the magnetically enhanced arc discharge [38] and it was demonstrated that the use of the magnetic-field-enhanced arc discharge is very promising for the production of the long SWNTs [39].

Experimental efforts to understand the arc plasma mechanism of synthesis concerned with anode erosion mechanism [152], current–voltage characteristics of the arc discharge [153], and cathode deposit mechanism [154]. In particular, it was demonstrated that anode erosion increases with anode radius decrease, current–voltage characteristics have typical V-shape, and radius of the cathode deposit increases with arc current as shown in Figure 6.18.

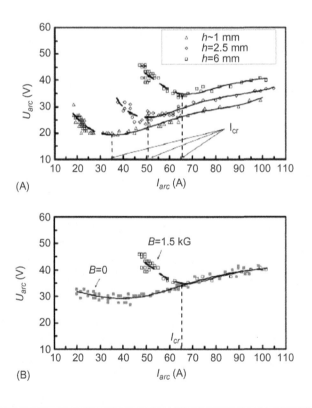

FIGURE 6.18

V–I characteristics of arc for different interelectrode gap sizes (*h*) for *p* = 300 Torr: (A) for different gaps and *B* = 1.5 kG and (B) comparison of two V–I characteristics with and without magnetic field for *h* = 6 mm.

Source: *Reprinted with permission from Ref. [153]. Copyright (2008) by American Institute of Physics.*

CNT synthesis is a relatively recent application of the anodic arc discharge and only few theoretical models related directly to this application were developed. A 1D model (in axial direction) of the SWNT formation was developed and the SWNT growth rate in the anodic arc discharge was calculated [76]. In that model, some simplified gas phase analysis was employed to calculate the nanotube growth rate. The axial velocity of the carbon outflow from the anode was estimated from the measured erosion rate and thus such model is not predictive. Moreover, the temperature of the anode was given *a priori*, while another work showed the anode temperature dependent on the gas pressure [79]. In all existing models, there is no coupling between the interelectrode plasma and electrode phenomena such as ablation and electron emission.

6.3.1.1 Model of the arc discharge

In order to obtain transparent solution while preserving main physical effects relevant to SWNT synthesis, we develop a global (integral) model of an anodic discharge shown schematically in Figure 6.19 [155].

The main features of the model are coupling between the interelectrode plasma and electrodes, current continuity at the electrodes, thermal regime of the electrodes, and the anode erosion rate. A steady-state operation of the arc discharge with carbon electrodes (anode diameter is about 6.35 mm, while the cathode diameter is about 12.5 mm) is considered. Typical interelectrode gap is in the range of about 2−5 mm. During the arcing period, carbon species are supplied by anode erosion which is determined by the anode temperature. In turn, anode temperature is affected by the heat flux from the interelectrode plasma which is controlled by pressure of the ablated species. On the other hand, the experiment indicates that erosion of the cathode is negligible during the arcing.

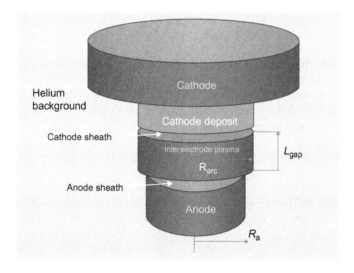

FIGURE 6.19

Schematics of the interelectrode gap.

Source: *Reprinted with permission from Ref. [155]. Copyright (2008) by American Institute of Physics.*

Ablated carbon species expand and interact with background gas (helium) at atmospheric pressure condition. Dynamic boundary of the arc (the arc radius) is therefore determined by the interaction of carbon vapor with the helium background.

In order to describe the plasma state in interelectrode gap of the arc discharge, we invoke the following model formulation. We start with energy balance of the interelectrode plasma [156]:

$$I_{arc}U_{pl} + U_c I_e = I_{arc}(2T_e + U_a) + U_{iz}I_{ion} + 3\frac{m_e}{m_i}(T_e - T_a)n_e\nu_e\pi R_{arc}^2 L_{gap} \tag{6.6}$$

where I_{arc} is the arc current, I_e is the electron current at the cathode, U_{pl} is the potential drop in the interelectrode gap, U_c is the cathode voltage, I_{ion} is the ion current at the cathode, T_e is the electron temperature, U_a is the anode voltage, U_{iz} is the ionization potential of carbon, R_{arc} is the arc radius, L_{gap} is the interelectrode gap length, n_e is the electron density in the interelectrode gap, ν_e is the electron collision frequency, and T_a is the neutral temperature. In this equation, the left-hand side term is Joule heating while right-hand side terms are heat losses to the anode, losses due to ionization, and heat transfer from electrons to neutral species. It is assumed that heavy particles (ions and neutrals) are in equilibrium and have the temperature which is equal to anode temperature.

Current continuity at the cathode implies that part of the current can be conducted by electrons emitted from the cathode so that the total arc current at the cathode consists of ion and electron current:

$$I_{arc} = I_{ion} + I_e \tag{6.7}$$

Balance of energy at the cathode is determined by heat flux from the interelectrode plasma and by the heat losses due to radiation and heat conduction [157]:

$$I_{ion}(U_c + U_{iz} - \varphi_w) = q_{con} + q_{rad} \tag{6.8}$$

where q_{rad} and q_{con} are heat losses due to radiation and conduction, respectively, and φ_w is the work function. According to Eq. (6.8), power deposited at the cathode is dissipated by thermal conduction through the cathode and by radiation, where $q_{con} = (T_c - T_0)\lambda/(\pi^{3/2}R_c)$ and $q_{rad} = \varepsilon\sigma T_c^4$.

Increase of the cathode surface temperature leads to thermoionic emission. Thermoionic electron current density is determined as follows:

$$j_e = AT_c^2 \exp\left(-\frac{e\varphi_w}{kT_c}\right) \tag{6.9}$$

where A is the constant dependent on cathode material and T_c is the cathode surface temperature. In the present model, the last equations (Eqs (6.7)–(6.9)) determine the solution for cathode surface temperature and cathode voltage. This can be illustrated in a more explicit manner. Firstly, by expressing cathode voltage from Eq. (6.8) and combining with Eq. (6.7), an explicit expression for the cathode voltage can be obtained in the following form:

$$U_c = \frac{\pi R_c^2 \varepsilon\sigma T_c^4 + (T_c - T_0)\lambda/\pi^{3/2}R_c}{(I_{arc} - I_e)} - U_i + \varphi_w \tag{6.10}$$

On the other hand, by combining the current continuity equation (Eq. (6.7)) with expressions for the electron and ion currents, we arrive at the following nonlinear equation for cathode temperature:

$$I_{arc} = \pi R_c^2 \left(AT_c^2 \exp\left(-\frac{e\varphi_w}{kT_c}\right) + 0.6 e n_e \left[\frac{T_e}{m_i}\right]^{0.5} \right) \tag{6.11}$$

It should be pointed out that the second term on the right-hand side of Eq. (6.11) is the ion current density. The ion current density is calculated based on assumption that the cathode sheath is collisionless and thus the Bohm condition at the cathode sheath edge can be used [158,159]. In addition, we want to note that electron density is the plasma parameter that couples cathode sheath model with the model of the interelectrode gap. Total arc current (used in Eq. (6.11)) is a given (known) parameter in this consideration. The interelectrode voltage, U_{pl}, can be calculated as $U_{pl} = I_{arc}L_{gap}/(\sigma\pi R_{arc}^2)$. We assume that interelectrode plasma reaches the local thermodynamic equilibrium (LTE) so that the plasma composition and ionization fraction of the gas can be calculated using Saha equation [160].

Anode sheath is established to provide current continuity at the anode. It will be shown below that anode voltage is negative under considered condition, i.e., anode sheath leads to decrease of the electron flux that reaches the cathode. Thus, anode potential drop is calculated as follows:

$$U_a = -T_e \ln(I_{th}/I_{arc}) \tag{6.12}$$

Heating of the anode by electrons leads to anode temperature increase. Anode temperature is determined by the heat diffusion equation in the anode body:

$$\frac{\partial T}{\partial t} = a\Delta T \tag{6.13}$$

where a is the thermal diffusivity. The boundary conditions at the anode surface for this equation take into account heat conduction as well as strong erosion:

$$-\lambda\frac{\partial T}{\partial z}(z = Z_a) = q_a - \Delta H \cdot \Gamma - c_p(T_a - T_0) \cdot \Gamma \tag{6.14}$$

where q_a is the anode heat flux density, Γ is the ablation flux (kg/m^2 s), T_0 is the initial anode temperature, c_p is the specific heat of carbon, and ΔH is the heat of vaporization of anode material (which is carbon in our case). The heat flux to the anode can be calculated as follows [161]:

$$q_a = \frac{I_{arc}}{\pi R_a^2}(2T_e + U_a + \varphi_w) \tag{6.15}$$

Anode erosion is calculated based on the Langmuir model [162]. It should be pointed out that a more accurate kinetic model may be used for ablation rate calculations (see Chapter 1). However, in our case, the arc discharge is diffuse and therefore vapor pressure near the anode is relatively small. As a result, Langmuir model predictions are close to those predicted by the kinetic model and, as such, the Langmuir model turns out to be satisfactory in this situation.

One of important characteristics of the interelectrode plasma in arc discharge is the arc radius. Being that this is an integral model of discharge that does not take into account spatial variation of plasma parameters, arc radius must be given as an input parameter.

Experimental study indicated that arc radius depends on arc current [154]. Since cathode deposit radius is determined by the radius of the arc column [154], the data shown in Figure 6.20 suggest that the arc radius varies with the arc current and can be described as follows:

$$R_{arc} = R_a(\alpha I_{arc}) \quad I_{arc} \geq 50 \text{ A} \tag{6.16}$$

$$R_{arc} = R_a \quad I_{arc} < 50 \text{ A} \tag{6.17}$$

where $\alpha = 0.02$ is a coefficient obtained from experiments [154]. The relations (6.16) and (6.17) were used to obtain solution of the system of equations (6.6)–(6.15).

The total discharge current consists of electron and ion currents. For the energy balance at the cathode, it is important to know the ion current fraction (I_{ion}/I_{arc}). The dependence of (I_{ion}/I_{arc}) on the discharge current is shown in Figure 6.21. On can see that (I_{ion}/I_{arc}) initially decreases and then slightly increases with arc current and also increases with helium pressure.

Cathode voltage has strongly nonmonotonic dependence on the arc-discharge current as shown in Figure 6.22. Initially cathode voltage decreases with arc current increase until about 50 A, reaches the minimum, and then increases. Nonmonotonic trend is also displayed for some other arc parameters. In particular, anode sheath voltage as well as plasma voltage (potential drop in the interelectrode gap) initially increases with arc current and then decreases as shown in Figure 6.22. In addition it is shown that all voltages depend on the helium pressure. Higher helium pressure leads to higher plasma density (see below) resulting in higher ion current fraction as shown in Figure 6.21. Electron temperature and electron density initially increase with arc current as it is shown in Figures 6.23 and 6.24. This dependence can be explained by increase of the power deposition into the plasma (Joule heating) with increase of the arc current. When arc-discharge current increases above the about 50 A, arc radius increases leading to increase of the plasma volume. In turn, this leads to decrease in the electron temperature, plasma density, and ionization fraction in

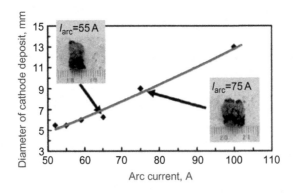

FIGURE 6.20

Dependence of the cathode deposit diameter on arc current ($p = 500$ Torr). Insert shows photographs of cathode deposit for $I_{arc} = 55$ A (on left) and 75 A (on right).

the interelectrode gap. According to our calculations, the interelectrode plasma is characterized by ionization degree of about 0.002−0.004.

Similarly, cathode temperature has nonmonotonic dependence on the arc current as shown in Figure 6.25. On the other hand, anode surface temperature increases monotonically with arc current as plotted in Figure 6.25. Such dependence can be explained by monotonic increase of the power deposition into the anode with arc current increase. Monotonic increase of the anode surface temperature leads to anode ablation rate increase as displayed in Figure 6.26. Experimental data is also shown for comparison. One can see that general trend is captured by the model while the model predicts relatively moderate increase of the ablation rate in comparison with experiment.

Calculated voltage−current (V−I) characteristic of the arc discharge is shown in Figure 6.27. One can see that calculated arc voltage (square symbols in Figure 6.27) initially decreases with arc current, reaches the minimum, and then increases. Such trend is generally in agreement with experimental data as shown in Figure 6.27 for comparison. It should be pointed out that nonmonotonic behavior of the arc voltage displayed in Figure 6.27 is the result of cathode voltage dependence on the arc current as described above and therefore is a direct consequence of model condition that arc radius increases with arc current (for $I > 50$ A). To illustrate the effect of the assumption regarding the arc radius on the arc voltage, the calculations were performed for constant arc radius (which is equal to the anode radius). These results are plotted in Figure 6.27. It can be seen that in this case, the arc voltage decreases monotonically with arc current over the entire range of arc currents.

It should be pointed out that nonmonotonic dependence of arc-discharge parameters on the arc current is a direct outcome of considered condition that the arc radius changes with arc current. Thus the main issue of the present model is implementation of experimentally observed variation of arc radius with arc current which determines features of all calculated results. In fact it was concluded that nonmonotonic behavior of the arc voltage can be only reproduced by considering arc radius increase with arc current. In this model, we have assumed linear variation of the arc radius

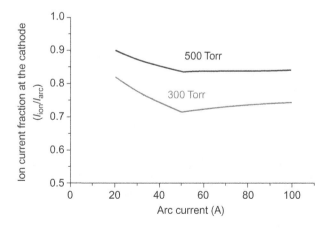

FIGURE 6.21

Ion current fraction at the cathode vs. arc current with gas pressure as a parameter.

Source: *Reprinted with permission from Ref. [155]. Copyright (2008) by American Institute of Physics.*

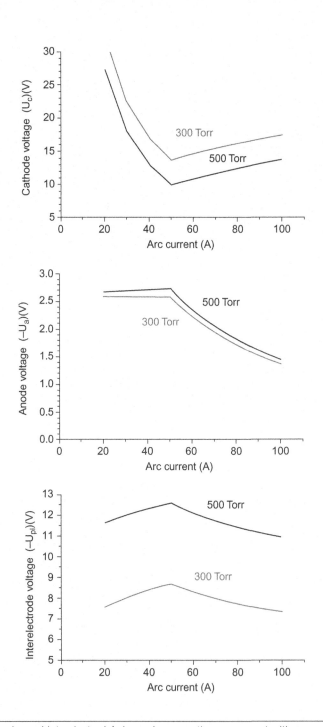

FIGURE 6.22

Voltages (cathode, anode, and interelectrode) dependence on the arc current with gas pressure as a parameter.

Source: *Reprinted with permission from Ref. [155]. Copyright (2008) by American Institute of Physics.*

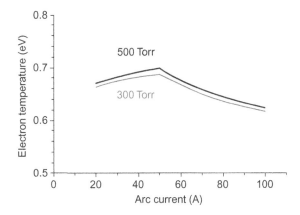

FIGURE 6.23

Electron temperature dependence on the arc current with gas pressure as a parameter.

Source: *Reprinted with permission from Ref. [155]. Copyright (2008) by American Institute of Physics.*

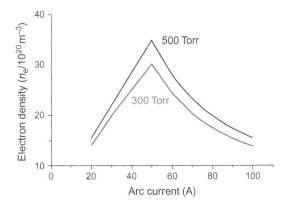

FIGURE 6.24

Electron density dependence on the arc current with gas pressure as a parameter.

Source: *Reprinted with permission from Ref. [155]. Copyright (2008) by American Institute of Physics.*

with arc current in accordance with experimental observations. As a result, calculated results display sharp nonsmooth variation of properties with the turning point at about 50 A.

Calculations show that both cathode voltage and cathode temperature exhibit nonmonotonic behavior with arc current increase. In general, in arc discharges, the cathode voltage is determined by the amount of energy deposition into the cathode required to provide electron emission. Initially arc-discharge current increase leads to higher cathode temperature resulting in higher electron emission current. As a consequence of electron current increase, the cathode voltage contributing to cathode heating decreases. When the arc current is higher than about 50 A, the arc radius increases leading to decrease in electron density. In turn, this results in ion current density decrease at the

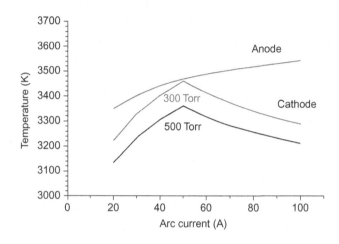

FIGURE 6.25

Cathode and anode temperatures vs. the arc current with gas pressure as a parameter.

Source: *Reprinted with permission from Ref. [155]. Copyright (2008) by American Institute of Physics.*

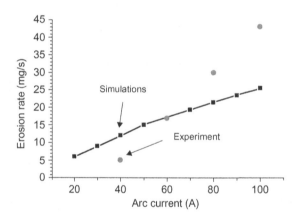

FIGURE 6.26

Anode erosion rate vs. the arc current and comparison with experimental data [153].

cathode. Thus, electron current increase is required to maintain the arc current. On the other hand, further increase of the arc current requires higher electron current to support the current continuity at the cathode and thus leads to increase of the cathode voltage.

This model predicts that the cathode temperature is relatively high although significantly smaller than the anode temperature as shown in Figure 6.25. However, from experiment it is known that cathode erosion is negligible [153,154]. Recall that the cathode is in direct contact with arc plasma only during the initial stage of the discharge. During the continuous arcing (after about 30 s), part of the anode material is deposited on the cathode and forms the so-called cathode deposit. This deposit

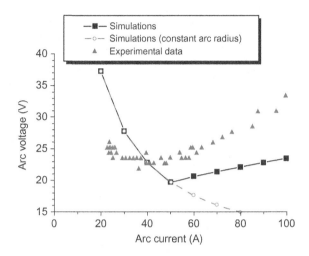

FIGURE 6.27

Arc voltage dependence on the arc current and comparison with experimental data [153]. The calculated arc voltage dependence on arc current based on the assumption about constant arc radius (open circle) is shown for comparison.

Source: *Reprinted with permission from Ref. [155]. Copyright (2008) by American Institute of Physics.*

material is a porous carbon structure with properties dependent on the arc parameters [154]. Thus during the continuous arcing, cathode deposit is in contact with arc plasma protecting the cathode from the erosion. This conclusion is in agreement with experimental evidence [153,154].

6.3.2 Experimental studies of the arc-discharge plasmas for nanoparticle synthesis

Plasma diagnostics is an essential tool for the *in situ* studies of nanostructures formation during the synthesis processes. Combining with the postsynthesis characterizations of the nanostructures, it can establish the important correlations between external arc-discharge parameters and intrinsic nanostructure properties, thus allowing ultimate control of the synthesis process.

Several experimental techniques can be utilized to investigate the plasma and nanostructures in arc discharge. We will start with description of the Langmuir probe. In addition, a nondestructive optical spectrometry technique will be described. UV–Vis emission spectra of arc in different locations under various magnetic conditions will be analyzed to provide an *in situ* investigation for transformation processes of evaporated carbon and catalyst species into growing carbon nanostructures. Based on the arc spectra of carbon diatomic Swan bands, vibrational temperature in arc is determined.

6.3.2.1 Langmuir probe diagnostics

The applicability of Langmuir probe technique for highly collision plasma of atmospheric anodic arc producing SWCNT remains the subject of active ongoing investigation (see Chapter 2). The complication of use of Langmuir probes in nanostructures producing arc is the very fast condensed contamination of the probe with the synthesized carbon nanostructures leading to the uncontrollable increasing of the collecting probe area and short circuit of the probe with surrounding bodies. In this respect, probe having a fast-moving shutter, providing exposure times to the plasma environment in the millisecond range, was recently shown to be the effective technique for plasma diagnostics [163].

Three modifications of single electrostatic probes were used for arc plasma studies as shown in Figure 6.28. The first two types of regular metallic probes for measurements were relatively far from the discharge (the distance to arc center $r > 15$ mm). In this case, probe contamination with the synthesized nanostructures was relatively low, and therefore shutter was not necessary. The circular probe oriented perpendicular to radial plasma flow expanding for the gap is shown in Figure 6.28A, while another surrounding probe used large-area prolonged cylinder covering the arc axis, oriented coaxially to the axis and equipped with opening for arc video recording as shown in Figure 6.28B. The single electrostatic probe with shutter was put in vicinity of the interelectrode gap ($r = 8$ mm). The probe was equipped with electrically controlled shutter providing exposure time to the plasma environment in milliseconds range. Fast shutter was utilized to limit the material contamination to the probe and to prevent its fast deposition by various carbon species synthesized in arc. Such deposition may lead to uncontrollable growth of collecting area and produce inadequate data. The probe consists of a copper foil collector with 3 mm width installed inside the ceramic tube which has outside diameter of 6.3 mm. The probe opening is about 5×3 mm^2 as shown in Figure 6.28C. The cylindrical shutter made of molybdenum sheet was closely fit to the outside surface of the ceramic tube and was able to slide along it, so that the collector was exposed

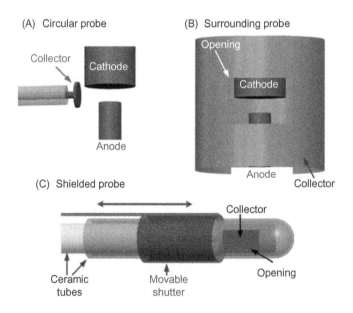

FIGURE 6.28

Schematic diagrams of three types of electrostatic probes.

Source: *Reprinted with permission from Ref. [150]. Copyright (2011) by American Institute of Physics.*

to the plasma solely during the period of time when shutter was open. Shutter was controlled using electrically driven solenoid. The collector surface containing nanostructures deposited on it during the exposition to plasma flux was analyzed using SEM. A 1 kHz voltage sweeping voltage was applied to the probes for measurements of V−I characteristics. A wide range of gas pressures were considered, while the main focus is concerned with relatively high-pressure range of several hundred Torr corresponding to the synthesis conditions of SWCNT and graphene.

The V−I characteristic of circular probe obtained at residual pressure of about 0.1 Torr is shown in Figure 6.29. The V−I characteristics demonstrated the ratio of saturation currents to positively (I_p) and negatively (I_n) biased probe of about 100, which is in agreement with conventional collisionless probe theory predicting this ratio to be about $(M/m)^{0.5}$, where M and m stand for the mass of ion and electron, respectively. In this case, the value is about 85 for helium or 140 for carbon.

The changes of I_p and I_n with the increase of helium pressure are presented in Figure 6.30A. Currents were measured at 50 ms after arc ignition by surrounding probe shown in Figure 6.30B. It was observed that I_p and I_n remained approximately constant before the gas pressure increased to 10 Torr, but after this critical pressure, both decreased with pressure significantly. In addition, it was observed that the ratio of I_p and I_n changed dramatically with increase of He pressure. According to Figure 6.30B, the value of I_p/I_n decreased from about 100 at 0.1 Torr to about 1−4 for pressures of about several hundred Torr. The similar ratios of I_p/I_n of about 3−5 were observed for arcs produced in argon at the pressure of several hundred Torr. This effect can be explained by the increase of measured current at the negatively biased probe above the level of ion saturation current due to secondary electron emission from the probe surface. The V−I characteristic of the probe in high helium pressure conditions is shown in Figure 6.30.

It is seen that the ratio of I_p/I_n was about 4 suggesting significant deviation from conventional collisionless probe theory prediction as shown in Fig. 6.31.

One possible pathway leading to large current in the negative bias is the collection of nanoparticles by probe [164]. In order to assess this effect, surface of the probe in negative and positive polarities was investigated.

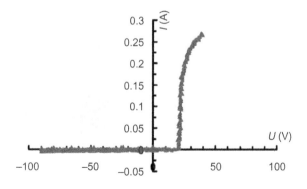

FIGURE 6.29

V−I characteristics of single probe at 0.1 Torr.

TEM images of surface morphology of the probe surface after interaction with arc plasma and negative voltage of 80 V applied were shown in Figure 6.32. It was observed that a large amount of nanostructures containing entangled bundles of CNT, amorphous carbon, and catalyst particles were deposited on the probe surface. The probe surface immersed in arc plasma under applied voltage bias of about $+80$ V are displayed in Figure 6.33. One can see that the deposit of carbon nanostructures are alike regardless the polarity of the bias.

In order to explain these results, we have invoked consideration of influence of the surrounding gas pressure on probe surface. Increase of I_n above the ion saturation current might be due to secondary electron emission from the probe by the Auger deexcitation of long-living excited background gas atoms (He and Ar were used in this work) on the probe surface [163].

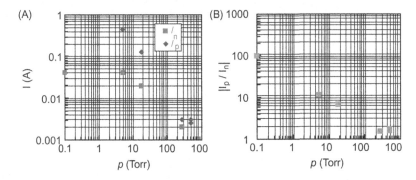

FIGURE 6.30

Saturation currents and their ratios as function of helium pressure.

Source: *Reprinted with permission from [150]. Copyright (2011) by American Institute of Physics.*

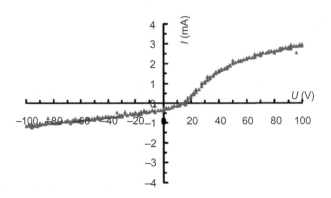

FIGURE 6.31

Typical V−I characteristic of probe taken at 500 ms after arc ignition. Helium pressure is 300 Torr.

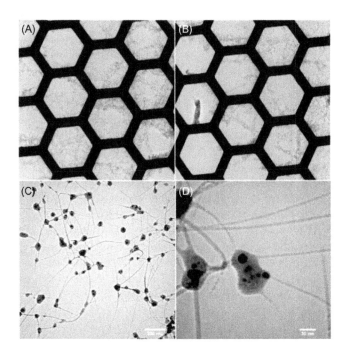

FIGURE 6.32

TEM images of carbon nanostructures captured by the probe with −80 V bias.

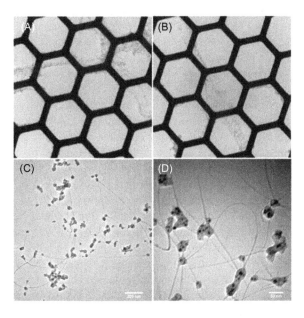

FIGURE 6.33

TEM images of carbon nanostructures captured by the probe with +80 V bias.

It should be emphasized that the secondary electron current in the conditions measured here may be significantly higher than ion saturation current even for relatively low secondary electron emission coefficients of about $10^{-3}-10^{-4}$, because the excited atom density can significantly exceed the plasma density ($\sim10^{15}-10^{17}$ cm^{-3} for excited atom density compared to less than 10^{13} cm^{-3} for plasma density) [165].

Note that if the secondary electron emission dominates the ion saturation current, the shape of V−I curve for potentials is more negative than plasma potential as well as the value of I_n will be significantly deviated from that measured by the conventional one utilized in collisionless case. In this case, the plasma electron temperature and plasma density no longer can be precisely determined using the standard expressions using the slope of V−I curve and the ion saturation current as described in Chapter 2. In contrary, the measurements of V−I curve for potentials below the floating potential can be used for determination of the metastable He* density. Using the diffusion theory, plasma density was estimated to be about 10^{18} m^{-3} [150].

In summary it should be pointed out that the measured V−I characteristic of single Langmuir probe in high-pressure arc shows unusually low ratio of saturation current on positively biased probe to that on negatively biased, which is about 1−4. This result was explained by additional electron current with secondary electron emission from the probe due to the deexcitation of excited background gas atoms at the negatively biased probe surface.

6.3.2.2 Analysis of emission spectra from arc plasmas

The *in situ* investigation of the carbon and catalyst precursor species is crucial for the understanding of the nucleation process and for mastering control over the nanoparticle synthesis. Arc plasma consists of many excited states and hence arc emission spectral measurement can offer a good way to investigate the temporal evolution and dynamics of nucleation and growth processes. Furthermore, the analysis of arc emission spectra is a noninvasion approach that provides accurate measurement compared to postsynthesis methods. There are few recent reports on the optical emission spectra diagnostics of the plasma during the CNT synthesis [163,166−168]. In this section, the effect of arc parameters on the nucleation of various precursor species in synthesis of SWCNT and graphene will be described.

Carbon nanostructures synthesis was carried out in a cylindrical stainless steel chamber with 254 mm length and 152 mm diameter as described in Section 6.2. A black copolymer cylinder with 120 mm length and 50 mm diameter was amounted outside a quartz view port of the reaction chamber [166]. Two straight holes of 1 mm diameter with 10 mm interval on the surface were drilled through the cylinder to fit optical fibers of spectrometer. According to the distance between the head of optical fiber and the arc center, one can calculate the effective spot radius in arc discharge, which equals 1.2 mm in this experimental setup. The emission spectra of the arc were measured using a spectrometer (EPP2000-UVN-SR model, StellarNet Inc.), which has the optical resolution of 1 nm with 14 μm slit and the diffracted light was recorded by a 2048 pixel CCD. For the synthesis of SWCNT without magnetic field, the first spot (N1) shown in Figure 6.34B was located in the center to analyze arc spectra in highest temperature region, while the other spot (N2) with 10 mm horizontal distance to N1 corresponding to the arc edge, which is considered the growth region (see Section 6.2). In the case of a magnetically enhanced synthesis, the two spots for optical fibers were designed differently due to the jet-shaped arc plasmas. The spot marked as B1 in Figure 6.34B was in the middle of plasmas jet, for the purpose to investigate the various species delivered by plasmas jet. The spot B2 was drilled at the edge of cylinder with 10 mm distance to B1 in order to capture the arc spectra in the tail of plasmas jet near the Mo substrate for the synthesis and collection of SWCNT and graphene.

After the arc ignition carbon, nickel, and yttrium powder are evaporated. To capture presence of various species the time-series emission spectra with interval of 190 ms were collected through the arc vapor mixture for 3 s starting from arc ignition under various magnetic fields. Figure 6.35 shows time-series emission spectra captured from spot N1 (see Figure 6.34). According to Figure 6.35, one can see that the arc emission spectra have the range from 200 to 850 nm. Main features appear at about 1.5 s after arc ignition. Typical arc emission spectra captured after 1.5 s from the arc ignition are shown in Figure 6.36. One can see that the emission spectra are dominated by carbon diatomic Swan bands (C_2), which is the transition from upper electronic lever of $d^3\Pi_g$ to lower lever of $a^3\Pi_u$. The prominent Swan band sequences of $\Delta v = -2, -1, 0, 1,$ and 2 marked in spectra could be noticed. The emission lines of nickel (Ni I) around the wavelength of 352.5 nm and yttrium atom (Y I) at 643.5 nm are also displayed in Figure 6.36. The major peaks of spectral bands and their possible transactions during the synthesis processes are listed in Table 6.2.

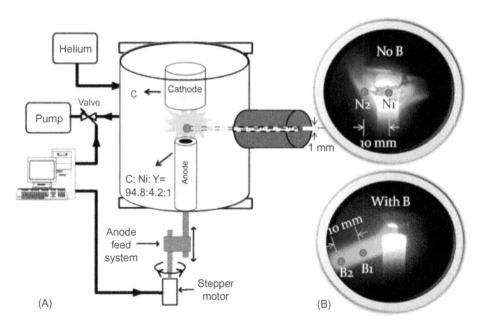

FIGURE 6.34

Schematic diagram of the synthesis system and spectrometer setup (A) and snapshots of arc without and with magnetic field marked with the four spots for arc emission spectra analysis (B).

Source: Reprinted with permission from Ref. [168]. Copyright (2012) by American Institute of Physics.

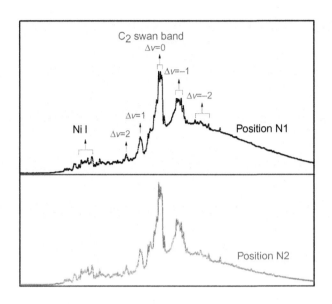

FIGURE 6.35

Time-series emission spectra from various locations and different conditions.

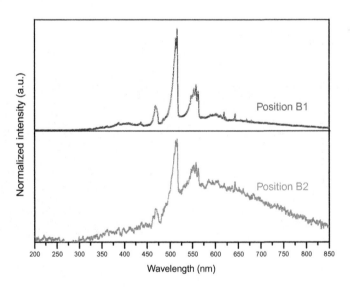

FIGURE 6.36

The typical arc emission spectra in different arc spots under various magnetic conditions.

Table 6.2 The List of Identified Emission Lines from Arc Plasma and Their Possible Transitions

Emission (nm)	Species	Transition (Band Head)
352.5	Ni I	$3d^9\ (^2D)4s \rightarrow 3d^9\ (^2D)4p$
361.9	Ni I	$3d^9\ (^2D)4s \rightarrow 3d^9\ (^2D)4p$
438.2	C_2	$d^3\Pi_g \rightarrow a^3\Pi_u$ (2-0)
468.5	C_2	$d^3\Pi_g \rightarrow a^3\Pi_u$ (4-3)
509.7	C_2	$d^3\Pi_g \rightarrow a^3\Pi_u$ (2-2)
512.9	C_2	$d^3\Pi_g \rightarrow a^3\Pi_u$ (1-1)
516.5	C_2	$d^3\Pi_g \rightarrow a^3\Pi_u$ (0-0)
547.0	C_2	$d^3\Pi_g \rightarrow a^3\Pi_u$ (4-5)
550.2	C_2	$d^3\Pi_g \rightarrow a^3\Pi_u$ (3-4)
558.5	C_2	$d^3\Pi_g \rightarrow a^3\Pi_u$ (1-2)
563.5	C_2	$d^3\Pi_g \rightarrow a^3\Pi_u$ (0-1)
605.9	C_2	$d^3\Pi_g \rightarrow a^3\Pi_u$ (2-4)
612.2	C_2	$d^3\Pi_g \rightarrow a^3\Pi_u$ (1-3)
619.1	C_2	$d^3\Pi_g \rightarrow a^3\Pi_u$ (0-2)
643.5	Y I	

According to the band peaks of well-defined species in arc emission spectra and assuming LTE, the Boltzmann plot method can be employed to determine the vibrational temperature [169,170].

The intensity of a spectral line transition from initial state (n) to state (m) is defined as

$$I_{nm} = N_n hc\sigma_{nm}A_{nm} \tag{6.18}$$

where N_n is the density of molecules in initial state, h is Planck constant, c is speed of light, σ_{nm} is wave number emitted in the transition and A_{nm} is the transition probability, which can be deduced as

$$A_{nm} = \frac{64\pi^4\sigma_{nm}^3}{3h}\frac{S_{nm}}{g_n} \tag{6.19}$$

where S_{nm} is defined as band strength between the two energy levels and g_n is statistical weight of initial energy level.

The Swan bands C_2 which dominate in the spectra shown in Figure 6.36 arise from transitions between the electronic states of $d^3\Pi_g$ and $a^3\Pi_u$, containing well-defined vibrational heads in the $\Delta v = -2, -1, 0, 1,$ and 2 sequences. Vibrational temperature can be determined according to the vibrational fine structures of C_2 spectra, where the thermal equilibrium is assumed among vibrational states. Define v' and v'' as the vibrational quantum numbers of the upper and lower vibrational levels, respectively. Hence, taking into account the two equations above, it follows that the relative emission coefficient of two vibrational lines between the $d^3\Pi_g$ and $a^3\Pi_u$ electronic states of the C_2 molecules is given by

$$I_{v'v''} = D\sigma_{v'vv''}^4 S_{v'vv''} \exp(-G(v')/k_B T_{vib}) \tag{6.20}$$

where D is a constant and $G(v')$ is the vibrational energy of the upper state.

After taking logarithms, Eq. (6.20) can be deduced as

$$\ln\left(\frac{I_{v'v''}}{\sigma^4_{v'v''}S_{v'v''}}\right) = \text{const} - \frac{G(v')}{k_B T_{\text{vib}}}$$

(6.21)

According to Eq. (6.19), left-hand side term can be expressed as a function of $G(v')$. Therefore, the vibrational temperature is determined by the slope of the straight line obtained from the linear regression. The parameters of Swan bands C_2 for temperature determination can be found in Refs [168,171]. Figure 6.37 shows the Boltzmann plots of four spots in arc under various conditions. Considering the standard errors of linear regression, the vibrational temperature of N1, N2, B1, and B2 is calculated as $(6.95 \pm 1.01) \times 10^3$ K, $(4.19 \pm 0.60) \times 10^3$ K, $(5.20 \pm 0.40) \times 10^3$ K, and $(3.62 \pm 0.46) \times 10^3$ K, respectively.

Below this measured temperature is compared with that predicted by the 2D simulations described in Section 6.3.4.

The calculated distributions of temperature and electron density in the discharge are shown in Figure 6.38. The highest numbers of temperature and density, 7020 K and 7.5×10^{20} m^{-3} respectively, were observed in the arc center. The disk-like radial distribution of temperature is due to convection. The subplot in Figure 6.38 shows calculated temperature (black curve) along the radial direction from arc center and measured temperature (blue points with vertical error bar) at spots of

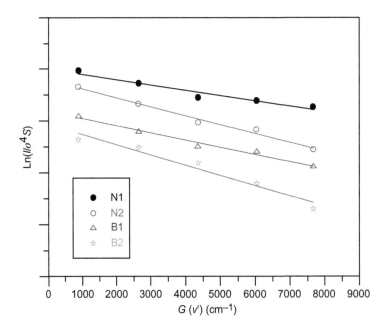

FIGURE 6.37

Data of Boltzmann plots calculated from the emission intensities of the C_2 Swan bands for upper vibrational levels of $G(v) = 0-4$.

N1 and N2 without magnetic field. Note in the context, the effective radius of spots is 1.2 mm, which is also marked as horizontal error bar shown in Figure 6.38.

Among various species detected in arc emission spectra shown in Figure 6.36, the band emission originating from C_2 dominates arc spectra and it is also considered as the intermediate product of carbon vapor nucleation and the precursor of carbon nanostructures growth. The analysis of C_2 band is of particular interest since it can provide an estimation of the plume temperature, which is essential to determine the growth regions of carbon nanostructures. According to the Boltzmann plots shown in Figure 6.37, the vibrational temperature of arc center (N1) without magnetic field is

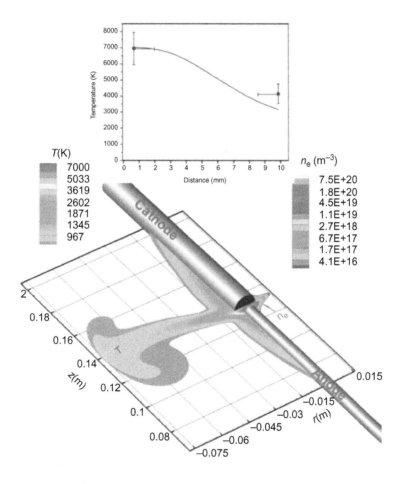

FIGURE 6.38

The temperature and electron density distribution contours inside the discharge chamber. The subplot shows simulated temperature along the radial direction from arc center and measured temperature at spots of N1 and N2 without magnetic field. (For interpretation of the references to color in this figure legend, the reader is referred to the web version of this book.)

Source: Reprinted with permission from Ref. [168]. Copyright (2012) by American Institute of Physics.

about 6950 K, where the mixture of carbon and catalyst powder can be evaporated to gaseous status by tremendous heat. Once the vapor mixture flows to lower temperature zone, i.e., arc plume boundary (N2), the temperature reduces to about 4000 K. The carbon atoms can nucleate with catalyst particles for carbon–catalyst alloy and then precipitate to form SWCNT. This step is considered as the essential step during the synthesis processes of SWCNT. Regarding the synthesis of SWCNT and graphene with magnetic field, the magnetically enhanced arc is confined by the Lorentz force, which generates the plasma jet (shown in Figure 6.17) and makes effective delivery of carbon particles and heat flux. The vibrational temperature in spot B1 is around 5200 K and much higher than that in spot N2. At the tail of plasma jet (B2), Mo sheet was placed on the side of permanent magnetic serving as growth substrate for graphene in terms of surface-catalyzed mechanism. The plasma jet can provide heat flux to Mo sheet continuously, keeping the vibrational temperature of spot B2 around 3600 K.

Since the density of species can be estimated from the intensity of emission, the intensity ratio of carbon and catalyst in arc emission spectra is appropriate to investigate the growth condition of carbon nanostructures. The ratio of carbon diatoms and nickel atoms in temporal evolution is displayed in Figure 6.39. The emission intensities of C_2 and Ni I are selected from the wavelength of 516.5 and 352.5 nm, respectively. In the case of SWCNT synthesis without magnetic field, it can be concluded that steady-state condition is reached since the values of $I(C_2)/I(Ni\ I)$ in the center (N1) and boundary (N2) of arc are approximately constant. However, in the case of a magnetically

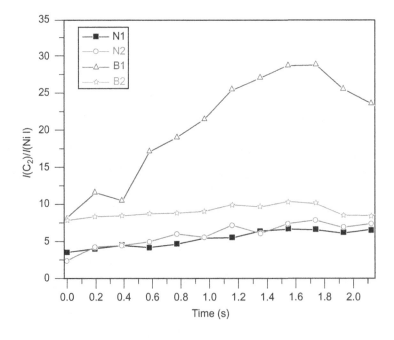

FIGURE 6.39

The intensity ratio of carbon diatoms (516.5 nm) and nickel atom (352.5 nm) as a function of synthesis time.

enhanced arc, $I(C_2)/I(\text{Ni I})$ in the middle of plasma jet (B1) is much larger. This is due to the fact that nickel particles (which are ferromagnetic) are attracted to the magnet and as such nickel density is depleted. Based on the TEM, Raman, and emission spectra measurements, it can be concluded that the proper density ratio of carbon diatoms and nickel atoms is around 5−7 for the synthesis of SWCNT and graphene.

In summary, analysis of arc spectra could provide a great deal of information about the transformation and nucleation processes of different species. Simulated temperature distribution was found to be in agreement with experimental data. Based on the TEM, Raman, and emission spectra measurements, it was concluded that the proper density ratio of carbon diatoms and nickel atoms is around 5−7 for the synthesis of SWCNT and graphene.

6.3.3 Two-dimensional simulation of atmospheric arc plasmas

The purpose of the multidimensional simulation of the arc plasma is to obtain plasma parameters of the discharge relevant for CNT synthesis, which are very difficult to measure. To be useful, the numerical models should combine various phenomena such as arc operation, electrode heating, sublimation, flow expansion, species diffusion, plasma generation, and finally, nanostructure growth.

In order to obtain the temperature and species distribution, the numerical simulation of carbon arc discharge was performed [172] and will be described in this section. Electrode heating and ablation rate were coupled with flow expansion to evaluate the instantaneous mass rate of ablation self-consistently. Conservative form of Navier−Stokes equations with electromagnetic source and energy equation are solved using SIMPLER algorithm [173]. Species diffusion is solved separately for C, Ni, and Y to obtain the respective mass fractions inside the fluid domain. Ionization fractions are calculated for the individual species using Saha equation with LTE plasma assumption. Momentum and energy equations are solved using finite-volume discretization and SIMPLER algorithm to obtain velocity distribution. Power law scheme is used to obtain the fluxes on the cell faces. Energy equation is then solved to obtain the temperature distribution. Using the equation of state, overall density distribution is obtained. This order is repeated until convergence is achieved at any time step. This procedure is repeated at all time steps to obtain the transient results. Further details of modeling, boundary conditions, and the simulation are found in Ref. [172].

Axisymmetric formulation is adapted here. Formulation can be divided into five major areas: arc, sublimation, flow expansion, species transport, and ionization. The domain and boundary conditions are shown in Figure 6.40.

Current continuity in electric potential form is solved to obtain the potential field and then current j is obtained. Where, σ is the electrical conductivity and φ is the electric potential.

$$\nabla \cdot (-\sigma \nabla \varphi) = 0 \tag{6.22}$$

Electrical conductivity of weakly ionized plasma in DC field is obtained using the Chapmann−Enskog equation

$$\sigma = \frac{e^2}{m_e} \frac{n_e}{(\nu_{e,a} + \nu_{e,i})} \tag{6.23}$$

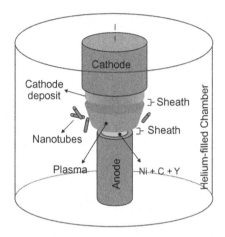

FIGURE 6.40

Schematics of the arc-discharge geometry and simulation domain.

Source: *Reprinted with permission from Ref. [172]. Copyright (2012) by American Institute of Physics.*

where, e, n_e, and m_e are unit charge, number density, and mass of electron, respectively. $\nu_{e,a}$ and $\nu_{e,i}$ are collision frequency of electron−neutrals and electron−ions, respectively, as given by:

$$\nu_{e,a} = \frac{4}{3}\left(\frac{8K_BT}{\pi m_e}\right)^{1/2} n_a Q_m \tag{6.24}$$

where K_B, T, n_a, and n_i are Boltzmann constant, temperature, neutrals density, and ion density, respectively. Q_m is momentum transfer cross section for electrons and neutrals collision varying with temperature [161]. Coulomb logarithm, $\ln(\Lambda)$ of Eq. (6.25) is given in Eq. (6.26) for $T_i = T_e$ and $n_i = n_e$:

$$\nu_{e,i} = \frac{4\sqrt{2\pi}}{3} n_i \left(\frac{e^2}{K_BT}\right)^2 \left(\frac{K_BT}{m_e}\right)^{1/2} \ln(\Lambda) \tag{6.25}$$

$$\ln(\Lambda) = \ln\left(\frac{4K_BT}{\gamma^2 e^2 k_e}\right) - 2\ln\left(\sqrt{2}\right) \tag{6.26}$$

where $k_e^2 = (4\pi n_e e^2)/(K_BT)$ and $\ln(\gamma) = 0.577$ [174]. The momentum transfer cross section Q_m is species dependent. The collision frequencies of all the electron−neutral interactions should be added. Here the collision cross section of helium is considered for all neutrals. In general, the collision cross sections of metallic species are about two orders of magnitude greater than that of noble gases. In the present case, the evaporated material vapor density is one to two orders of magnitude less than that of He.

The azimuthal component of self-induced magnetic field is obtained using Ampere's law. B_θ is azimuthal component of magnetic field, μ_0 is permittivity of vacuum, r is spatial coordinate in radial direction, j_z is axial component of current flux, and R is radius of chamber.

$$B_\theta = \frac{\mu_0}{r} \int_0^R j_z \, r \, \mathrm{d}r \qquad (6.27)$$

Initially, arc concentrates near the axial region and the catalyst-filled core intensively evaporates. Due to this intensive evaporation, the gap between electrodes increases near the catalyst core, and hence, the arc shifts toward the enclosed carbon shell. This transition in arc position is expected to alternate between the catalyst core and enclosed carbon shell throughout the experiment. Simulation of this transition is complicated and not necessary to obtain the overall evaporation rate. Hence, it is assumed that anode is made of single uniform compound material and arc is distributed uniformly throughout the anode tip.

Sublimation is calculated using Langmuir evaporation model (for details see Chapter 1).

$$\Gamma = p_{\mathrm{sat}}(2\pi RT/M)^{-1/2} \qquad (6.28)$$

Γ is evaporation mass flux rate, p_{sat} is the saturation pressure, R and M are gas constant and molecular weight, respectively. Equation (6.28) is applied at all radial locations along the anode tip and plasma interface to account for the variation in surface temperature and the subsequent sublimation rate. It has to be noted here that, Langmuir model predicts higher ablation rate compared to model based on Knudsen layer kinetics (see Chapter 1) as the former does not consider the influence of background pressure on the ablation rate.

In order to find the temperature and density of fluid inside the chamber, standard Navier–Stokes equations are solved. Conservative form is used to account for the variations in density:

$$\frac{\partial \rho}{\partial t} + \nabla \cdot (\rho \, \vec{u}) = 0 \qquad (6.29)$$

$$\frac{\partial(\rho \vec{u})}{\partial t} + \nabla \cdot (\rho \vec{u} \, \vec{u}) = -\nabla p + \nabla \cdot (\mu \nabla \vec{u}) + \rho \vec{g} + \vec{j} \times \vec{B} \qquad (6.30)$$

$$\frac{\partial(\rho h)}{\partial t} + \nabla \cdot (\rho \vec{u} h) = \frac{Dp}{Dt} + \nabla \cdot \left(k \nabla \left(\frac{h}{C_{\mathrm{p,m}}} \right) \right) + \frac{\vec{j} \cdot \vec{j}}{\sigma} + \frac{5}{2} \frac{k_{\mathrm{B}}}{e} \vec{j} \cdot \nabla \left(\frac{h}{C_{\mathrm{p,m}}} \right) \qquad (6.31)$$

where u is the velocity, μ is the viscosity, and h is the enthalpy. All the species are combined to obtain overall density and then treated as a single fluid. Velocity is zero on all solid fluid interfaces, due to no-slip, other than at the anode tip. Mass averaged velocity of the sublimated vapor at the interface is given as the boundary condition for axial velocity at the anode tip. The mass averaged velocity is obtained by dividing the evaporation mass rate with sum of vapor density and local density of fluid existing near the interface. Radial velocity and normal gradient of axial velocity are zero along the axis. Vent condition is specified on the whole bottom periphery of the chamber to maintain constant pressure inside the chamber.

Temperature on chamber walls is maintained constant at 350 K and normal gradient along the axis is considered to be zero. The chamber wall temperature 350 K is observed from the experiments. Heat flux boundary condition at the anode tip and plasma interface is given by Eqs (6.14) and (6.15).

Mass diffusion equation (Eq. (6.32)) is employed to find the distribution of individual species inside the chamber. c_1 and D_1 are the mass fraction and diffusion coefficient of species 1, respectively.

$$\frac{\partial(\rho c_1)}{\partial t} + \nabla \cdot (\rho \vec{u} c_1) = \nabla \cdot (D_1 \nabla(\rho c_1)) \tag{6.32}$$

Binary diffusion coefficients with hard sphere model [175] are used to account for the influence of temperature on the diffusion:

$$D_{AB} = \frac{2.63 \times 10^{-7}}{p(\sigma_{AB})^2} \left(\frac{T^3(M_A + M_B)}{2M_A M_B}\right)^{1/2} \tag{6.33}$$

where D_{AB} is diffusion coefficient of species A diffusing into species B. M_A and M_B are molecular weights of species A and B. σ_{AB} is rigid sphere collision diameter.

Assuming that plasma is in LTE, Saha equation is used to obtain the ionization fractions of individual species (see Chapter 1):

$$\frac{n_e n_{i,1}}{n_{0,1}} = \left(\frac{2\pi m_e k_B T_e}{h^2}\right)^{3/2} \exp\left(-\frac{E_1}{k_B T_e}\right) \tag{6.34}$$

where, $n_{i,1}$ and $n_{0,1}$ are number density of ions and neutrals of species 1, respectively. E_1 and h are ionization energy of species 1 and Planck's constant. The set of Saha equations with the ionization energies corresponding to each species is solved subjected to charge neutrality condition.

Figure 6.41 shows the direct comparison of temperature contours from the simulation (left-hand side) with the photo of experiments. Total current is 60 A and the electrodes are separated

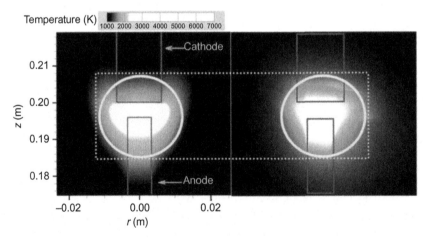

FIGURE 6.41

Comparison of temperature contours (left side) with the direct image of the arc-discharge experiment with 4 mm electrode gap and total current of 60 A.

Source: *Reprinted with permission from Ref. [172]. Copyright (2012) by American Institute of Physics.*

by 4 mm. In the contour plot, the temperature of plasma at the cathode periphery level is greater than 4000 K. This 1:1 comparison is not accurate in terms of plasma emission and has only qualitative character. The similarity in the shape of the temperature contour plot and photo image can be noted here.

The current flux and self-induced magnetic field are shown in Figure 6.42A. The current flux is uniform near the anode tip and self-adjusts to a lower value toward the cathode tip. The maximum value of self-induced magnetic field is 0.0034 T. Pressure and density are shown in Figure 6.42B. Pressure of the gas inside chamber is 68,280 Pa throughout the chamber except in the arc region. Pressure along the axis near the anode tip is 68,480 Pa and decreases to 68,350 Pa in the mid arc region and then increases to 68,390 Pa toward cathode tip, due to flow stagnation. The density is low in the arc region due to high temperatures. Streamlines are shown in Figure 6.42C. The material evaporates from the anode with a velocity of 95 m/s and accelerates to 176 m/s in the mid arc region due to heat addition. The vapor then deflects away from the axis by cathode and velocity decreases gradually due to expansion. Vapor velocity is 20 m/s at a distance of 20 mm from the

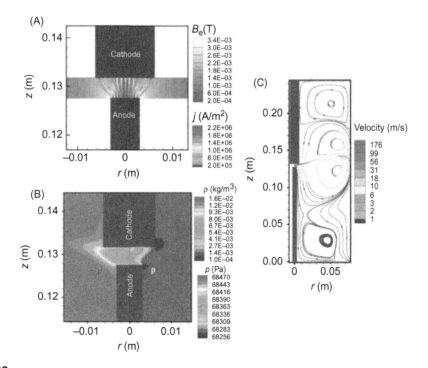

FIGURE 6.42

Electromagnetic and hydrodynamic parameters in the chamber for an arc discharge with 4 mm electrode gap and total current of 60 A in 68 kPa background pressure. (A) Current flux lines with magnitude and self-induced magnetic field. (B) Mass density of the mixture (left side) and pressure (right side). (C) Streamlines with velocity magnitude.

Source: *Reprinted with permission from Ref. [172]. Copyright (2012) by American Institute of Physics.*

axis. In addition, the deflected vapor separates into two recirculation regions after hitting the chamber wall. Some of the gas leaves the chamber through the vent to maintain constant pressure.

Figure 6.43A and B shows the density distribution of neutral species C, Ni, Y, and their ions in the arc-discharge chamber. The highest values of density existing at the anode tip are 4.2×10^{22}, 3.2×10^{21}, and 7.8×10^{20} m^{-3} for C, Ni, and Y, respectively. As expected, the quantity of carbon

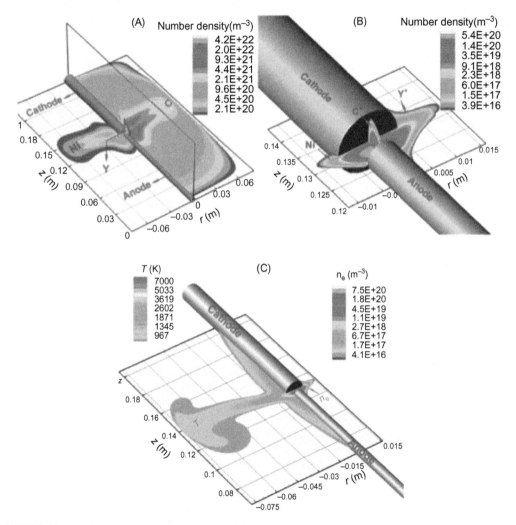

FIGURE 6.43

Typical distribution of plasma parameters inside the arc-discharge chamber for $I_{arc} = 60$ A, electrode gap = 4 mm, and background helium pressure = 68 kPa. (A) Number density of C, Ni, and Y. (B) Number density of C^+, Ni^+, and Y^+. (C) Number density of electrons and the temperature. *Note:* All of the parameters coexist.

Source: *Reprinted with permission from Ref. [172]. Copyright (2012) by American Institute of Physics.*

in the plasma exceeds that of the catalyst. The transport of species below the arc is mainly due to diffusion, whereas the diffusion and convection ensure the transport above the arc. The diffusion coefficient increases with the temperature; since the temperature in the fluid below the anode tip is high due to conduction from the anode lateral surface, the downward diffusion of species is observed in this area. Transport in radial direction is mainly due to convection. The highest densities of ions are 1.9×10^{20}, 5.5×10^{20}, and 2.6×10^{20} m^{-3} respectively for C^+, Ni^+, and Y^+. It is interesting to note that number density of C^+ is lowest though it has highest neutral density. This is due to its high ionization potential. Out of the three species, Y has the lowest ionization potential and hence more Y^+ are observed outside the arc, where temperatures are not sufficient to ionize C and Ni. Helium background gas has even higher ionization potential, so its ionization is negligible. Figure 6.43C illustrates the calculated distributions of temperature and electron density in the discharge with the highest numbers, 7020 K and 7.5×10^{20} m^{-3} respectively, observed in the interelectrode gap. The disk-like radial distribution of temperature is due to convection.

The flow parameter distribution for $I = 100$ A of arc current is shown in Figure 6.44. Anode evaporation rate increases due to higher energy deposition. As a result, the species density and temperature increase inside the chamber. Figure 6.44A shows the increased spreading of neutrals in all directions. The peak density of neutrals is 6.56×10^{22}, 5.02×10^{21}, and 1.21×10^{21} m^{-3} for C, Ni, and Y, respectively. The peak densities of C^+, Ni^+, and Y^+ (Figure 6.44B)) are 7.31×10^{20}, 9.89×10^{20}, and 4.25×10^{20} m^{-3}, respectively. A slight reduction in the thickness of the disk-like structure is noted from Figure 6.44C, which is due to the increased flow speed. However, the peak temperature in the core region is 8640 K, which is greater than that observed for $I = 60$ A case. The peak density of electrons, for this case, is 1.6×10^{21} m^{-3}.

In the model described, an attempt was made to identify and outline the probable location of nanotube growth, directly from the simulation results. Majority of MWNTs are found in the soft core of the cathode deposit while SWNTs are found in the collaret, lateral surface of the cathode, upper wall of the chamber, and in the web suspended between cathode and chamber walls. It was also observed that, the cathode deposit has negligible amount of Ni, while it is high inside the soot deposited on the electrode lateral surfaces and in the web. The temperature distribution in Figure 6.43C also shows that temperature inside the arc region is high for Ni clusters to form. Three major theories were suggested for the growth of nanotubes: open-ended model [81], two-step growth model [84], and root-growth model [176]. Either one or all of the three mechanisms may contribute for the growth. Nevertheless, the root-growth model alone is considered here, due to the presence of large Ni clusters outside the arc region. It was shown analytically that growth of nanotubes is terminated due to the solidification of Ni clusters [177]. By considering the solidification point of 1730 K and condensation point of 3180 K, the region of nanotube growth can be outlined using the isothermal lines, directly from the simulation.

The probable growth region in vapor is shown in Figure 6.45 for $I = 20$, 60, and 100 A. The outlined region also shows the possibility of nanotube growth on the walls of the electrodes. It can be deduced now that, the clusters grown in this region will be transported away due to convection and buoyancy, and deposited on the chamber walls. The size of growth region decreases with the increase of arc current, which is mainly due to the increases in the flow velocity as a result of increased anode evaporation rate. However, the production rate of nanoparticles does not decrease as the growth depends on the local density of contributing species which increase with the arc current as shown in Figure 6.44. Hence, there exists an optimum value of current for which production rate is the highest.

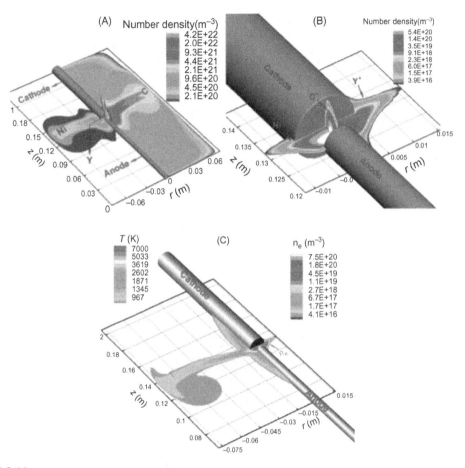

FIGURE 6.44

Typical distribution of plasma parameters inside the arc-discharge chamber for $I_{arc} = 100$ A, electrode gap = 4 mm, and background helium pressure = 68 kPa. (A) Number density of C, Ni, and Y. (B) Number density of C^+, Ni^+, and Y^+. (C) Number density of electrons and the temperature. *Note*: Legend limits synchronized with those in Figure 6.42.

Source: *Reprinted with permission from Ref. [172]. Copyright (2012) by American Institute of Physics.*

6.3.4 Model of the CNT synthesis in arc-discharge plasmas

Inside the growth region, vapor flux contributing to the growth consists of neutrals and ions of metal and carbon. The nanoparticles are negatively charged due to high mobility of electrons, creating a sheath of thickness close to Debye length, and hence ions are attracted toward the nanoparticle as shown in Figure 6.46. Nanoparticle of diameter d_{np} is surrounded by Debye sphere of radius $d_{np}/2 + \lambda_{De}$ consisting of nonneutral plasma. Ions enter the sheath with Bohm speed, v_B at the

FIGURE 6.45

Nanoparticle growth region and number density of C and Ni for $I_{arc} = 20$, 60, and 100 A. The size and configuration of growth region decreases with the increase of arc current.

Source: Reprinted with permission from Ref. [178]. Copyright (2012) by Institute of Physics.

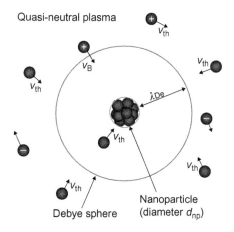

FIGURE 6.46

Nanoparticle growth in plasma. The cluster is negatively charged. The positive ions are focused on to the cluster from Debye sheath edge with Bohm velocity. Neutrals approach the cluster with thermal velocity. *Note*: Not to scale.

Source: Reprinted with permission from Ref. [178]. Copyright (2012) by Institute of Physics.

Debye sphere edge, and create a focused flux that contributes to the growth of nanoparticle [178]. Neutrals, on the other hand, attach to nanoparticle with thermal velocity v_{th}. However, the focusing of ion flux vanishes if the collisions with the neutrals of background gas are dominant inside the sheath, which occurs for smaller mean free paths λ_{mf}, i.e., $\lambda_{De} > \lambda_{mf}$. Besides, the nanoparticle travels at the flow speed macroscopically, which has to be considered to account for the variation in the contributing factors for nanoparticle growth [179]. The flux balance model used in Ref. [177] was for neutral vapor in the exhaust of rocket nozzles. Coming to the specific case of root-growth method of nanotubes, the effect of surface diffusion has to be considered for estimating the flux of carbon atoms. Quasi-steady form of continuum surface diffusion model was used previously for nanotube growth in stationary plasmas with constant properties [177].

Let us describe the mathematical model and prediction related to the growth of CNTs in arc plasmas. The growth of catalyst clusters can be obtained by first finding the critical cluster size using Gibbs free energy of formation of a spherical cluster of radius r given by Eq. (6.35):

$$\Delta G = 4\pi r^3 \sigma_\tau - 4\pi r^3 \frac{k_B T}{V_a} \ln(S) \tag{6.35}$$

where σ_τ is surface energy (surface tension of isotropic materials), V_a is the atomic volume, and S is the saturation ratio given by the ratio of oncoming vapor pressure p_v and equilibrium pressure of the spherical cluster p_{sat}. Equation (6.35) has maxima at $r_c = 2\sigma V_a/[k_B T \ln(S)]$ with a maximal energy of $\Delta G_c = (4/3)\pi r_c^3 \sigma$. Here, r_c is the radius of critical nucleate. The clusters with $r \geq r_c$ are stable and grows larger with further addition of vapor atoms. Furthermore, the probable number of critical clusters may be given using Boltzmann like distribution function, $n_{r_c} = n_a \exp[-\Delta G_c/(k_B T)]$ as given in Ref. [180].

The growth model for the cluster moving along a random path l is given by [177]:

$$\frac{dN}{dl} = 4\pi (r + r_a)^2 \frac{Q_{Ni}\beta}{v_l} \tag{6.36}$$

where $r + r_a \approx r = (3/4\pi)1/3N^{1/3}(2r_a)$, N is the number of atoms in the cluster, r_a is the radius of atom, r is the radius of nanoparticle, and v_l is the velocity of the cluster. The coefficient β is the ratio of the vapor atoms contributing to the growth of cluster to those arriving on to the cluster's surface. β is given as $(R_{arr} - R_{evap})/R_{arr}$. Since arrival rate (R_{arr}) and evaporation rate (R_{evap}) are proportional to pressure of the vapor atoms and equilibrium pressure of the cluster, by assuming bulk vapor and cluster exist at the same temperature, $R_{evap}/R_{arr} = S$. The equilibrium pressure p_{sat} of the spherical cluster surface can be calculated from its value corresponding to a flat surface (p^*) using Kelvin's equation, $p_{sat} = p^* \exp[2\sigma V_a/(rRT)]$. Here, r is radius of the sphere and R is gas constant. The particle flux Q_{Ni} is obtained by adding neutral and ion fluxes as

$$Q_{Ni} = \frac{nv_{th}}{4}\left[1 + \frac{n_{Ni^+}}{n}\frac{4v_B}{v_{th}}\left(1 + \frac{\lambda_{De}}{r}\right)^2\right] \tag{6.37}$$

Finally, Q_{Ni} from Eq. (6.37) is substituted in Eq. (6.35) to obtain the equation for catalyst cluster growth:

$$\frac{dN}{dl} = \pi\left(\frac{3}{4\pi}\right)^{2/3}(2r_a)^2 N^{2/3}\left(\frac{8k_B T}{\pi m_a}\right)^{1/2}\frac{n\beta}{v_l}\eta \tag{6.38}$$

where $\eta = [1 + (n_{Ni^+}/n)(4v_B/v_{th})(1 + (\lambda_{De}/r))^2]$. The value of η gives the ratio of total flux to the neutral vapor flux.

The effective flux of carbon atoms j_C directly contributing to the growth of nanotube is estimated using quasi-steady approach of continuum surface [176]:

$$D_s \frac{d^2 j_C}{dx^2} + Q_C - \frac{j_C}{\tau_a} = 0 \tag{6.39}$$

where x is the length coordinate along nanotube and Q_C is the rate of carbon flux from the bulk vapor arriving on the nanotube surface, D_s is the surface diffusion coefficient, and τ_a is the time required to absorb carbon atoms. Q_C, D_s, and τ_a are estimated using the following system of equations [176]:

$$Q_C = \frac{n_C v_{th}}{4} \left[1 + \frac{n_{C^+}}{n_C} \frac{4v_E}{v_{th}} \left(1 + \frac{\lambda_{De}}{r_C}\right)\right] \tag{6.40}$$

$$D_s = a_0^2 v \exp\left(\frac{-\delta E_D}{k_B T}\right) \tag{6.41}$$

$$\tau_a = \frac{1}{v} \exp\left(\frac{E_a}{k_B T}\right) \tag{6.42}$$

where a_0 is the inter atomic distance for carbon 0.14 nm, v is the vibrational frequency of the atoms $= 3 \times 10^{13}$ (value based on thermal vibrations), δE_D is the activation energy for surface diffusion of carbon ($0.3 - 1.8$ eV), and E_a is adsorption energy ($1.8 - 3.5$ eV) [176].

Equation (6.39) is solved analytically using the following boundary conditions specified at the root ($x = 0$) and closed end ($x = L$) to obtain the flux distribution:

$$\frac{dj_C}{dx}\bigg|_{x=0} = 0 \text{ and } D_s \frac{dj_C}{dx}\bigg|_{x=L} = kj_C$$

$$j_C(x) = Q_C \tau_a \left[\frac{\lambda_D \cosh(x/\lambda_D)}{\sinh(L/\lambda_D) + (k\lambda_D/D_s)\cosh(L/\lambda_D)} + 1\right] \tag{6.43}$$

where L is the instantaneous length of nanotube, $\lambda_D = \sqrt{D_s \tau_a}$ is diffusion length, and $k = a_0/\tau_{inc}$ is the kinetic constant of incorporation. The incorporation time τ_{inc} can be calculated as [176]:

$$\tau_{inc} \approx \frac{1}{v} \exp\left(\frac{\delta E_{inc}}{k_B T}\right) \tag{6.44}$$

Now, the increase in the length of the nanotube moving along a path l inside the chamber can be estimated using the flux balance as

$$\frac{dL}{dl} v_1 = -\Omega D_s \frac{dj_C}{dl}\bigg|_{x=L} \tag{6.45}$$

where Ω is area of one carbon atom in the SWCNT.

Let us now discuss some predictions by the model described. The growth region in plasma is outlined using the isotherms 3180 (inner line) and 1730 K corresponding to the condensation and solidification points of Ni. The outlined region for an arc current of 60 A is shown in Figure 6.46. Now, the path of nanoparticle in the chamber has to be traced in order to extract the vapor density and temperature local to the particle, which contributes to its growth. Since the particle size varies from nanometer to micrometer, they can be conveniently assumed to follow the streamlines of the flow. Three typical streamlines are shown in Figure 6.47. The following calculations are performed along streamline-3, which originates inside the arc core and passes through the growth region. The temperature and species distribution used for these calculations were obtained from detailed simulation of arc discharge.

Electrode heating and sublimation rate were coupled with flow expansion to evaluate the instantaneous mass rate of ablation self-consistently. 2D electric field was considered to simulate the arc. Conservative form of Navier−Stokes equations with electromagnetic source and energy equation were solved. Species diffusion was solved separately for C, Ni, and Y to obtain the respective mass fractions inside the fluid domain. Ionization fractions were calculated for the individual species using Saha equation with LTE plasma assumption. Further details of modeling, boundary conditions, and the simulation are found in Refs [170,176]. For 60 A of arc current, the evaporation

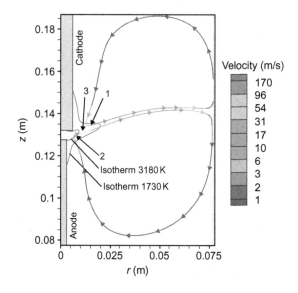

FIGURE 6.47

Nanoparticle growth region and streamlines in the discharge chamber. Particle growth region in plasma is marked by the isotherms. The inner isotherm corresponds to 3180 K and outer isotherm corresponds to 1730 K. The streamlines 1, 2, and 3 are colored with the magnitude of velocity. Nanoparticles are assumed to follow the streamlines and growth calculations are performed along these paths within the growth region. Modeling and simulation details to obtain the streamlines and temperature distribution are found in Refs [170,176].

Source: *Reprinted with permission from Ref. [178]. Copyright (2012) by Institute of Physics.*

model has predicted almost double that of experiment value. Nevertheless, the trend shown the simulation for $10-100$ A arc currents was consistent with the experiment values.

Growth calculations performed for $I = 60$ A and 4 mm interelectrode gap with a background pressure of 68 kPa are shown in Figure 6.48. Mass fractions of C, Ni, and Y in the anode are 0.66, 0.25, and 0.09, respectively. Calculations are performed along streamline-3 shown in Figure 6.47. The mean free path, λ_{mf} is calculated based on neutral gas density. Here, He neutrals alone are considered as their density is more than three orders of magnitude compared to Ni. Figure 6.48A shows the effect of plasma on the growth of Ni cluster. The abscissa represents the distance along the streamline-3 starting from the isotherm corresponding to 3180 K and ending at the isotherm corresponding to 1730 K. The coefficient η shown on the left ordinate reflects the ratio of ion flux to neutral flux $(\eta - 1)$ contributing to the growth.

The ion density decreases as the particle moves away from the arc and hence η also decreases. Debye sheath thickness, λ_{De} shown on right ordinate, increases with the reduction in plasma density. The mean free path λ_{mf} decreases due to reduction in the temperature. At $l = 6$ mm, the ion flux to the Ni cluster ceases completely as $\lambda_{De} > \lambda_{mf}$. The resultant growth is shown in Figure 6.48B. Though the flux is high in the region up to $l = 3$ mm, cluster growth is negligibly small due to high rate of evaporation, R_{evap}. From $l = 3$ to 6 mm, steep increase in the particle diameter d_{np} is observed due to the dominance of ion flux and beyond this point, the growth rate is low due to the cessation of ion flux. The final size of the Ni cluster is around 9.3 nm. It has to be noted here that, SWCNT starts growing on the Ni cluster simultaneously, which will reduce the Ni vapor flux to the cluster. The maximum reduction in the area may be 50%. It remains relatively unknown, exactly when the nanotube starts growing on the Ni cluster. Hence, on an average, only 75% of the cluster area is considered to receive Ni vapor flux.

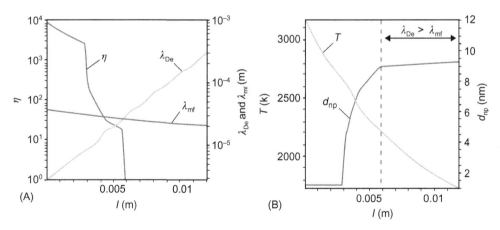

FIGURE 6.48

Catalyst cluster growth calculations for $I = 60$ A. (A) Effect of plasma on catalyst nanoparticle growth and (B) diameter of the catalyst (Ni) nanoparticle. It is assumed that the ion flux contribution for the growth ceases completely for Debye length greater than mean free path of vapor.

Arc-discharge experiment with 60 A current was carried out to measure the Ni particle size. The details are found in Ref. [126]. The diameter distribution of Ni particles in a sample taken at a distance of 20 mm from the arc axis is shown in Figure 6.49. Diameter varies from 2 to 12 nm and the highest number of particles is of 6 nm. The average size of the particles is 7.5 nm. The size of the Ni cluster obtained from the present simulation at this location is 9.2 nm. As mentioned earlier, the rate of evaporation predicted by the simulation was higher compared to experiment. The accurate values of evaporation rate may decrease the Ni concentration in the growth region leading to a reduction in the particle size. But, the flow speed also decreases simultaneously, due to low evaporation rate. The slower flow causes the particle to stay for a longer time in the growth region and eventually increases the particle size.

Figure 6.50A shows that the ratio of total (ions + neutrals) flux to the neutral vapor flux is about 1 throughout the growth region. This means the ion flux no longer contributes to the growth of nanotube, which is evident from Figure 6.50B.

The ratio of ion density to neutrals density of carbon is negligibly small (less than 10^{-11}) even when the Debye sheath increases the collecting area. The reason for low ion flux is high ionization potential of carbon. Figure 6.50C shows the length of SWCNT as it travels along the streamline in the temperature range 3180–1730 K. The nanotube grows up to 3.6 μm long mostly due to neutrals flux.

The nanotube growth model, used here, cannot specify the diameter; however, the Ni cluster growth model can be used to specify the range of diameters. In fact it was shown that nanotubes grow as bundles as well on a single catalyst particle [21].

The SWCNT length distribution from the experiments conducted with 70–80 A arc current is shown in Figure 6.51. More details are found in Ref. [39]. The length distribution shows that 50% of the nanotubes are under 0.6 μm length while 90% are less than 1.3 μm. Also, the SEM image from Ref. [39] shows the nanotube grown up to 3.04 μm long. The growth calculations performed for 70 A and 80 A arc current are shown in Figure 6.52. The SWCNT length is 2.2 μm and

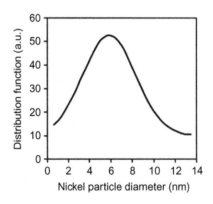

FIGURE 6.49

Diameter distribution of Ni particles measured at a distance of 20 mm from the electrode axis. Arc current was 60 A.

2.02 µm respectively for the arc currents 70 A (Figure 6.52A)) and 80 A (Figure 6.52B). The length obtained using the growth model agreed well with experimental data.

It should be pointed out that the growth of nanoparticles can be improved by (1) increasing the size of growth region, (2) increasing the density of contributing species (Ni, Ni^+, C, and C^+), and (3) decreasing the velocity of nanoparticle ($\approx$fluid velocity). For a given configuration, contributing species density and velocity are interdependent as velocity is directly proportional to the evaporation rate. Though ion flux has a significant effect on the growth rate, density of C^+ may not increase without increasing the temperature above 4500 K, at which nanotubes cannot grow [176].

Keeping this in view, parametric studies were carried out by varying the arc current, background pressure, and interelectrode gap. The results are listed in Table 6.3. Columns 2, 3, and 4 represent

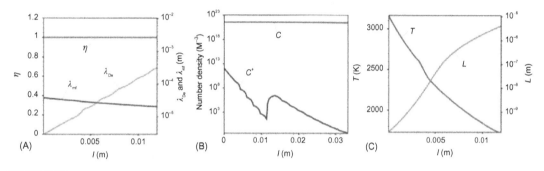

FIGURE 6.50

SWCNT growth calculations for $I = 60$ A. (A) Effect of plasma on SWCNT growth. (B) Density of carbon neutrals and ions. (C) Length of SWCNT.

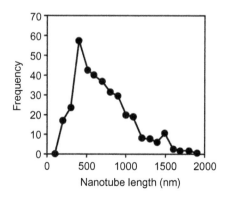

FIGURE 6.51

CNT length distribution from experiments with 70–80 A arc current. Further details of the experiment are found in Ref. [39].

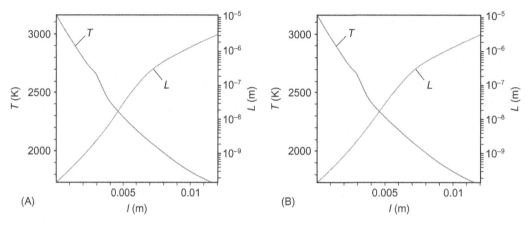

FIGURE 6.52

SWCNT growth calculations for (A) $I = 70$ A and (B) $I = 80$ A arc current.

Source: *Reprinted with permission from Ref. [178]. Copyright (2012) by Institute of Physics.*

Table 6.3 Parametric Study Cases for Nanoparticle Growth in Arc Discharge

Case No.	I (A)	Gap (mm)	p (kPa)	L (mm)	d_{np} (nm)	L (μm)
0	60	4	68	12	9.25	3.6
1	20	4	68	15	20	51
2	100	4	68	11	7.5	2
3	60	4	33	12.5	8	2.35
4	60	4	132	11	10.3	5.1
5	60	2	68	9.5	6.9	0.04
6	60	6	68	14	11.7	7

arc current, interelectrode gap, and background pressure, respectively. The length of path traversed by nanoparticle in the growth region, l, is shown in column 5. The size of Ni cluster, d_{np}, and length of SWCNT, L, are given in columns 6 and 7, respectively.

The nanoparticle sizes, d_{np} and L, decrease with the increase of arc current. This is mainly due to the increase in the flow velocity along the growth region. The velocity increases by 40 m/s compared to case-0. Though the density of species Ni, Ni$^+$, C, and C$^+$ is high due to the increased evaporation, the particles quickly move out of the growth region. On the other hand, reduction in arc current has the exactly opposite effect on the growth of nanoparticles. Reduced background pressure (case-3) and interelectrode gap (case-5) accelerate the flow and decrease the species density as well. Hence the particle sizes also decrease. Increase in background pressure and electrode gap decelerates the flow and also results in the increase of species density along the streamline. The diameter of Ni cluster increases by 1.1 and 2.5 nm, and the length of SWCNT increases by 1.5 and 3.4 μm respectively for the cases 4 and 6 in comparison to case-0. The increment of pressure and interelectrode gap has limitations in terms of transformation to MWCNT growth mode and arc stability. High pressure causes the arrival rate of vapor atoms to exceed the incorporation rate, which may result in the formation of MWCNTs.

Homework problems

1. *Calculate charge evolution* of a 10 μm long carbon nanotube. Consider 1.5 nm diameter nanotube in carbon plasma electron density of about 10^{17} m^{-3} and electron temperature of about 1 eV. Consider collisionless sheath around nanotube and neglect electron emission from the nanotube.
2. *Compute SWCNT residence time* in the 2 mm plasma region having density of about 10^{18} m^{-3} and flow velocity of about 10^3 m/s. Consider electric field of about 1 V/m. Use conditions from problem #1 to calculate SWCNT charge in steady state.
3. *Taking into account results presented in Section 6.3.2 compute the heat flux* to anode and cathode in the case of 100 A arc discharge. Consider pressure of about 300 Torr, the anode diameter of about 6.35 mm, and the cathode diameter of about 12.5 mm. Both cathode and anode are made out of carbon.
4. *Compute the conductivity* of the weakly ionized 1 eV carbon plasma as a function of pressure. Consider that ionization degree is about 0.001.
5. *Calculate the binary diffusion coefficient* of carbon into helium as a function of temperature (1000−3000 K) for 1 atm arc.

References

[1] R. Feynman, in: Alan Chodos (Ed.), American Physics Society, Feynman's Classic CalTech Lecture, 1959.
[2] I. Levchenko, K. Ostrikov, M. Keidar, S. Xu, Deterministic nanoassembly: neutral or plasma route? Appl. Phys. Lett. 89 (2006) 033109.
[3] K. Ostrikov, Plasma Nanoscience, Wiley-VCH, 2008.
[4] S. Iijima, Helical microtubules of graphitic carbon, Nature 354 (1991) 56−58.
[5] K.S. Novoselov, A.K. Geim, S.V. Morozov, D. Jiang, Y. Zhang, S.V. Dubonos, et al., Electric field effect in atomically thin carbon films, Science 306 (5696) (2004) 666−669.
[6] M.S. Dresselhaus, et al., Physics of carbon nanotubes, Carbon 33 (7) (1995) 883−891.
[7] M. Baxendale, The physics and applications of carbon nanotubes, J. Mater. Sci. Mater. Electron. 14 (10-12) (2003) 657−659.
[8] C. Lee, X.D. Wei, J.W. Kysar, J. Hone, Measurement of the elastic properties and intrinsic strength of monolayer graphene, Science 321 (5887) (2008) 385−388.
[9] S. Hong, S. Myung, Nanotube electronics—a flexible approach to mobility, Nat. Nanotechnol. 2 (4) (2007) 207−208.
[10] A.K. Geim, K.S. Novoselov, The rise of graphene, Nat. Mater. 6 (3) (2007) 183−191.
[11] S.S. Chen, Q.Z. Wu, C. Mishra, J.Y. Kang, H.J. Zhang, K.J. Cho, et al., Thermal conductivity of isotopically modified graphene, Nat. Mater. 11 (3) (2012) 203−207.

[12] S.B. Lee, A.S. Teh, K.B.K. Teo, M. Chhowalla, D.G. Hasko, W.I. Milne, et al., Fabrication of carbon nanotube lateral field emitters, Nanotechnology 14 (2003) 192−195.

[13] A. Buldum, J.P. Lu, Electron field emission properties of closed carbon nanotubes, Phys. Rev. Lett. 91 (2003) 236801/1−236801/4.

[14] M. Chhowalla, K.B.K. Teo, C. Ducati, N.L. Rupesinghe, G.A.J. Amaratunga, A.C. Ferrari, et al., Growth process conditions of vertically aligned carbon nanotubes using plasma enhanced chemical vapor deposition, J. Appl. Phys. 90 (2001) 5308−5317.

[15] H. Gao, X.B. Wu, Ji.T. Li, G.T. Wu, J.Y. Lin, K. Wu, et al., Hydrogen adsorption of open-tipped insufficiently graphitized multiwalled carbon nanotubes, Appl. Phys. Lett. 83 (2003) 3389−3391.

[16] J. Kong, N.R. Franklin, C. Zhou, M.G. Chapline, S. Peng, K. Cho, et al., Nanotube molecular wires as chemical sensors, Science 287 (2000) 622−625.

[17] L. Schadler, S. Giannaris, P. Ajayan, Load transfer in carbon nanotube epoxy composites, Appl. Phys. Lett. 73 (1998) 3842−3844.

[18] T. Tersoff, Energies of fullerenes, Phys. Rev. B 46 (1992) 15546−15549.

[19] Z. Huang, J. Xu, Z. Ren, J. Wang, M. Siegal, P. Provencio, Growth of highly oriented carbon nanotubes by plasma-enhanced hot filament chemical vapor deposition, *Appl. Phys. Lett.* 73 (1998) 3845.

[20] Y.H. Wang, M.J. Kim, H.W. Shan, C. Kittrell, H. Fan, L.M. Ericson, et al., Continued Growth of Single-Walled Carbon Nanotubes, Nano. Lett. 5 (2005) 997.

[21] C. Journet, W.K. Maser, P. Bernier, A. Loiseau, M.L. de la Chapelle, S. Lefrant, et al., Large-scale production of single-walled carbon nanotubes by the electric-arc technique, Nature 388 (1997) 756−758.

[22] A.P. Moravsky, E.M. Wexler, R.O. Loufty, in: M. Meyyappan (Ed.), Carbon Nanotubes: Science and Applications, CRC Press, Boca Raton, FL, 2004.

[23] H. Takikawa, M. Yatsuki, T. Sakakibara, S. Itoh, Carbon nanotubes in cathodic vacuum arc discharge, J. Phys. D Appl. Phys. 33 (2000) 826−830.

[24] M. Meyyappan (Ed.), CarbonNanotubes: Science and Applications, CRC Press, Boca Raton, FL, 2004.

[25] S. Farhat, M.L. De La Chapelle, A. Loiseau, C.D. Scott, S. Lefrant, C. Journet, et al., Diameter control of single-walled carbon nanotubes using argon−helium mixture gases, J. Chem. Phys. 115 (2001) 6752−6759.

[26] E.I. Waldorff, A.M. Waas, P.P. Friedmann, M. Keidar, Characterization of carbon nanotubes produced by arc discharge: effect of the background pressure, J. Appl. Phys. 95 (2004) 2749−2754.

[27] J.W.G. Wildoer, L.C. Venema, A.G. Rinzler, R.E. Smalley, C. Dekker, Electronic structure of atomically resolved carbon nanotubes, Nature 391 (1998) 59−62.

[28] C. Bower, O. Zhou, W. Zhu, D.J. Werder, S. Jin, Nucleation and growth of carbon nanotubes by microwave plasma chemical vapor deposition, Appl. Phys. Lett. 77 (2000) 2767−2769.

[29] J. Kunstmann, A. Quandt, I. Boustani, An approach to control the radius and the chirality of nanotubes, Nanotechnology 18 (2007) 155703/1−155703/3.

[30] Y.Y. Ando, X. Zhao, K. Hirahara, K. Suenaga, S. Bandow, S. Iijima, Mass production of single-wall carbon nanotubes by the arc plasma jet method, Chem. Phys. Lett. 323 (2000) 580−585.

[31] T. Zhao, Y. Liu, Large scale and high purity synthesis of single-walled carbon nanotubes by arc discharge at controlled temperatures, Carbon 42 (2004) 2765−2768.

[32] X. Lv, F. Du, Y. Ma, Q. Wu, Y. Chen, Synthesis of high quality single-walled carbon nanotubes at large scale by electric arc using metal compounds, Carbon 43 (2005) 2020−2022.

[33] D. Tang, L. Sun, J. Zhou, W. Zhou, S. Xie, Two possible emission mechanisms involved in the arc discharge method of carbon nanotube preparation, Carbon 43 (2005) 2812−2816.

[34] M. Yao, B. Liu, Y. Zou, L. Wang, D. Li, T. Cui, et al., Synthesis of single-wall carbon nanotubes and long nanotube ribbons with Ho/Ni as catalyst by arc discharge, Carbon 43 (2005) 2894−2901.

[35] Y. Okawa, S. Kitamura, Y. Iseki, An experimental study on carbon nanotube cathodes for electrodynamic tether propulsion, in: Proc. Ninth Spacecraft Charging Technology Conference, Japan, 2005.

[36] I. Stepanek, G. Maurin, P. Bernier, J. Gavillet, A. Loiseau, R. Edwards, et al., Nano-mechanical cutting and opening of single wall carbon nanotubes, Chem. Phys. Lett. 331 (2000) 125−131.

[37] K.G. Kostov, J.J. Barroso, Numerical simulation of magnetic-field-enhanced plasma immersion ion implantation in cylindrical geometry, IEEE Trans. Plasma Sci. 34 (2006) 1127−1135.

[38] K. Anazawa, K. Shimotani, C. Manabe, H. Watanabe, M. Shimizu, High-purity carbon nanotubes synthesis method by an arc discharging in magnetic field, Appl. Phys. Lett. 81 (2002) 739−741.

[39] M. Keidar, I. Levchenko, T. Arbel, M. Alexander, A.M. Waas, K. Ostrikov, Increasing the length of single-wall carbon nanotubes in a magnetically enhanced arc discharge, Appl. Phys. Lett. 92 (2008) 043129/1−043129/3.

[40] L.X. Zheng, M.J. O'Connell, S.K. Doorn, X.Z. Liao, Y.H. Zhao, E.A. Akhadov, et al., Ultralong single-wall carbon nanotubes, Nat. Mater. 3 (2004) 673−676.

[41] Z.L. Wang, D.W. Tang, X.B. Li, X.H. Zheng, W.G. Zhang, L.X. Zheng, et al., Length-dependent thermal conductivity of an individual single-wall carbon nanotube, Appl. Phys. Lett. 91 (2007) 123119/1−123119/3.

[42] K.S. Novoselov, A.K. Geim, S.V. Morozov, D. Jiang, Y. Zhang, S.V. Dubonos, et al., Electric field effect in atomically thin carbon films, Science 306 (2004) 666.

[43] S. Kim, J. Nah, I. Jo, D. Shahrjerdi, L. Colombo, Z. Yao, et al., Realization of a high mobility dual-gated graphene field-effect transistor with Al$_2$O$_3$ dielectric, Appl. Phys. Lett. 94 (2009) 062107.

[44] M.P. Levendorf, C.S. Ruiz-Vargas, S. Garg, J. Park, Transfer-free batch fabrication of single layer graphene transistors, Nano. Lett. 9 (2009) 4479.

[45] X. Li, W. Cai, J. An, S. Kim, J. Nah, D. Yang, et al., Large-area synthesis of high-quality and uniform graphene films on copper foils, Science 324 (2009) 1312.

[46] R.F. Service, Carbon sheets an atom thick give rise to graphene dreams, Science 324 (2009) 875.

[47] J.H. Chen, M. Ishigami, C. Jang, D.R. Hines, M.S. Fuhrer, E.D. Williams, Printed graphene circuits, Adv. Mater. 19 (2007) 3623.

[48] A.K. Geim, Graphene: status and prospects, Science 324 (2009) 1530.

[49] Y.-M. Lin, K.A. Jenkins, A. Valdes-Garcia, J.P. Small, D.B. Farmer, P. Avouris, Operation of graphene transistors at gigahertz frequencies, Nano. Lett. 9 (2009) 422.

[50] D.H. Kim, J.H. Ahn, W.M. Choi, H.S. Kim, T.H. Kim, J. Song, et al., Stretchable and foldable silicon integrated circuits, Science 320 (2008) 507.

[51] T. Sekitani, Y. Noguchi, K. Hata, T. Fukushima, T. Aida, T. Someya, A rubberlike stretchable active matrix using elastic conductors, Science 321 (2008) 1468.

[52] K.S. Kim, Y. Zhao, H. Jang, S.Y. Lee, J.M. Kim, K.S. Kim, et al., Large-scale pattern growth of graphene films for stretchable transparent electrodes, Nature 457 (2009) 706.

[53] G. Eda, G. Fanchini, M. Chhowalla, Large-area ultrathin films of reduced graphene oxide as a transparent and flexible electronic material, Nat. Nanotechnol. 3 (2008) 270.

[54] K.S. Novselov, A.K. Geim, S.V. Morozov, D. Jiang, M.I. Katsnelson, I.V. Grigorieva, et al., Two-dimensional gas of massless Dirac fermions in graphene, Nature 438 (2005) 197.

[55] K.S. Novoselov, D. Jiang, F. Schedin, T.J. Booth, V.V. Khotkevich, S.V. Morozov, et al., Two-dimensional atomic crystals, Proc. Natl. Acad. Sci. U.S.A. 102 (2005) 10451.

[56] http://www.nobelprize.org/nobel_prizes/physics/laureates/2010/.

[57] C. Berger, Z. Song, X. Li, X. Wu, N. Brown, C. Naud, et al., Electronic confinement and coherence in patterned, Science 312 (2006) 1191.

[58] T. Ohta, F. El Gabaly, A. Bostwick, J.L. McChesney, K.V. Emtsev, A.K. Schmid, et al., Morphology of graphene thin film growth on SiC(0001), New J. Phys. 10 (2008) 023034.

[59] A.N. Obraztsov, Making graphene on a large scale, Nat. Nanotech. 4 (2009) 212.

[60] S. Park, R.S. Ruoff, Chemical methods for the production of graphenes, Nat. Nanotech. 4 (2009) 217.

[61] E.V. Rut'kov, A.Y. Tontegode, A study of the carbon adlayer on iridium, Surf. Sci. 161 (1985) 373.

[62] C. Berger, Z. Song, T. Li, X. Li, A.Y. Ogbazghi, R. Feng, et al., Ultrathin epitaxial graphite: 2D electron gas properties and a route toward graphene-based nanoelectronics, J. Phys. Chem. B 108 (2004) 19912.

[63] http://www.darpa.mil/Our_Work/MTO/Programs/Carbon_Electronics_for_RF_Applications_(CERA).aspx.

[64] J.D. Wood, S.W. Schmucker, A.S. Lyons, E. Pop, J.W. Lyding, Effects of polycrystalline Cu substrate on graphene growth by chemical vapor deposition, Nano. Lett. 11 (2011) 4547.

[65] Z. Luo, Y. Lu, D.W. Singer, M.E. Berck, L.A. Somers, B.R. Goldsmith, et al., Effect of substrate roughness and feedstock concentration on growth of wafer-scale graphene at atmospheric pressure, Chem. Mater. 23 (2011) 1441.

[66] A. Reina, X. Jia, J. Ho, D. Nezich, H. Son, V. Bulovic, et al., Large area, few-layer graphene films on arbitrary substrates by chemical vapor deposition, Nano. Lett. 9 (2009) 30.

[67] R.F. Bunshah, Handbook of Deposition Technologies for Films and Coatings: Science, Technology and Applications, Noyes Publications, Park Ridge, NJ, 1994.

[68] R.F. Bunshah, Handbook of Hard Coatings: Deposition Technologies, Properties and Applications, Noyes Publications, Park Ridge, NJ, 2001.

[69] J.L. Qi, W.T. Zheng, X.H. Zheng, X. Wang, H.W. Tian, Relatively low temperature synthesis of graphene by radio frequency plasma enhanced chemical vapor deposition, Appl. Surf. Sci. 257 (2011) 6531.

[70] O. Volotskova, I. Levchenko, A. Shashurin, Y. Raitses, K. Ostrikov, M. Keidar, Single-step synthesis and magnetic separation of graphene and carbon nanotubes in arc discharge plasmas, Nanoscale 2 (10) (2010) 2281−2285.

[71] T. Terasawa, K. Saiki, Growth of graphene on Cu by plasma enhanced chemical vapor deposition, Carbon 50 (2012) 869.

[72] J.W. McKelliget, J. Szekely, A mathematical model of the cathode region of a high intensity carbon arc, J. Phys. D Appl. Phys. 16 (1983) 1007.

[73] I.I. Beilis, Application of vacuum arc cathode spot model to graphite cathode, IEEE Trans. Plasma Sci. 27 (1999) 821.

[74] A. Lefort, M.J. Parizet, S.E. El-Fassi, M. Abbaoui, Erosion of graphite electrodes, J. Phys. D Appl. Phys. 26 (1993) 1239.

[75] J.F. Bilodeau, J. Pousse, A. Gleizes, A mathematical-model of the carbon-arc reactor for fullerene synthesis, Plasma Chem. Plasma Proc. 18 (1998) 285.

[76] I. Hinkov, S. Farhat, C.D. Scott, Influence of the gas pressure on single-wall carbon nanotube formation, Carbon 43 (2005) 2453.

[77] I. Hinkov, J. Grand, M.L. de la Chapelle, S. Farhat, C.D. Scott, P. Nikolaev, et al., Effect of temperature on carbon nanotube diameter and bundle arrangement: Microscopic and macroscopic analysis, J. Appl. Phys. 95 (2004) 2029.

[78] S. Farhat, C.D. Scott, Review of the arc process modeling for fullerene and nanotube production, J. Nanosci. Nanotechnol. 6 (2006) 1189.

[79] M. Keidar, A.M. Waas, On the conditions of carbon nanotube growth in the arc discharge, Nanotechnology 15 (2004) 1571.

[80] E.G. Gamaly, T.W. Ebbesen, Mechanism of carbon nanotube formation in the arc discharge, Phys. Rev. B 52 (1995) 2083.

[81] S. Iijima, P.M. Ajayan, T. Ichihashi, Growth model for carbon nanotubes, Phys. Rev. Lett. 69 (1992) 3100.

[82] Y. Saito, T. Yoshikawa, M. Inagaki, Growth and structure of graphitic tubules and polyhedral particles in arc-discharge, Chem. Phys. Lett. 204 (1993) 277.

[83] D.T. Colbert, R.E. Smalley, Electric effects in nanotube growth, Carbon 33 (1995) 921.

[84] D. Zhou, L. Chow, Complex structure of carbon nanotubes and their implications for formation mechanism, J. Appl. Phys. 93 (2003) 9972.

[85] P.M. Ajayan, T.W. Ebbesen, T. Ichihashi, S. Iijima, K. Tanigaki, H. Hiura, Opening carbon nanotubes with oxygen and implications for filling, Nature 362 (1993) 522−525.

[86] R.S. Wagner, W.C. Ellis, Vapor−liquid−solid mechanism of single crystal growth, Appl. Phys. Lett. 5 (1964) 89−90.

[87] F. Ding, A. Rosen, K. Bolton, The role of the catalytic particle temperature gradient for SWNT growth from small particles, Chem. Phys. Lett. 393 (4-6) (2004) 309−313.

[88] M. Keidar, I. Levchenko, T. Arbel, M. Alexander, A.M. Waas, K.K. Ostrikov, Magnetic-field-enhanced synthesis of single-wall carbon nanotubes in arc discharge, J. Appl. Phys. 103 (9) (2008) 094318.

[89] W.H. Chiang, R.M. Sankaran, Linking catalyst composition to chirality distributions of as-grown single-walled carbon nanotubes by tuning Ni_xFe_{1-x} nanoparticles, Nat. Mater. 8 (11) (2009) 882−886.

[90] E.I. Waldorff, A.M. Waas, P.P. Friedmann, M. Keidar, Characterization of carbon nanotubes produced by arc discharge: effect of the background pressure, J. Appl. Phys. 95 (2004) 2749.

[91] M. Keidar, Y. Raitses, A. Knapp, A.M. Waas, Current-driven ignition of single-wall carbon nanotubes, Carbon 44 (2006) 1022.

[92] A.G. Ostrogorsky, C. Marin, Heat transfer during production of carbon nanotubes by the electric-arc process, Heat Mass Transfer 42 (2006) 470−477.

[93] M. Kundrapu, Modeling and Simulation of Ablation-Controlled Plasmas, Ph.D Dissertation, George Washington University, 2011.

[94] O. Volotskova, A. Shashurin, M. Keidar, Y. Raitses, V. Demidov, S. Adams, Ignition and temperature behavior of a single-wall carbon nanotube sample, Nanotechnology 21 (095705) (2010).

[95] T. Mieno, M. Takeguchi, Thermal motion of carbon clusters and production of carbon nanotubes by gravity-free arc discharge, J. Appl. Phys. 99 (2006) 104301.

[96] D. Winske, M.E. Jones, Effect of finite charging time on particulates in RF discharges, IEEE Trans. Plasma Sci. 23 (1995) 188.

[97] M. Keidar, I. Beilis, R.L. Boxman, S. Goldsmith, Non-stationary macroparticle charging in an arc plasma jet, IEEE Trans. Plasma Sci. 23 (1995) 902.

[98] M. Keidar, A.M. Waas, Y. Raitses, E. Waldorff, Modeling of anodic arc discharge and condition for single wall nanotubes growth, J. Nanosci. Nanotechnol. 6 (2006) 1309.

[99] I. Levchenko, K. Ostrikov, E. Tam, Uniformity of postprocessing of dense nanotube arrays by neutral and ion fluxes, Appl. Phys. Lett. 89 (2006) 223108.

[100] M.A. Lieberman, A.J. Lichtenberg, Principles of Plasma Discharges and Materials Processing, Wiley, New York, NY, 1994.

[101] M. Keidar, O.R. Monteiro, A. Anders, I.D. Boyd, Magnetic field effect on the sheath thickness in plasma immersion ion implantation, Appl. Phys. Lett. 81 (2002) 1183.

[102] I. Levchenko, K. Ostrikov, Nanostructures of various dimensionalities from plasma and neutral fluxes, J. Phys. D 40 (2007) 2308.

[103] I. Levchenko, A.E. Rider, K. Ostrikov, Control of core-shell structure and elemental composition of binary quantum dots, Appl. Phys. Lett. 90 (2007) 193110.

[104] J. Gavillet, J. Thibault, O. Stephan, H. Amara, A. Loiseau, C. Bichara, et al., Nucleation and growth of single-walled nanotubes: the role of metallic catalysts, J. Nanosci. Nanotech. 4 (2004) 346.

[105] R. Sen, S. Suzuki, H. Kataura, Y. Achiba, Growth of single-walled carbon nanotubes from the condensed phase, Chem. Phys. Lett. 349 (2001) 383.

[106] I. Levchenko, O. Baranov, Simulation of island behavior in discontinuous film growth, Vacuum 72 (2003) 205.

[107] I. Levchenko, K. Ostrikov, J.D. Long, S. Xu, Plasma-assisted self-sharpening of platelet-structured single-crystalline carbon nanocones, Appl. Phys. Lett. 91 (2007) 113115.

[108] R.H. Baughman, A.A. Zakhidov, W.A. De Heer, Carbon nanotubes—the route toward applications, Science 297 (2002) 2.

[109] K. Ostrikov, Colloquium: reactive plasmas as a versatile nanofabrication tool, Rev. Mod. Phys. 77 (2005) 489.

[110] Z. Han, B. Tay, C. Tan, M. Shakerzadeh, K. Ostrikov, Electrowetting control of cassie-to-wenzel transitions in superhydrophobic carbon nanotube-based nanocomposites, ACS Nano. 3 (10) (2009) 3031−3036.

[111] N. Li, X. Wang, F. Ren, G.L. Haller, L.D. Pfefferle, Diameter tuning of single-walled carbon nanotubes with reaction temperature using a Co monometallic catalyst, J. Phys. Chem. C 113 (2009) 10070−10078.

[112] C. Journet, W.K. Maser, P. Bernier, A. Loiseau, M. Lamy de la Chapelle, S. Lefrant, et al., Large-scale production of single-walled carbon nanotubes by the electric-arc technique, Nature 388 (1997) 756.

[113] A.P. Moravsky, E.M. Wexler, R.O. Loufty, Growth of carbon nanotubes by arc discharge and laser ablation, in: M. Meyyappan (Ed.), Carbon Nanotubes: Science and Applications, CRC, Boca Raton, FL, 2004, pp. 65−97.

[114] I. Stepanek, G. Maurin, P. Bernier, J. Gavillet, A. Loiseau, R. Edwards, et al., Nano-mechanical cutting and opening of single wall carbon nanotubes, Chem. Phys. Lett. 331 (2000) 125.

[115] O. Łabedz, H. Langea, A. Huczko, J. Borysiuk, M. Szybowicz, M. Bystrzejewski, Influence of carbon structure of the anode on the synthesis of single-walled carbon nanotubes in a carbon arc plasma, Carbon 47 (2009) 2847−2854.

[116] E. Waldorff, A.M. Waas, P.P. Friedmann, M. Keidar, Characterization of carbon nanotubes produced by arc discharge: effect of the background pressure, J. Appl. Phys. 95 (5) (2004) 2749.

[117] M. Keidar, Factors affecting synthesis of single-wall carbon nanotubes in arc discharge, J. Phys. D Appl. Phys. 40 (2007) 2388−2393.

[118] S. Karmakar, N.V. Kulkarni, A.B. Nawale, N.P. Lalla, R. Mishra, V.G. Sathe, A novel approach towards selective bulk synthesis of few-layer graphenes in an electric arc, J. Phys. D Appl. Phys. 42 (2009) 115201.

[119] T.M.G. Mohiuddin, A. Lombardo, R.R. Nair, A. Bonetti, G. Savini, R. Jalil, et al., Uniaxial strain in graphene by Raman Spectroscopy: G peak splitting, Gruneisen parameters and sample orientation, Phys. Rev. B 79 (2009) 205433.

[120] M.S. Dresselhaus, G. Dresselhaus, P.C. Eklund, C_{60}-related tubilus and spherulus, Science of Fullerence and Carbon Nanotubes, Academic Press, San Diego, CA, 1996, pp. 756−864

[121] B. Gao, Y. Zhang, J. Zhang, J. Kong, Z. Liu, Systematic comparison of the Raman spectra of metallic and semiconducting SWNTs, J. Phys. Chem. C 112 (2008) 8319−8323.

[122] M.E. Itkis, D.E. Perea, S. Niyogi, S.M. Rickard, M.A. Hamon, H. Hu, et al., Purity evaluation of as-prepared single-walled carbon nanotube soot by use of solution-phase near-IR spectroscopy, Nano. Lett. 3 (3) (2003) 309.

[123] M.S. Arnold, A.A. Green, J.F. Hulvat, S.I. Stupp, M.C. Hersam, Sorting carbon nanotubes by electronic structure using density differentiation, Nat. Nanotech. 1 (2006) 60−65.

[124] J.A. Fagan, B. Landi, J. Simpson, I. Mandelbaum, B.J. Bauer, V. Bajpai, et al., Comparative measures of single-wall carbon nanotube dispersion, J. Phys. Chem. B 110 (2006) 801.

[125] D.V. Kosynkin, A.L. Higginbotham, A. Sinitskii, J.R. Lomeda, A. Dimiev, B.K. Price, et al., Longitudinal unzipping of carbon nanotubes to form graphene nanoribbons, Nat. Lett. 458 (2009) 07782.

[126] J. Li, O. Volotskova, A. Shashurin, M. Keidar, Controlling diameter distribution of catalyst nanoparticles Arc discharge, J. Nanosci. Nanotechnol. 11 (11) (2011) 10047−10052.

[127] M. Keidar, I.I. Beilis, R.L. Boxman, S. Goldsmith, 2-D Expansion of the low-density interelectrode vacuum arc plasma jet in an axial magnetic field, J. Phys. D Appl. Phys. 29 (1996) 1973.

[128] I. Levchenko, O. Volotskova, A. Shashurin, Y. Raitses, K. Ostrikov, M. Keidar, The large-scale production of graphene flakes using magnetically-enhanced arc discharge between carbon electrodes, Carbon 48 (15) (2010) 4570−4574.

[129] A. Moisala, A.G. Nasibulin, E.I. Kauppinen, The role of metal nanoparticles in the catalytic production of single-walled carbon nanotubes—a review, J. Phys. Condens. Matter. 15 (2003) S3011.

[130] C. Thomsen, S. Reich, Double resonant raman scattering in graphite, Phys. Rev. Lett. 85 (2000) 5214.

[131] A. Reina, X.T. Jia, J. Ho, D. Nezich, H.B. Son, V. Bulovic, et al., Large area, few-layer graphene films on arbitrary substrates by chemical vapor deposition, Nano. Lett. 9 (1) (2009) 30−35.

[132] Y.Y. Ando, X. Zhao, K. Hirahara, K. Suenaga, S. Bandow, S. Iijima, Mass production of single wall carbon nanotubes by arc discharge method, Chem. Phys. Lett. 323 (2000) 580.

[133] T. Zhao, Y. Liu, Temperature and catalyst effects on the production of amorphous carbon nanotubes by a modified arc discharge, Carbon 42 (2004) 2765.

[134] I. Levchenko, K. Ostrikov, M. Keidar, S. Xu, Deterministic nanoassembly: Neutral or plasma route? Appl. Phys. Lett. 89 (2006) 033109.

[135] X. Lu, F. Du, Y. Ma, Q. Wu, Y. Chen, Synthesis of high quality single-walled carbon nanotubes at large scale by electric arc using metal compounds, Carbon 43 (2005) 2020.

[136] D. Tang, L. Sun, J. Zhou, W. Zhou, S. Xie, Two possible emission mechanisms involved in the arc discharge method of carbon nanotube preparation, Carbon 43 (2005) 2812.

[137] M. Yao, B. Liu, Y. Zou, L. Wang, D. Li, T. Cui, et al., Synthesis of single-wall carbon nanotubes and long nanotube ribbons with Ho/Ni as catalyst by arc discharge, Carbon 43 (2005) 2894.

[138] M. Keidar, A.M. Waas, On the conditions of carbon nanotube growth in the arc discharge, Nanotechnology 15 (2004) 1571.

[139] K.S. Novoselov, D. Jiang, F. Schedin, T.J. Booth, V.V. Khotkevich, S.V. Morozov, et al., Two-dimensional atomic crystals, Proc. Natl. Acad. Sci. *U.S.A.* 102 (2005) 10451.

[140] E.V. Rut'kov, A.Y. Tontegode, A study of the carbon adlayer on iridium, Surf. Sci. 161 (1985) 373.

[141] C. Berger, Z. Song, T. Li, X. Li, A.Y. Ogbazghi, R. Feng, et al., Ultrathin epitaxial graphite: 2D electron gas properties and a route toward graphene-based nanoelectronics, J. Phys. Chem. B 108 (2004) 19912.

[142] T. Ohta, F. Gabaly, A. Bostwick, J.L. McChesney, K.V. Emtsev, A.K. Schmid, et al., Morphology of graphene thin film growth on SiC(0001), New J. Phys. 10 (2008) 023034.

[143] R.B. Little, Mechanistic aspects of carbon nanotube nucleation and growth, J. Cluster Sci. 14 (2003) 135.

[144] I. Levchenko, O. Volotskova, A. Shashurin, K. Ostrikov, M. Keidar, The large scale production of graphene flakes using magnetically-enhanced arc discharge between carbon electrodes, Carbon 48 (2010) 4570.

[145] M. Keidar, Y. Raitses, A. Knapp, A.M. Waas, Current-driven ignition of single-wall carbon nanotubes, Carbon 44 (2006) 1022.

[146] I. Hinkov, J. Grand, M.L. de la Chapelle, S. Farhata, J.B. Clement, C.D. Scott, et al., Effect of temperature on carbon nanotube diameter and bundle arrangement: Microscopic and macroscopic analysis, J. Appl. Phys. 95 (2004) 2029.

[147] S.M. Bachilo, M.S. Strano, C. Kittrell, R.H. Hauge, R.E. Smalley, R.B. Weisman, Raman-active modes of single-walled carbon nanotubes derived from the gas-phase decomposition of CO (HiPco process), Science 298 (2002) 2361.

[148] M. Keidar, A. Shashurin, J. Li, O. Volotskova, M. Kundrapu, T. Zhuang, Arc plasma synthesis of carbon nanostructures: where is the frontier? J. Phys. D Appl. Phys. 44 (174006) (2011).

[149] M.T.C. Fang, J.L. Zhang, J.D. Yan, Langmuir probe technique was applied for plasma parameter measurements, IEEE Trans. Plasma Sci. 33 (4) (2005) 1431−1442.

[150] A. Shashurin, J. Li, T. Zhuang, M. Keidar, I.I. Belis, Application of electrostatic Langmuir probe to atmospheric arc plasmas producing nanostructures, Phys. Plasmas 18 (073505) (2011).

[151] M. Cau, N. Dorval1, B. Attal-Trétout, J.L. Cochon, A. Foutel-Richard, A. Loiseau, et al., Formation of carbon nanotubes: In situ optical analysis using laser-induced incandescence and laser-induced fluorescence, Phys. Rev. B 81 (2010) 165416.

[152] A. Fetterman, Y. Raitses, M. Keidar, Enhanced ablation of small anodes in a carbon nanotube arc plasmas, Carbon 46 (2008) 1322.

[153] A. Shashurin, M. Keidar, I.I. Beilis, Voltage-current characteristics of an anodic arc producing carbon nanotubes, J. Appl. Phys. 104 (2008) 063311.

[154] A. Shashurin, M. Keidar, Factors affecting the size and deposition rate of the cathode deposit in an anodic arc used to produce carbon nanotubes, Carbon 46 (13) (2008) 1826.

[155] M. Keidar, I.I. Beilis, Modeling of atmospheric-pressure anodic carbon arc producing carbon nanotubes, J. Appl. Phys. 106 (2009) 103304.

[156] I.I. Beilis, R.L. Boxman, S. Goldsmith, Interelectrode plasma evolution in a hot refractory anode vacuum arc: Theory and comparison with experiment, Phys. Plasmas 9 (7) (2002) 3159.

[157] I.I. Beilis, Theoretical modeling of cathode spot phenomena, in: R.L. Boxman, P.J. Martin, D.M. Sanders (Eds.), Handbook of Vacuum arc Science and Technology, Noyes Publication, Park Ridge, NJ, 1995, pp. 208−256.

[158] M. Keidar, I.I. Beilis, Transition from plasma to space-charge sheath near the electrode in electrical discharges, IEEE Trans. Plasma Sci. 33 (2005) 1481.

[159] A. Porwitzky, M. Keidar, I.D. Boyd, On the mechanism of energy transfer in the plasma-propellant interaction, J. Propell. Explos. Pyrot. 32 (2007) 385.

[160] M. Keidar, I.D. Boyd, I.I. Beilis, Electrical discharge in the Teflon cavity of a co-axial pulsed plasma thruster, IEEE Trans. Plasma Sci. 28 (2000) 376.

[161] I.I. Beilis, Anode spot vacuum arc model: Graphite anode, IEEE Trans. Compon. Packag. Technol. 23 (2) (2000) 334.

[162] I.I. Beilis, Wall interactions with plasma generated by vacuum arcs and targets irradiated by intense laser beams, Plasma Sources Sci. Technol. 18 (2009) 014015.

[163] X.L. Chen, J. Mazumder, Emission-spectroscopy during excimer laser ablation of graphite, Appl. Phys. Lett. 57 (21) (1990) 2178−2180.

[164] M. Keidar, I.I. Beilis, On a model of nanoparticle collection by an electrical probe, IEEE Trans. Plasma Sci. 38 (11) (2010) 3249−3251.

[165] O. Sakai, Y. Kishimoto, K. Tachibana, Integrated coaxial-hollow micro dielectric-barrier-discharges for a large-area plasma source operating at around atmospheric pressure, J. Phys. D Appl. Phys. 38 (2005) 431.

[166] A. Huczko, H. Lange, M. Bystrzejewski, P. Baranowski, Y. Ando, X. Zhao, et al., Formation of SWCNTs in arc plasma: effect of graphitization of Fe-doped anode and optical emission studies, J. Nanosci. Nanotechnol. 6 (5) (2006) 1319−1324.

[167] M. Cau, N. Dorval, B. Attal-Tretout, J.L. Cochon, A. Foutel-Richard, A. Loiseau, et al., Formation of carbon nanotubes: *in situ* optical analysis using laser-induced incandescence and laser-induced fluorescence, Phys. Rev. B 81 (16) (2010) 165416.

[168] J. Li, M. Kundrapu, A. Shashurin, M. Keidar, Emission spectra analysis of arc plasma for synthesis of carbon nanostructures in various magnetic conditions, J. Appl. Phys. 112 (2012) 024329.

[169] J.M. Williamson, C.A. DeJoseph, Determination of gas temperature in an open-air atmospheric pressure plasma torch from resolved plasma emission, J. Appl. Phys. 93 (4) (2003) 1893–1898.

[170] L.L. Danylewych, R.W. Nicholls, Intensity measurements on the C_2 ($d^3\Pi_g - a^3\Pi_u$) Swan band system. I. Intercept and partial band methods, Proc. Royal Soc. London A 339 (1617) (1974) 197–212.

[171] J.G. Phillips, Perturbations in the Swan system of the C_2 molecule, J. Mol. Spectrosc. 28 (2) (1968) 233–242.

[172] M. Kundrapu, M. Keidar, Numerical simulation of carbon arc discharge for nanoparticle synthesis, Phys. Plasmas 19 (7) (2012) 073510.

[173] S.V. Patankar, Numerical Heat Transfer and Fluid Flow, Hemisphere, New York, NY, 1980.

[174] Y. Itikawa, Effective collision frequency of electrons in atmospheric gases, Planet Space Sci. 19 (1971) 993.

[175] J.O. Hirschfelder, C.F. Curtiss, R.B. Bird, Molecular Theory of Gases and Liquids, Wiley, New York, NY, 1964.

[176] J. Gavillet, A. Loiseau, C. Journet, F. Willaime, F. Ducastelle, J.-C. Charlier, Root-growth mechanism for single-wall carbon nanotubes, Phys. Rev. Lett. 87 (275504) (2001).

[177] O.A. Louchev, H. Kanda, A. Rosen, K. Bolton, Thermal physics in carbon nanotube growth kinetics, J. Chem. Phys. 121 (2004) 446.

[178] M. Kundrapu, J. Li, A. Shashurin, M. Keidar, Model of the carbon nanotube synthesis in arc discharge plasmas, J. Phys. D Appl. Phys. 45 (2012) 315305.

[179] W. Knauer, Formation of large metal clusters by surface nucleation, J. Appl. Phys. 62 (1987) 841.

[180] F. Abraham, Homogeneous Nucleation Theory: The Pretransition Theory of Vapor Condensation, Academic Press, New York, NY, 1974.

[181] O. Volotskova, J. Fagan, J.Y. Huh, F. Phelan Jr., A. Shashurin, M. Keidar, Tailored distribution of single-wall carbon nanotubes from Arc plasma synthesis using magnetic fields, ASC NANO 4 (9) (2010) 5187–5192.

Plasma Medicine

7.1 Plasmas for biomedical applications

7.1.1 Introduction

Plasma medicine is an emerging field combining plasma physics and engineering, medicine, bioengineering to use plasmas for therapeutic applications.

Recent progress in atmospheric plasmas led to creation of cold plasmas with ion temperatures close to room temperature [1]. Cold nonthermal atmospheric plasmas can have tremendous applications in biomedical technology. In particular, plasma treatment can potentially offer a minimally invasive surgery that allows specific cell removal without influencing the whole tissue. Conventional laser surgery is based on thermal interaction and leads to accidental cell death, i.e., necrosis, and may cause permanent tissue damage. In contrast, nonthermal plasma interaction with tissue may allow specific cell removal without necrosis. These interactions include cell detachment without affecting cell viability, controllable cell death, etc. It can be also used for cosmetic methods of regenerating the reticular architecture of the dermis. The aim of plasma interaction with tissue is not to denature the tissue but rather to operate below the threshold of thermal damage and to induce chemically specific response or modification. Presence of the plasma can promote chemical reactions that would have the desired effect. Chemical reactions can be promoted by tuning the pressure, gas composition, and energy. Thus, it is important to find conditions that produce effect on tissue without thermal treatment. Overall plasma treatment offers advantages that were never considered in most advanced laser surgery [2]. Due to the myriad of potential applications for cold plasma in biomedical science, it is critical to understand the mechanism regulating the plasma interaction with tissue at the cellular and the molecular level.

It should be pointed out that earlier applications of plasma in medicine relied mainly on the thermal effects of plasma [3]. Heat and high temperature have been utilized in medicine for a long time for the purpose of tissue removal, sterilization, and cauterization [3]. One of the successful applications of thermal plasma is the argon plasma coagulation (APC) in which highly conductive plasma allows passing a current through the tissue [4]. APC is being used to cut tissue and, in particular, in endoscopic applications [5].

Stoffels et al. [2,6] studied the plasma needle and demonstrated the promising potential of the cold plasma in biomedical applications. It was concluded that plasma can interact with organic materials without causing thermal/electric damage to the surface, although this conclusion was not supported by direct measurements. In recent years, several new devices have been presented that were able to generate a cold plasma [7]. It is becoming clear that low-temperature plasmas will play an increasing role in biomedical applications. This understanding motivates development of a variety of reliable and user-friendly plasma sources. Plasma jets have to meet many requirements such as low-temperature, stable operation at atmospheric pressure, and no risk of arcing. Electrosurgical medical devices based on pulsed nonequilibrium plasma in saline solutions were developed [8]. Laroussi and Lu [9] described operation of a cold plasma plume using helium as carrier gas. They demonstrated that the plasma plume can be touched by bare hands and can come in contact with skin and dental gums without causing any heating or painful sensation. The device that later received the name "plasma pencil" was further characterized by Laroussi et al. [10]. A nonequilibrium plasma plume with lengths of 4 and 11 cm was generated by Lu et al. [11,12] Marriotti [13] shows that atmospheric plasma jets are highly nonequilibrium with strong departure between temperatures of electrons and heavy species. A similar plasma source was described by Kolb et al. [14]. They demonstrated that yeast grown on agar can be eradicated with a treatment lasting only a few seconds. Fridman et al. [3,15] demonstrated that cold plasmas can promote blood coagulation and tissue sterilization. It was shown previously that thermal plasma treatment is very beneficial in terms of blood coagulation and sterilization, but it induces significant damage. On the contrary, non-thermal plasmas can lead to the same result without any side effects associated with thermal plasmas [12,13]. It was demonstrated that plasmas efficiently eliminate a mix of "skin flora" [16]—a mix of bacteria collected from cadaver skin containing *Staphylococcus*, *Streptococcus*, and yeast. Sterilization under plasma treatment occurs in the experiments generally after 4 s of the treatment in most cases and 5−6 s in a few cases [13]. There is some controversy in the literature with respect to the mechanism of plasma−cell interaction. Some authors are of the opinion that ion species play the most important role in plasma−cell interactions by triggering intracellular biochemistry [17]. On the other hand, the same and other authors suggested that neutral species play the primary role in some plasma−cell interaction pathways [18]. It was suggested that various plasma effects are highly selective and that different species can have either "plasma killing" (such as O) or "plasma healing" (such as NO) effects [19]. The roles of other species such as O_3 and OH are not clear.

Thus recent studies of cold atmospheric plasmas (CAPs) have shown great potential for the use in biomedical applications. Their distinguished physical and chemical properties are defined by the uniqueness of the nonthermal, nonequilibrium plasmas. Depending on their configuration, they can be used in the following areas: wound healing, skin diseases, hospital hygiene, sterilization, antifungal treatments, dental care, cosmetics targeted cell/tissue removal, and cancer treatments [19−23].

In addition, studies of the impact of cold plasma on cell motility have been conducted [24]. It was shown that cold plasma leads to decrease in the cell motility. This effect will be considered in detail in the next section.

Some limited research has been performed with respect to application of cold plasma for cancer therapy. These studies were limited to skin cells and only simple cell responses to the cold plasma treatment were reported [25,26]. It was shown that cold plasma treatment demonstrates a selective effect by causing apoptosis primarily in cancer cells. More recent work presents preliminary results on the *in vivo* treatment of U87 glioma-bearing mice [27]. It was demonstrated [22] that cold plasma treatment of U87-luc glioma tumor leads to a decrease of tumor bioluminescence intensity and tumor volume. It was demonstrated [21] that cold plasma application (a) selectively eradicates cancer cells *in vitro* a lesser effect on normal cells and (b) significantly reduces tumor size *in vivo*. It is shown that ROS metabolism and oxidative stress responsive genes are deregulated. The development of cold plasma tumor ablation has the potential to transform cancer treatment technologies by utilizing another state of matter.

In summary, cold plasma has demonstrated intriguing potential for cancer treatment. Further progress is associated with creating new devices that can enhance the established CAP effect and understanding the underlying mechanism of plasma action on cells.

The variety of different effects of plasma can be explained by their complex chemical composition and variations in the way that CAP is generated. CAP is a cocktail containing variety of reactive oxygen species (ROS), reactive nitrogen species (RNS), charge particles, UV, etc. This variety leads to the variety of effects mentioned above. In general, the CAP sources can be classified into three major groups according to the principal mechanism of generation and application:

a. *Direct plasmas* employ living tissue or organs as one of the electrodes, and thus, living tissue directly participates in the active discharge plasma processes. Some current may flow through living tissue in the form of small conduction current, displacement current, or both. Conduction current has to be limited to avoid any thermal effects or electrical stimulation of the muscles. The dielectric barrier discharges (DBD) are typical example of direct plasma sources as shown in Figure 7.1A [18].
b. *Indirect plasmas* are produced between two electrodes and are then transported to the area of application entrained in a gas flow. There is a great variety of different configurations of indirect plasma sources exist in the size, type of gas, and power. They range from very narrow "plasma needles" to larger "plasma torches" as shown in Figure 7.1B–F [28,29].
c. *Hybrid plasmas* that combine the production technique of direct plasma with the current-free property of indirect plasma, which is achieved by introducing a grounded wire mesh electrode, which has much smaller electrical resistance than the skin—so that practically all the current passes through the wire mesh. One of the best examples is the plasma dispenser "HandPlaSter" (Max-Plank Institute, Germany) shown in Figure 7.1E [30].

7.1.2 Cold atmospheric plasmas

CAP is a newly developed technology that recently attracted a lot of attention because of their tremendous potential in the areas of nanotechnology, bioengineering, and medicine [24,31–35]. As an example, of CAP device we will consider atmospheric plasma jet that is widely used nowadays in biomedical applications.

The experimental data on atmospheric jet parameters are very scanty and overall understanding of phenomena involved is still lacking. Below we describe most recent efforts in understanding the physics of the CAPs.

The potential for new diagnostics of these plasma jets has been recently demonstrated by microwave scattering diagnostics of small size atmospheric plasmas, which has been successfully applied for laser-induced avalanche ionization in air, resonance-enhanced multiphoton ionization in argon, and CAP jet [36–38].

7.1.2.1 *State of the art modeling of the cold plasma jets*

The state of the art numerical modeling of CAPs for medical applications is mainly limited to the study of the "plasma needle" described by Stoffels [2]. The first numerical study of this device was

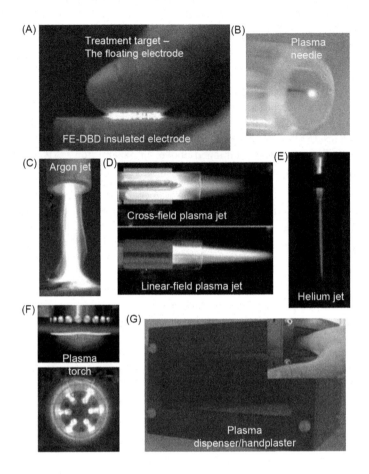

FIGURE 7.1

Plasma medicine devices. (A) Direct plasma device, (B—F) indirect plasma devices, (G) hybrid plasma device.

performed by Brok et al. [39]. The device was studied using a time-dependent, 2D fluid model based on the diffusion equation. The code included air chemistry and modeled a large number of helium and nitrogen species. The code considered a wide range of chemical reactions, including excitations, ionization, disassociation, and recombination. To account for deviation from the Maxwellian distribution, the authors used a Boltzmann solver to tabulate the transport coefficients and reaction rates as a function of electron energy. Similar approach was taken by Sakiyama and Graves [40—42]. The authors also used a 2D fluid model implementing the fine element method in order to accurately capture the needle geometry. An even more limited set exists for codes utilizing a particle-based approach. One example of this approach is a work by Shi et al. [43]. Here the authors used the particle-in-cell method coupled with Monte Carlo collisions to model a pulse-induced discharge between two parallel plates. This model accounted for the avalanche ionization;

however, air chemistry was neglected. Feasibility of particle-in-cell modeling was performed by Hong et al. [44] by comparison with fluid simulations. It was found that electron energy distribution is strongly nonequilibrium especially near the sheath region. Hybrid approach for the atmospheric pressure discharge was recently presented by Iza et al. [45]. Kinetic approach is used to simulate pulse-on phase of plasma while fluid approach is used to simulate multiple pulses. Only very recently models of the atmospheric plasma jets were developed [46]. A numerical study of ionization waves propagating through a circuitous capillary channel and impinging upon a target, in the context of remote delivery of plasma species for biomedical applications, has been conducted [47]. Unlike the plasma bullets in open configurations, the ionization wave fronts in the capillary channel are followed by an extended tail of high electron temperature and ionization up to several centimeters long [45].

7.1.2.2 Characterization of cold atmospheric plasma jet

The typical cold plasma jet device is shown schematically in Figure 7.2A. It is a Pyrex syringe through which a helium flow is supplied. The gun is equipped with a pair of high-voltage (HV) electrodes—a central electrode (which is isolated from direct contact with plasma by ceramics) and an outer ring electrode, as shown schematically in Figure 7.2A. The electrodes are connected to the HV resonant transformer (voltage U up to 10 kV, frequency $\sim$30 kHz). A typical photograph of the plasma jet is shown in Figure 7.2B for $U \sim 5$ kV and helium jet corresponding to the flow rate $v_{fl} = 17$ l/min without plasma (measured at $U = 0$). The visible well-collimated plasma jet has a typical length of $\sim$5 cm. The length of the plasma jet varied with gas flow. In particular, the increasing of the helium feeding results in jet elongation, while further increase in the flow rate caused jet shortening and appearance of a turbulent tail at the jet's end. Maximal plasma jet length was observed at conditions presented in Figure 7.2A. The increase in the HV amplitude applied to the electrodes resulted in increase in plasma jet intensity but did not affect the plasma jet diameter.

In the following, we describe experimental technique that was developed recently to study CAP jets.

The microwave scattering on the plasma jet was studied to resolve the temporal evolution of plasma density in the jet [48]. The experimental microwave system is schematically presented in Section 2.16. Two microwave horns with centers of exit sections located at $(x,y,z) = (6 \text{ cm},0,0)$ and $(0,15 \text{ cm},0)$ were used for radiation and detection of microwave signal. Microwave radiation was linearly polarized along the z-axis (12.6 GHz).

The output signal of the microwave system can be written as follows for the case of dielectric and conducting scatterers:

$$U = \begin{cases} A\sigma V, & \text{for conductor} \\ A\varepsilon_0(\varepsilon - 1)\omega V, & \text{for dielectric} \end{cases} \tag{7.1}$$

The coefficient A was determined to be $\sim$11 V $\times$ W/cm [31] using the dielectric scatterers made of Teflon ($e = 2.1$), alumina ($e = 9.2$), polyethylene ($e = 2.25$), and quartz ($e = 3.8$) [49−51]. Then, plasma conductivity and plasmas were determined based on Eq. (7.1), known coefficient A, and plasma volume. The temporal evolutions of average plasma density are shown on Figure 2.17 for two driven voltage amplitudes, 2.7 and 3.8 kV. It was observed that after discharge initiation, the plasma density reaches $\sim 5-10 \times 10^{13}$ cm^{-3} and then decays with characteristic times of few ms governed by electron attachment. The second peak of plasma density appears with a certain delay

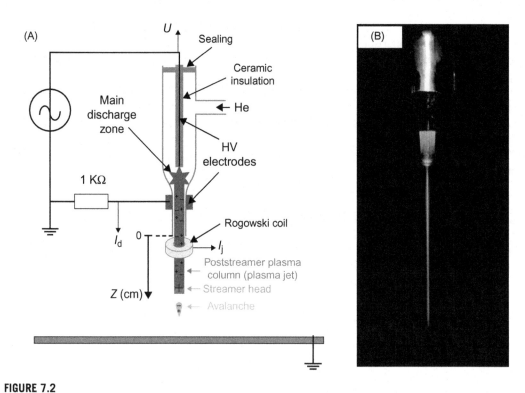

FIGURE 7.2

(A) Schematic presentation of experimental setup and (B) typical image of plasma jet for $U = 3.8$ kV.

Source: *Reprinted with permission from [118]. Copyright (2010) by American Institute of Physics.*

after decay of the first discharge (∼1 ms for $U_{HV} = 2.7$ kV) and indicates the presence of the second breakdown event. This additional ionization of the channel can be provided by ongoing increase of discharge driver voltage (see voltage waveform in Figure 2.17).

A typical microwave scattering signal from the jet, ICCD images of the jet (taken at times indicated by the rectangular bars, exposition time = 500 ns, the plasma bullet is moving down), the jet (I_j), and discharge (I_d) currents are all shown in Figure 7.3 ($t = 0$ is chosen at the moment of zero voltage as shown in Figure 7.3A and C). The ICCD images in Figure 7.3B(a and b) indicate that intensively radiating "plasma bullets" were ejected from the syringe. They had velocities of ∼2 × 10⁶ cm/s and existed for ∼2 ms (from t∼1 ms up to t∼3 ms) as indicated in Figure 7.3A and C by the red bar. Such high velocities of "plasma bullets" can be explained by a streamer model, which we will invoke here following previous works [31]. If in fact a streamer model is applicable, the ICCD photos in Figure 7.3B (a and b) are images of the propagation of a positively charged head of streamer photoionizing the gas in front of it and causing avalanche directed toward the head.

The microwave scattering signal (see Figure 7.3A) consists of two sequential peaks. The first peak occurs at the same time that the ICCD camera registered the presence of the streamer.

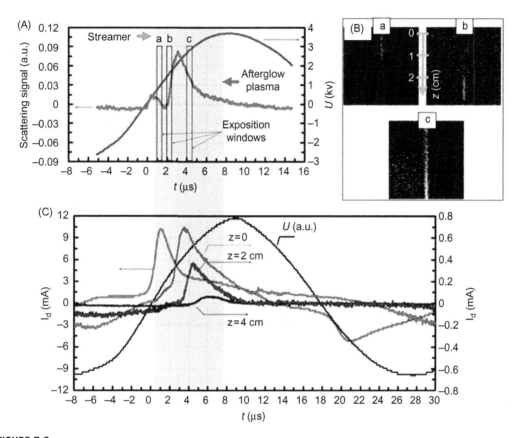

FIGURE 7.3

(A) Scattering signal from the jet and voltage waveform. Gray bars a, b, and c indicate time windows, when ICCD images were taken. (B) ICCD images of jet at a, b, and c: "plasma bullet" velocity is ~2 × 10⁶ cm/s. (C) Discharge and jet currents at different z positions. Red bar indicates time when streamer exists. During this time, the first peak in scattering signal and I_d peak were observed. Blue bar indicates time after streamer passing when signals from plasma were registered. Scattering signal and jet current indicate the presence of the afterglow plasma column with a characteristic decay time of ~3–5 μs. (For interpretation of the references to color in this figure legend, the reader is referred to the web version of this book.)

Source: *Reprinted with permission from [119]. Copyright (2009) by American Institute of Physics.*

Note that the first peak starts even before the head comes out from the syringe, and this may be caused by the electron avalanche developing in front of the streamer head or by scattering from the plasma in the syringe. The second peak on the microwave scattering signal (indicated by blue bar in Figure 7.3A) and image (c) in Figure 7.3B (integrated over 100 shots of 500 ns exposition time each)) was registered after the streamer had already gone. Thus we may conclude that an afterglow plasma column remains after streamer passing, having a decay time of ~3–5 ms. This fast decay of the column plasma may be explained by three body and dissociative attachment of electrons to oxygen molecules present in ambient air.

Streamers observed by high-speed cameras in atmospheric jets (see Figure 7.3) were characterized by a new method of stopping ("scattering") by means of external DC potential [52]. Interaction of the streamer with the DC potential created by the ring located at $z = 3$ cm and dependence of streamer length L as a function of the ring potential U_r are shown in Figure 7.4. One can see that application of higher positive potential to the ring electrode led to significant perturbation of the streamer propagation, namely to shortening of the streamer. Basing on this experimental evidence, the term "streamer scattering on the external DC potential" was introduced to denote this interaction.

As the electric field around the streamer head is governed by difference $U_h - U_r$, where U_h is potential of streamer head, one can consider the condition $U_r = U_h$ as sufficient to stop propagation of the streamer [50,53]. Figure 7.5 presents the temporal evolution of streamer head potential $U_h(t)$ for the amplitude of driven high voltages $U_{HV} = 2.6$, 3.1, and 3.8 kV, superimposed with the temporal evolution of voltage applied to the discharge electrodes. The discharge current is also shown. It can be seen that in all cases, U_h was close to the voltage applied to the electrodes (within $10-15\%$) and followed its temporal evolution. Thus, the experimental evidence does not support the model of the electrically insulated streamer head. On the contrary, the experiment indicates that the electrode potential is transferred to the streamer head along the streamer column to which it is attached with no significant voltage drop. Measurement of the streamer head potential allow to determine number of key streamer properties such as head charge ($1-2 \times 10^8$ electrons), electrical field

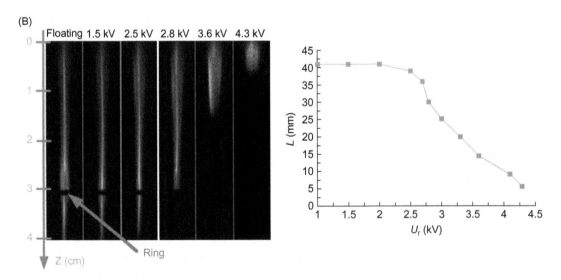

FIGURE 7.4

Interaction (scattering) of the streamer on the DC potential created by the ring located at $z = 3$ cm and dependence of streamer length L vs. ring potential U_r (exposure window $t = 2-7$ ms, $U_{HV} = 3.1$ kV).

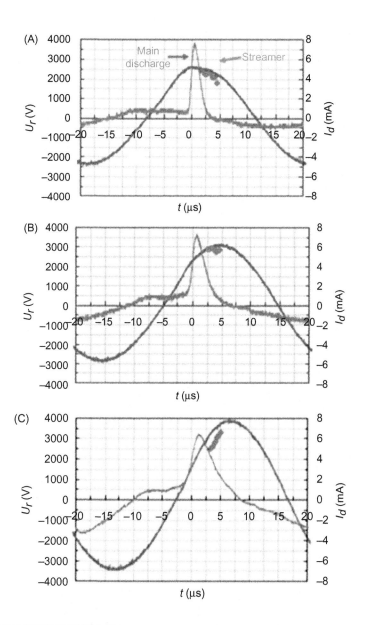

FIGURE 7.5

Temporal evolution of discharge current (green), discharge voltage (blue) and streamer head potential (red) for U_{HV} = 2.6, 3.1, and 3.8. Potential of streamer head is close to potential of central electrode and following its temporal behavior in all cases. This indicates that voltage drops potential of central electrode is transferred to the streamer head without significant drops. (For interpretation of the references to color in this figure legend, the reader is referred to the web version of this book.)

Source: *Reprinted with permission from [52]. Copyright (2012) by Institute of Physics.*

Table 7.1 Typical Parameters of the Streamer on the Growth Stage

Head Charge, Electrons	Streamer Length, cm	Speed of Ionization Front, cm/s	Streamer Radius, cm	Characteristic Electrical Field in Head Vicinity, V/cm	Average Conductivity in the Streamer, Ω^{-1} cm^{-1}	Average Plasma Density in the Streamer, cm^{-3}
$1-2 \times 10^8$	4–5	$1.5-2 \times 10^6$	3×10^{-2}	$\sim 10^5$	$3-7 \times 10^{-3}$	$2-6 \times 10^{13}$

He flow = 11.5 l/min and U_{HV} = 2.6–3.8 kV.

in the head vicinity (~100 kV/cm), average conductivity ($3-7 \times 10^{-3}$ W^{-1} cm^{-1}), and plasma density of the streamer column ($2-6 \times 10^{13}$ cm^{-3}). All measured parameters are listed in Table 7.1.

In addition to measurement techniques presented in this section, optical emission spectroscopy (OES) is also widely used for characterization of the atmospheric pressure plasmas. Plasma composition and temperature at various modes (translational, rotational, vibrational) can be measured using this technique. Composition of the CAP jet is shown in Figure 2.13.

7.2 Cold plasma interaction with cells

Among others, CAP treatment leads to modification in cell migration as will be described below. In particular, these results suggest that by controlling cell migration, it is possible to stabilize wounds. As an immediate consequence, this would lead to mitigation of tissue damage after injury. It should be pointed out that wound healing can be delayed in the elderly and in patients with diabetes. These delays can be due to problems with any or all of the three phases of wound healing cited above. Ulcers form when wound healing is impaired and can lead to amputation and loss of life if not treated. Delayed reepithelialization or poorly adherent epithelial tissue fails to provide the needed barrier and contributes to delayed healing. Persistent inflammation in the wound bed will also cause delayed healing. Finally, poor reestablishment of circulation caused by delayed angiogenesis also causes challenges to wound healing, particularly in diabetics and the elderly. Paradoxically, some patients, particularly young adults, have robust wound healing responses which also results in complications. When wound healing takes place too quickly, reduced epithelial stability and increased scar formation result. If the epithelial barrier is not tightly adhere to the underlying connective tissue, the newly formed epithelial sheet over the wound site will rip off due to scratching and/or normal wear and tear, and the adjacent epithelial cells will have to initiate reepithelialization once again. This exposes the underlying tissue once again to microbes and increases the risks of infection. Scars are thick disordered accumulations of collagen and connective tissue elements produced by the fibroblasts during the proliferative phase of wound healing. During the remodeling stage of healing, these scars have to be remodeled to allow proper regrowth of nerves and blood vessels. Remodeling involves the selective secretion of proteases that degrade collagen. Excessive scar formation is more than just a cosmetic problem. It can slow down recovery by preventing angiogenesis and the regrowth of nerves. However, for injuries to exposed sites on the face, scar formation is indeed a serious cosmetic concern [54]. Proper treatment of traumatic and chronic injuries demands that we monitor the healing process in each patient and step in to apply treatments to enhance healing when it is delayed or to slow it down when it is taking place too rapidly. Thus *cell migration reduction* triggered by the cold plasma treatment becomes very important. While treatments are available for the purpose of improving wound healing, more are needed. Although several recent studies have begun to sort out the cellular and subcellular events altered when cold plasmas interact with living cells, these studies have yet to address the dose–response (the relationship between duration of plasma jet interaction with living tissue and a change in the migration rates) and permanency of the migration effects induced in cells by cold plasmas.

In addition, migration of the tissue cells plays an important role in many physiological and pathological processes, including embryonic development, angiogenesis, and the metastasis of tumor cells. The migration of the cells based on the interaction (mechanical, electrical, and chemical) between the cell adhesion receptors and the extracellular matrix. The most important family of cell adhesion proteins are integrins. As their name suggests, integrins are cell adhesion receptors that integrate signals received from the extracellular matrix with intracellular elements of the cytoskeleton to regulate mechanical signal transduction. In so doing, integrins function to maintain cell adhesion, tissue integrity, cell migration, and differentiation [21,55]. These issues will be discussed in the following sections.

7.2.1 **Cell migration**

Cell migration is a broad term that describes those processes that involve the translation of cells from one location to another. This may occur on glass/plastic (common *in vitro* setups), or within complex, multicellular organisms. Cells migrate in response to multiple situations they encounter during their lives. Some examples include morphogenetic events that require the mobilization of precursors to generate new structures/layers/organs, sometimes at distant locations (during embryogenesis, organogenesis, and regeneration); or the presence of environment cues that inform the cells of the need for their movement to accomplish a larger goal (e.g., wound healing or the immune response).

There are different modes of cell migration depending on the cell type and the context in which it is migrating. Cells can move as single entities, and the specifics of their motility depend on several factors, e.g., adhesion strength and the type of substratum (including extracellular matrix ligands and other cells), external migratory signals and cues, mechanical pliability, dimensionality, and the organization of the cellular cytoskeleton. Cell migration can be characterized by velocity, which is the measure of speed at which cells change their location.

The unique chemical and physical properties of CAPs enable their numerous recent applications in biomedicine [56]. A wide range of cold plasmas applications have been investigated including sterilization, the preparation of polymer materials for medical procedures, wound healing, tissue or cellular removal, and dental drills [35,57]. One of the recent research trends is the investigation of CAPs interaction with living tissue at the cellular level [58–60]. Initial experiments on direct interaction of cold plasmas with living cells demonstrated immediate detachment of treated cells from the extracellular matrix [58]. Later it was demonstrated that cell detachment occurs for several different cell types including primary mouse fibroblast cells, PAM212 (skin) cancer cells, and

BEL-7402 liver cancer cells [58]. The effects of mild intensity and short duration cold plasma treatment below the threshold required for cell detachment have also been studied [61]. It was observed that the migration rate of primary mouse fibroblast cells slowed significantly after mild intensity plasma treatment [61]. It was found that integrin expression at the cell surface reduced at plasma treatment. Integrins (transmembrane proteins) are cell adhesion receptors having dual function: intracellular (integrin occupancy coordinates the assembly of cytoskeletal filaments and signaling complexes) and extracellular (engaging either extracellular matrix macromolecules or counter receptors on adjacent cell surfaces). Integrins function in maintaining cell adhesion, tissue integrity, cell migration, and differentiation [55,62,63].

The interaction of the CAP with a living tissue was examined on fibroblast cells and PAM212 cells (mouse squamous carcinoma cells) [64]. Uniform densities of cells in culture were grown on the surfaces of six-well tissue culture dishes using the methodology described in Ref. [65]. The schematic of the treatment is shown in Figure 7.6.

Experiments were conducted with different volumes of media covering the cell. Immediately following the plasma jet treatment time-lapse cell migration analysis was performed. Cells were maintained in normal cell culture media containing serum at 36°C and 5% CO_2. Cell velocity distributions were built and analyzed using SAS software [66].

Interaction of CAP jet with cells leads to media displacement away from the point of contact of jet with cells, and for the duration of treatment, the cells were not covered by media as shown in Figure 7.6. Immediately after jet interruption, the media flowed over the cells. Three distinct regions were observed as shown in Figure 7.7. Region 1 is the zone which stayed uncovered during

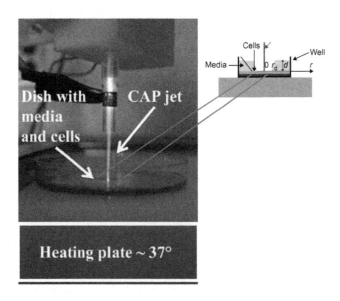

FIGURE 7.6

CAP jet treatment of the living tissue. The distance between nozzle and plate bottom was of around ~20 mm during all treatment. The cell culture was placed on the heating plate to maintain the liquid media temperature ~37°C. Inset shows schematics of media and cells in the vicinity of CAP action.

the treatment where the cells were subject to desiccation and after the treatment was occupied by immovable or "frozen" dead cells. Second region is the area which remains covered with a media at all times. This region was filled up with migrating alive cells. A third region which was at the interface between the two areas with alive and dead cells contained voids with no cells. Images taken in the center of region with frozen cells at $r = 0$ and at the edge $r = r_d$ are shown in Figure 7.7A and B immediately after 30 s treatment with plasma jet demonstrating all three regions. Note that no voids or frozen cells were found after treatment of the cell culture with He flow only without plasma at the same conditions as that on Figure 7.7C and D. After treatment with He flow without plasma, no voids were found in the cell culture as shown in Figure 7.7D.

Alive cells tracks are shown in Figure 7.8. These observations led to the conclusion that untreated cells display higher mobility.

The velocity of cells treated with these conditions was compared to that of untreated cells and cells treated with helium only (80 cells were tracked in each case) The velocity distributions are presented in Figure 7.9 (velocities are normalized to the average velocity of untreated cells). The cell velocity distribution after treatment with helium (no plasma) coincides well with the distribution exhibited by untreated cells. Treating with the plasma jet decreased the average cell migration rate and standard deviation by a factor of ~ 2 in comparison with both untreated cells and cells treated with helium (no plasma).

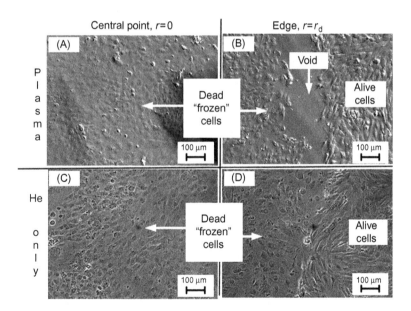

FIGURE 7.7

Photographs of the fibroblast cell culture immediately after treatment: (A) and (B) with CAP jet at the central point $r = 0$ (A) and at the edge between the areas with alive and dead cells $r = 3$ mm (B); (C) and (D) with He flow only.

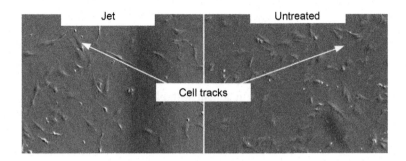

FIGURE 7.8

Cell tracks (in red) after 16 h of migration: treated (with plasma jet) cells and untreated cells.
(For interpretation of the references to color in this figure legend, the reader is referred to the web version
of this book.)

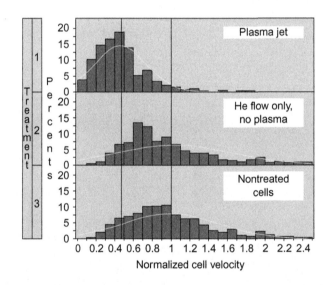

FIGURE 7.9

Velocity distribution functions for fibroblast cells treated with (1) plasma, (2) He flow only, and (3) nontreated
cells.

Source: *Reprinted with permission from [24]. Copyright (2008) by American Institute of Physics.*

The migration of cells within tissues plays an important role in many physiological and pathological processes, including embryonic development, wound repair, angiogenesis, and the metastasis. Various signaling pathways control these processes. Integrins are a major family of metazoan
cell-surface-adhesion receptors playing a key role in the signaling and mechanotransduction
(i.e., conversion mechanical stimulus into the chemical activity) mechanisms. Their functions
include maintaining cell adhesion, tissue integrity, cell migration, and differentiation. Integrins

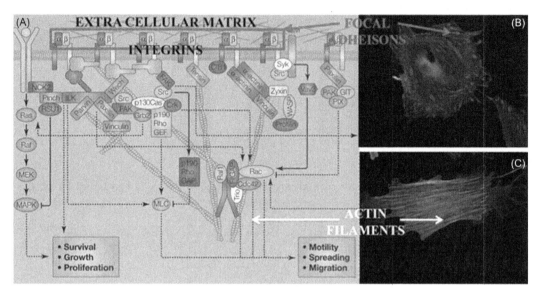

FIGURE 7.10

(A) Structure and functions of integrins [57]. (B and C) WTDF 3° (wild-type dermal fibroblast tertiary passage) cells were stained against vinculin (focal adhesions, red) and actin filaments (cytoskeleton, green). (For interpretation of the references to color in this figure legend, the reader is referred to the web version of this book.)

consists of two noncovalently associated subunits α and β subunits, each of which is a single-pass type I transmembrane protein. There are 24 types of α subunits and 8 types of β subunits. However, not all $\alpha\beta$ combinations exist, only 24 are possible as shown in Figure 7.10 [67–69].

An integrin expression study was performed in order to understand the effects of the plasma jet on fibroblast cells migration. Fibroblast cells 4 days after being placed in cell culture were used. Treatment with plasma jets consisted of consecutive exposure of seven adjacent positions. Each position was exposed to the plasma jet for 5 min, while the thickness of protecting media was kept ~2 mm during the entire treatment. Cells whose surface integrins were stained with the β1-integrins-PE-conjugated antibody expressed about two orders of magnitude higher fluorescence response—$I_{\beta1} = 440$. This increase of fluorescence intensity is caused by fluorescence of antibodies bound to the β1-integrins on the cell surface. Similar as above, stained cells had a higher $I_{\beta1}$ of 310. However, their mean fluorescence intensity was significantly less (~30%) than that for untreated cells (440). The histograms summarizing change of average fluorescence responses after cell treatment and their deviations are presented in Figure 7.11. To summarize, we observed that treatment of cells with the plasma jet resulted in a decrease of cell fluorescence responses: 25% and 10% for staining of β_1 and α_v-integrins, respectively. This clearly indicates that plasma treatment reduces expression of β_1 and α_v surface integrins. This suggests that one of the elementary cell components being affected by plasma treatment are surface integrins. Because surface integrins are responsible for cell adhesion and mediate cell migration, changes in integrin

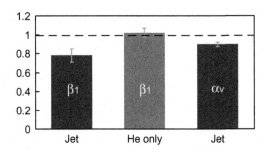

FIGURE 7.11

Average fluorescence treated with plasma cells (stained with β_1- and αv-integrins PE-conjugated antibodies) and treated with He only cells (stained with β_1-integrins antibody) normalized to fluorescence of untreated cells stained with corresponding antibody.

Source: Reprinted with permission from [60]. Copyright Wiley-VCH Verlag GmbH & Co. KGaA.

expression may be responsible for the effects observed experimentally, including the decrease in migration rate and cell detachment after plasma jet treatment.

In addition, it is important to determine the region that is affected by CAP during the treatment, i.e., to determine how localized CAP treatment could be [70]. A schematic diagram indicating how these treatments were made is shown in Figure 7.12A. Controls included cells treated with helium alone for the same times. Cells were treated with plasma immediately after plasma treatment with trypan blue. It was shown previously that plasma treatments can cause cell death as assessed using trypan blue. Trypan blue is a vital dye that is excluded from entering live cells; dead cells fill with the blue dye. Data shown in Figure 7.12B for cells treated with plasma or helium alone for the times tested indicate that trypan blue is excluded from cells and therefore these treatment times do not cause cell death.

When cell migration is assessed as a function of the duration of plasma treatment, it was found (as shown in Figure 7.12C) that 100 s significantly reduced cell migration rates when assessed from 0 to 16.5 h after plasma treatment. Treatment times up to 500 s did not further reduce cell migration significantly when assessed between 0 and 16.5 h; however, when migration rates were assessed between 16.5 and 33 h (see Figure 7.12D), it was found that 100 s significantly reduced cell migration rates.

In order to address the effect of plasma treatment on pH-level of media, treatments of media at standard experimental conditions described above was performed. One of the components of cell culture media is the acid−base indicator phenol red. At acidic pH, the color of phenol red is yellow; at basic pH, phenol red is pink, and at neutral pH, phenol red is red/orange. These features of phenol red, along with its lack of toxicity, make it an excellent indicator of pH in cell culture media. No change of media color was observed after the 60, 120, and 200 s treatments of media (data not shown) indicating that pH-level of media remained unchanged at plasma treatment.

The ability of CAP treatment to impact cell migration was further considered by evaluating cell migration rates as a function of the distance from the plasma-treated zone. A schematic of the experiment is shown in Figure 7.13A. Three locations of interest were considered: inside the

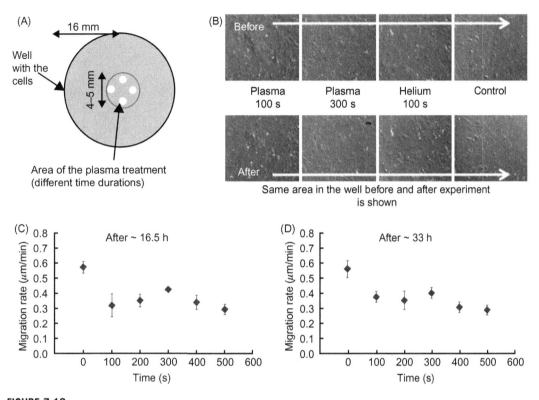

FIGURE 7.12

Reduction of the cell migration in WTDF 3. Temporal dynamics. (A) Schematic representation of the experiment is shown: reduction in migration rates dependence on the duration of the plasma treatment. Four points were taken inside the treated area and 40 cells analyzed. (B) "Alive-vs.-dead" testing of the treated cells with trypan blue stain with the same area in the well analyzed before and after treatment. Cell migration rates for ~0−16.5 h (C) and ~16.5−33 h (D) after treatment as function of treatment duration. Error bars indicate the standard deviations of data points.

plasma-treated area (red circle indicates the treated area, diameter of around 5 mm) and further equidistant at second and third locations outside, each location between two neighbor points was ~3−4 mm apart. The average velocity of 10 cells per each tracked location was calculated after 16.5 h. To increase the statistical significance of the results, data were taken symmetrically in six different locations thus giving 20 cells per point of interest. Experiments were repeated several times with various treatment times of 100 and 300 s. Thus, overall ~60 cells per each location were analyzed. Figure 7.13B shows spatial distribution of migration rates after 100 s of plasma treatment (data shown for 20 cells at each point after 16.5 h of tracking). Both the only helium and control velocity distributions do not show any significant changes. However, plasma-treated cells showed a localized reduction in their migration rates of around 30−40%. The standard deviation is indicated with vertical bars at the data points; it does not exceed 5%.

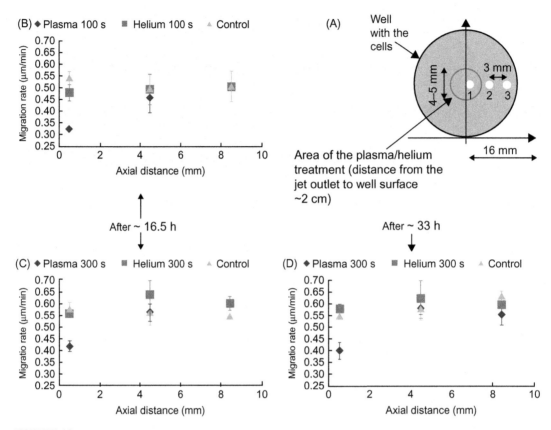

FIGURE 7.13

Localization of the cold plasma effect on the cell migration. (A) The experimental setup for spatial distribution of the cell migration rates for single well is shown. (B–D) The dependence of cell migration rates vs. distance from the center of treated zone: for cells tracked during 0–16.5 h after 100 and 300 s plasma treatment, respectively (B and C), for cells tracked during 16.5–33 h after 300 s plasma treatment (D). Cells treated with plasma, with helium only, and untreated cells are shown. Data is given with standard deviations. The reduction in migration rate is around 30% in the treated area after ~16.5 h (B and C), this trend remains after ~33 h (D) of cell tracking.

Spatial characteristics of the effect in cell migration induced by the plasmas treatment are shown in Figure 7.13B and C for cells tracked between 0 and 16.5 h after treatment (cells treated during 100 and 300 s) and in Figure 7.13D for cells tracked between 16.5 and 33 h after treatment (time duration of treatment 300 s).

Again for controlling purpose, the only helium treated cells and untreated cells are shown. Figure 7.13B and C shows that cells treated with plasma (100 and 300 s) have slowest migration rates in the treated area (~1 mm from center of the plate), while cells outside the treated area (5 and 8 mm from the center) migrate at the same rates as untreated cells. Also no reduction of cells migration was observed in the cells treated with the only helium. It was observed that reduction (around 30%) in cell migration is persistent for 33 h (see Figure 7.13D).

In conclusion, the reduction of the rates of the cells migration can be related to the changes in the integrin expression whose functions are related to cell motility and cell adhesions. In the following section, we present more detailed analysis of this effect.

7.2.2 Integrin activation by a CAP jet

Despite the progress made in studying of CAP interaction with the living tissue described above, questions remain. What causes the motility and cell adhesion to change in response to plasma treatment? [71].

The goal of this section is to understand the mechanism by which the CAP jet alters cell migration and influences adhesion [72]. Although the main focus here is on the study of the interaction between the CAP jet and fibroblasts, we also present data comparing fibroblast responses to those of epithelial cells.

Wild-type tertiary mouse fibroblast cells (WTDF) were cultured in the Dulbecco's Modified Eagle Medium (DMEM; Invitrogen Corp., Carlsbad, CA) enriched with 5% serum, 1% NEAA, 1% L-glutamine, and 1% Pen-Strep. Diluted cells (30% confluence) were plated in multiwell plates or double-well glass slides and treated with plasma on the third day in culture. Human corneal limbal epithelial cell line (HCLE, Dr. Ilene Gipson, Harvard Medical School and described in Ref. [73]) was grown in HCLE medium (500 ml Keratinocyte Serum Free Medium with added 1.2 ml Bovine Pituitary Supplement, 4 μl epidermal growth factor, 0.17 mM $CaCl_2$ and 5 ml Pen-Strep, Invitrogen Corp., Carlsbad, CA). HCLE cells were used on the second day of culture with confluence of around 30−40%. During plasma treatment, plates with cells were kept on a heating plate (Boekel Scientific, model 240000, Feasterville, PA) to maintain the media temperature at 37°C. Serum starvation experiments were performed on WTDF 3 cells and placing the cells in DMEM media with 1% serum ∼24 h before the plasma treatment. For studies using $MnCl_2$ to activate integrins, 0.5 mM MnCl2 was added to DMEM with 5% serum immediately after cells were subjected to cold plasma treatment.

Some cold plasma treatments were conducted on cells that had been placed in plates precoated with extracellular matrix proteins. Fibroblasts were placed onto plates precoated with either 10 mg/ml fibronectin (BD Bioscience, Pasadena, CA, cat#354008)/1% collagen type I (Advanced BioMatrix, San Diego, CA, Part#5005) or 10 ug/ml vitronectin (BD Bioscience, cat#354238). Plasma treatment was performed 8 h after replating.

In the described studies, dermal fibroblasts and epithelial cells were prepared and equivalent numbers plated onto culture plastic (for time-lapse studies) or glass slide (for immunostaining) and maintained at DMEM medium supplemented with 5% serum (fibroblasts) or HCLE medium (epithelial cells). Fresh media or fresh media with the $MnCl_2$ was added to cells following plasma treatment.

For experiments involving immunofluorescent staining of cells, cells were plated onto two well glass chamber slides for 24 h prior to CAP treatment. For cold plasma treatment, chamber slides containing cells were placed on the heating plate to maintain the temperature and treated with cold plasma as described above. Fresh media was added and cells returned to the 37°C CO_2 incubator for 3 h. Media was discarded and cells fixed in 4% paraformaldehyde (Prod#28908, Thermo

Scientific, Rockford, IL) in PBS for 10 min and then rinsed in PBS. Nonspecific staining of cells was blocked by incubating the cells for 15 min in blocking buffer (1% Bovine Serum Albumin (BSA) in 1X PBS). Cells were incubated in the appropriate dilution of each of the primary antibodies below for 1 h, followed by washing (15 min in PBS) and incubation in secondary antibodies. When the primary antibodies used were generated in different species, cells were stained with multiple antibodies simultaneously. For studies of active and total surface integrins, cells were used for immunostaining without permeabilization. However, for studies of vinculin, and the α_v receptor within focal adhesions, cells were permeabilized by incubating in 0.1% triton-$\times$100 for 10 min and then incubated in PBS. Permeabilization is needed because vinculin is an intracellular protein and antibodies to detect it cannot penetrate the cell unless it is permeabilized. Total β_1 integrin was detected with the hamster-derived CD 29 monoclonal antibody clone Ha2/5 (BD Bioscience, cat#555004) at a dilution of 1:200. For activated β_1 integrin, the rat-derived monoclonal antibody against activate CD 29, clone 9EG7 (BD Bioscience, cat#553715), was used at a dilution of 1:200. αv integrin was visualized using the rat-derived monoclonal antibody against CD 51 RMV7 (BD Bioscience, cat#552299) at a dilution of 1:200. Focal adhesions were visualized by staining for the focal adhesion protein vinculin using the mouse-derived monoclonal antibody clone VIIF9 (7F9) at a dilution of 1:200 (cat#MAB3574, Chemicon Int., Temecula, CA). Nuclei were visualized using DAPI, dilution 1:2000 (D21490; Invitrogen Corp., Carlsbad, CA). Images were acquired at $20\times$ magnification using a Nikon Fluorescent microscope equipped with a RT-Slider SPOT Camera (Melville, NY).

7.2.2.1 Dermal fibroblasts and human corneal epithelial cells reduce their migration in response to CAP

While it was described above that dermal fibroblasts migrate slower after CAP treatment, a difference between behavior of cell type will be described in this section. Presented in Figure 7.14A are relief-contrast images of the tracks taken (red lines) by migrating fibroblasts (WTDF 3) and epithelial (HCLE) cells with and without CAP treatment over a 16 h and 40 min time period.

The tracks of the cells after plasma treatment (fibroblast cells were treated for ~60 s, epithelial cells for ~100 s) are significantly shorter than untreated cells (control). Additional experiments were conducted to assess the impact of increasing plasma treatment times on cell migration rates of fibroblasts and epithelial cells, i.e., to determine the threshold of treatment (threshold here is determined as the treatment dose (duration of the treatment) after which the change becomes statistically nonsignificant (i.e., $P > 0.05$ between two neighbor points) and no other change occurs). In Figure 7.14B, cell migration rates are shown as a function of the duration of CAP treatment. Fibroblasts show a maximum drop in the migration rate at ~40 s; increasing treatment times ranging from 40 to 200 s do not alter cell migration rates. The impact of CAP on the migration of epithelial cells is more graded; migration rates of epithelial cells decrease with increasing treatment times from 10 to 100 s; treatment times between 100 and 200 s do not alter epithelial cell migration rates. The decrease in cell migration rates for both cell types is ~30–40%. It was shown early that helium treatment only does not affect the cell migration without additional conditions such as serum starvation or $MnCl_2$ activation. Figure 7.14C shows the persistence of the cell motion as a function of the duration of the plasma treatment. The persistence was measured as the ratio between displacement in a particular direction and the total displacements and, thus, represents the

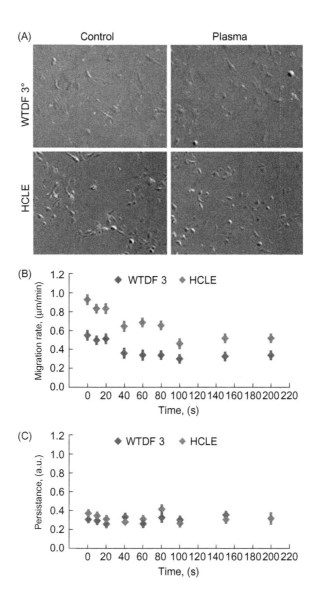

FIGURE 7.14

Temporal dynamics of the cells in response to CAP treatment. (A) Relief-contrast images of the WTDF 3 and HCLE cells with their tracks: control (no plasma treatment), plasma (WTDF 3–60 s, HCLE ~100 s). (B and C) The data is shown for ~16.5 h of tracking. Error bars indicate the standard error of mean for the presented data. At least ~80 cells were analyzed per each time point. The migration rates and the persistence of the cells are assessed as a function of the plasma treatment time: fibroblast cells are shown in blue and epithelial cells in red. (For interpretation of the references to color in this figure legend, the reader is referred to the web version of this book.)

Source: *Reprinted with permission from [72]. Copyright (2012) by Institute of Physics.*

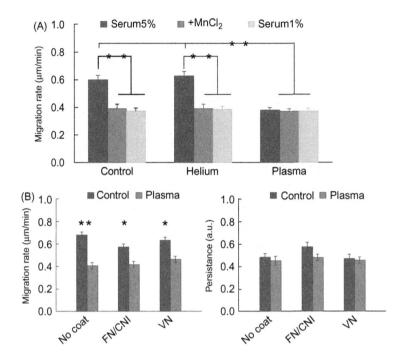

FIGURE 7.15

Integrin activation of the fibroblasts reduces their response to CAP. (A) The migration rates of the fibroblast cells controls (no CAP treatment and treated with helium only) and cells treated with plasma for ~100 s are shown under different conditions: blue—no additional pretreatments, red—treated with $MnCl_2$ and green—serum-starved cells. Data is shown for ~16.5 h of tracking. (B) The migration rates (μm/min) and persistence of the cells motion (a.u.) of the WTDF 3 cells control (blue)—no plasma treatment and cells treated with plasma for ~100 s (red) plated on a different precoated surfaces: no precoating, precoated with fibronectin/collagen type I and with vitronectin. The data are shown for the first 6 h of the cell tracking. The standard error of mean is shown with error bars. At least ~80 cells were analyzed per each conditional change. Double asterisk stands for the $P < 0.001$ and single asterisk for $P < 0.01$, otherwise there are no statistical significant difference in the data. (For interpretation of the references to color in this figure legend, the reader is referred to the web version of this book.)

Source: *Reprinted with permission from [72]. Copyright (2012) by Institute of Physics.*

directionality of the cells motion. Neither fibroblasts nor corneal epithelial cells show significant changes in persistence as a function of increasing plasma treatment time.

7.2.2.2 *Activating fibroblast integrins reduces their response to CAP*

One of the possible reasons for the decrease in cell migration rate after CAP treatment is the activation of integrins on cell surfaces. Integrins can be activated by adding $MnCl_2$ to their media [74]. This treatment shifts the integrins from a folded state to an unfolded state exposing their ligand binding sites. Serum starvation of fibroblasts increases their spreading which induces activation of

$\alpha_5\beta_1$ integrin [75]. Figure 7.15A shows the migration rates of fibroblasts under control (not treated, only helium treated for 100 s) and CAP treatment (100 s) followed by incubation in standard media, media supplemented with $MnCl_2$, and low serum media. In the case of no additional treatment, we found a 30–40% drop in the migration rates of the cells as expected. However, under conditions where integrins were activated, no change in migration rate was observed after CAP treatment. The decrease of cell migration rates for all three conditions is 30–40% of the optimal cell migration rate. These data suggest that CAP treatment activates integrins on cell surfaces.

Plating cells onto surfaces coated with integrin ligands activates integrins engaged by those ligands. β_1 family integrins bind to fibronectin (FN) and collagen (CN), whereas α_v-family integrins bind to vitronectin (VN). Next we assess whether plating cells onto FN/CNI- or VN-coated surfaces impacts their ability to respond to CAP. In Figure 7.15B, cell migration rates and persistence of migration for fibroblasts plated on uncoated surfaces and surfaces precoated with FN/CNI or VN are shown for untreated and for CAP (100 s) treated cells. Data are shown for just the first 6 h of cell tracking, i.e., right after cells were adhered to the surface. There are no significant differences in cell migration rates between cells plated onto uncoated compared to FN/CNI- and VN-coated surfaces for the control cells ($P > 0.05$). In addition, precoating surfaces with FN/CNI or VN did not alter their response to plasma treatment and a drop of ~30% in cell migration rate was seen for the three conditions studied. Statistical analyses of these data showed that there is greater statistical significance ($P < 0.001$) for fibroblasts plated on the uncoated surface rather than on the surface precoated with FN/CNI or VN ($P < 0.01$). No significant changes were found in the persistence of the cells motion. Note also that helium alone does not affect the cell migration [72].

7.2.2.3 Integrins activation by CAP

It was shown above that integrin activation by $MnCl_2$ and starvation attenuates the response of cells to CAP, but adhering cells to specific integrin ligands had no impact on cells ability to respond to CAP. To study directly the impact of CAP on integrin activation, control cells and cells treated with CAP were fixed and used for immunofluorescence studies without permeabilization to assess the ratio of activated to total β_1 integrin on their surface. The method requires the use of an antibody that recognizes total β_1 integrin regardless of conformation (Ha2/5) and another that recognizes only the active conformation of the β_1 integrin (9EG7). Figure 7.16 shows the results of immunofluorescence studies to assess the activation state of the β_1 integrin in the CAP treated fibroblasts. Untreated cells (control), cells treated with $MnCl_2$ only, CAP (100 s) treated cells, and cells treated with plasma (100 s) and $MnCl_2$ are shown in Figure 7.16A; Figure 7.16B shows an enlargement of the image shown in Figure 7.16A. Data show that activated β_1 integrin is present at significantly higher amounts in cells treated with $MnCl_2$, CAP, and both CAP and $MnCl_2$ confirming that β_1 integrin is activated by CAP treatment (see Figure 7.16C).

7.2.2.4 Focal adhesion of treated cells

Activated integrins accumulated within focal adhesions increasing their size and reducing cell migration rates. To determine whether focal adhesions of CAP treated cells were indeed larger, control cells and cells treated with CAP were fixed, permeabilized, and used for immunofluorescence to detect the presence of vinculin as well as αv integrin within focal adhesions. The data are presented in Figure 7.17A and quantified in Figure 7.17B. Vinculin is a focal adhesion protein and the amount of the protein engaged into the focal adhesion is proportional to the size of the focal

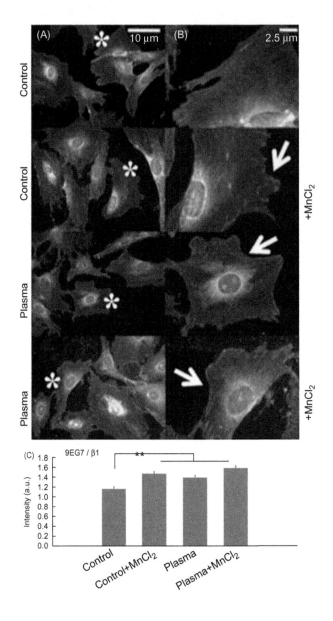

FIGURE 7.16

Activation of the β1 integrin with CAP. The immunofluorescence images of the fibroblast cells with different scales are shown: control (no treatments), control + MnCl₂ (treated with only MnCl₂), plasma (treated with CAP 100 s), plasma + MnCl₂ (treated with both plasma 100 s and MnCl₂). Total β1 integrin are shown in green, activated β1 (clone 9EG7)—in red and nuclei in blue. (A) Images were taken with the magnification 20 ×. Cells enlarged in Figure 3.3B are marked with the white asterisk. (B) Enlarged typical cells are shown. White arrows indicate the increase of the activated β1 integrin. (C) The data represents the ratio of 9EG7 and total β1 integrin for the peripheral part of the cells; error bars stand for with the standard error of mean, ∼20 cells were analyzed per each conditional change. The double asterisk shows the statistical significance ($P < 0.001$) in the increase of the intensity of the 9EG7 over the total β1 in the cells treated with MnCl₂, plasma, or both. (For interpretation of the references to color in this figure legend, the reader is referred to the web version of this book.)

Source: *Reprinted with permission from [72]. Copyright (2012) by Institute of Physics.*

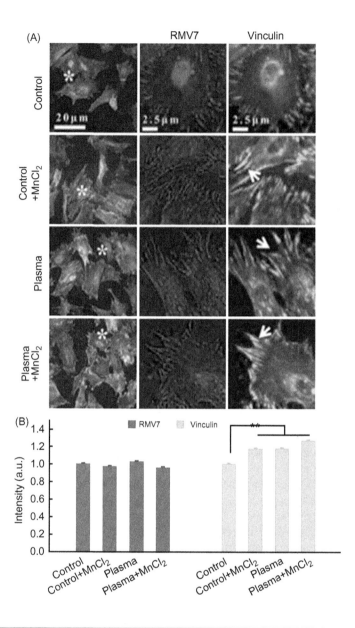

FIGURE 7.17

CAP treatment induces enlargement of the focal adhesions. (A) The immunofluorescence images of WTDF 3 cells (control—no treatments, control + MnCl$_2$—treated with MnCl$_2$ only, plasma—treated only with plasma 100 s and plasma + MnCl$_2$—treated with both plasma 100 s and MnCl$_2$) are shown. αv (RMV 7) integrin are shown in red, focal adhesion protein vinculin—in green and nuclei—in blue. The first column shows the images taken with the magnification 20 ×, the white asterisk indicates the cells enlarged in the second and third columns: the second shows only αv (red), the third one—only vinculin (green). The change in the size of the focal adhesions is shown with white arrows. (B) Statistical data of the pixels intensities (a.u.) of the RMV 7 and vinculin (normalized to the controls) are shown. The statistically significant increase ($\sim$20%) in the vinculin intensity is shown with the double asterisk ($P < 0.001$). The error bars indicate the error of the mean for the presented data. (For interpretation of the references to color in this figure legend, the reader is referred to the web version of this book.)

Source: Reprinted with permission from [72]. Copyright (2012) by Institute of Physics.

adhesion [76]. Thus the intensity of the fluorescence antibody will be proportional to the size of the focal adhesion. CAP treated cells, cells treated with $MnCl_2$ only, or cells treated with both CAP and $MnCl_2$ show an overall increase in vinculin intensity and thus have larger focal adhesions [76]. Whereas a statistically significant ($P < 0.001$) increase of ~20% is seen for vinculin intensity at the periphery of $MnCl_2$, CAP, or both treated cells. However, no significant changes were found in the expression of the total α_v integrin in ~3 h after treatments.

7.2.2.5 Relationship between integrins and focal adhesions

The results shown in Section 7.2.2.3 demonstrate that CAP activates $\beta1$ integrin, slows down cell migration, and increases focal adhesion size. The ability of CAP to alter cell migration rates is not specific to fibroblasts but extends to human corneal epithelial cells; both types of cell reduce their cell migration rates by maximum 30–40% after CAP treatment. The extent of reduction in cell migration rate after CAP in fibroblasts and epithelial cells was identical to the reduction in cell migration achieved by activating integrins using $MnCl_2$ or by serum starving cells.

In both cell types, CAP decreases cell migration as a function of plasma treatment time up to a threshold above which no further change in cell migration is seen. That threshold is reached at 40 s for fibroblasts and 100 s for epithelial cells. The difference in threshold between cell types suggests that epithelial cell adhesion is mediated by integrins that are more adhesive and more difficult to activate. Epithelial cells but not fibroblasts express $\alpha6\beta4$ integrin, which acts like cellular glue to mediate the tight adhesions needed between the epidermis and underlying dermis [71]. It is likely that the increased adhesion seen in epithelial cells underlies the higher threshold seen in CAP treatment time required to reduce epithelial cell migration by 30–40%. It is important to note that CAP treatment of ~40 s leads to discrimination between fibroblast and epithelial cells thus allowing differential treatment of various cells presented in tissue.

$MnCl_2$ treatment is known to activate integrins and to reduce cell migration rates; here it was shown that $MnCl_2$ alone reduces cell migration rates to the same extent as plasma treatment. The migration rates of plasma-treated cells and those of plasma-treated cells subsequently treated with $MnCl_2$ are similar. It is known that treatment with $MnCl_2$ leads to integrin activation, thus linking plasma treatment to integrin activation. If plasma treatment did not activate integrins, incubating cells with $MnCl_2$ would further reduce cell migration rates.

Serum starvation induces cell flattening and activates integrins via mechanotransduction of stress forces to the cells focal adhesions. The fact that plasma treatment does not alter the migration rates of serum-starved cells can be interpreted numerous ways. It is possible that the reduction in cell migration after plasma treatment requires nutrients lacking in media with 1% serum; this is unlikely since rapid cell migration demands energy and nutrients. It is also possible that cell migration in 1% media is reduced to a level that cannot be reduced further without a loss of cell viability. Although cell migration rates did not change after CAP treatment of serum-starved cells, it is remarkable that the most observed starved cells did not show an increase in cell death in response to CAP treatment.

Integrins are expressed on cell surfaces as $\alpha\beta$ heterodimers. There are two different classes of integrins on fibroblasts: those that adhere primarily to fibronectin (FN) and collagens and are primarily $\beta1$ containing integrins ($\alpha1\beta1$, $\alpha2\beta1$, $\alpha3\beta1$, $\alpha5\beta1$, $\alpha7\beta1$, $\alpha8\beta1$, $\alpha9\beta1$, and $\alpha11\beta1$) and those that adhere to vitronectin (VN), which include the αv containing integrins ($\alpha v\beta3$ and $\alpha v\beta5$) [77]. Epithelial cells are more complex having an additional integrin heterodimer ($\alpha6\beta4$) to mediate attachment to laminin [71,77].

Tissue culture plastic is charged. When cells are placed in dishes with media containing 5% serum, low concentrations of glycoproteins with charged sugar moieties on their surface that are present in the serum stick to the uncoated tissue culture plastic. FN and VN, two integrin ligands, are among the proteins present in serum that stick to tissue culture plastic. This yields a low concentration of those ligands on uncoated tissue culture plastic and supports cell adhesion and migration by both $\beta 1$ and αv integrins. When cells adhere to high concentrations of FN-collagen type I (FN/CNI), $\beta 1$ integrins cluster which induces their activation, adhering cells to VN clusters and induces activation of αv integrins. When cells migrate on FN/CNI- or VN-coated surfaces, they migrate using primarily $\beta 1$- or αv-integrins, respectively.

The purpose of the replating studies was to determine whether cells using primarily $\beta 1$ integrins to migrate respond to plasma treatment differently than cells using primarily αv integrins or cells using both $\beta 1$ and αv integrins on uncoated plastic. If plasma treatment acts by increasing $\beta 1$ integrin activity, then plating cells on FN/CNI would reduce or eliminate the impact of plasma treatment on cell migration rate because $\beta 1$ integrins would be active prior to plasma treatment; in addition, if plasma treatment acts by increasing $\beta 1$ integrin activity, plating cells on VN would enhance the plasma affect. Putting cells on VN reduces the cells' surface ligated and activated $\beta 1$ integrin compared to cells on uncoated plastic. There would be less active $\beta 1$ integrin present on cells on VN than on cells on uncoated plastic and therefore there would be an increase in the ability of plasma to alter cell migration if plasma operates by increasing $\beta 1$ integrin activation alone. As shown in Figure 7.15, there is a reduction in the ability of plasma to alter cell migration when cells are plated on FN/CNI compared to uncoated plastic and VN-coating dishes. These results show that plasma treatment operates primarily through $\beta 1$ integrins but that αv integrins are also involved because plating cells on VN did not increase the ability of plasma treatment to reduce cell migration.

The data obtained for $\beta 1$ integrin activation using immunofluorescence (see Figure 7.16) confirms that $MnCl_2$ treatment, as expected, activates $\beta 1$ integrin but data were only statistically significant for $\beta 1$ integrin at the cell periphery where the ratio of active to total integrin increased from 1.1 to 1.4; it was not observed an increase in $\beta 1$ integrin activation at the cell center. The failure to demonstrate $\beta 1$ integrin activation at the center of cells likely reflects limitations of 9EG7, which binds to its activation-state epitope more readily when the integrin is both ligand bound and active [78]. Because integrin activation at the cell center of control cells treated with $MnCl_2$ was not noticed, we focused our assessment of whether $\beta 1$ integrin was activated by plasma treatment on changes seen at the cell periphery. The data presented in Figure 7.16 show that the ratio of active to inactive is 1.1 without plasma treatment, whereas after plasma treatment it is 1.38. This value is statistically significant; treating cells with plasma followed by $MnCl_2$ increases the ratio of active to inactive $\beta 1$ integrin to 1.5. It is also worth noting that there is a statistically significant difference in the presence of the activated $\beta 1$ integrin between cells treated with only plasma and only $MnCl_2$ and between only plasma and $MnCl_2$ and both. This interesting observation can be subject of the future studies, since it is expressing different mechanisms involved in the integrin activation.

Integrin activation is one of the conditions required for the formation of focal adhesions [79,80]. The activation of integrins leads to an increase of the sizes of focal adhesions. The data presented in Figure 7.17 show that plasma-treated cells and cells treated with $MnCl_2$ both show an increase in the size of the vinculin positive focal adhesions measured in pixel intensities at the cell periphery of $\sim 20\%$. Treating cells with plasma followed by $MnCl_2$ further increased the sizes of the focal adhesions, but the increase was small. Ligated αv integrin also localizes to focal adhesions. After plasma treatment, focal adhesions increase in size, and activated β1 integrin accumulates within these larger focal adhesions and yet, αv integrin does not increase within focal adhesions. If the ratio of αv integrin to vinculin is determined, it is clear that both plasma and $MnCl_2$ treatments reduce the localization of αv integrin within focal adhesions. Consider together the data with the cell migration rate presented in Figure 7.16B and the β1 integrin activation data in Figure 7.17. These data suggest that plasma treatment increases β1 integrin activation leading to increased accumulation of β1 integrin and reduced accumulation of αv integrin within focal adhesions. These events increase focal adhesion size. The increased size of focal adhesions resists disassembly during cell migration and impedes cell migration rates. Interestingly, the migration rate reduction due to CAP treatment and $MnCl_2$ is about ∼30−40%; the low migration rate value seen after these treatments may represent some universal lower threshold for cell migration rate.

7.2.2.6 Thermodynamic model of the CAP effect on the cell membrane

Let us discuss a possible physical mechanism leading to integrin activation and, thus, decrease in the cell motility as described in the previous sections. A mechanism based on the thermodynamic model of mechanotransduction developed by Shemesh et al. [81] will be described here. The model uses the thermodynamic argument that the pulling force leads to self-assembly of molecules into an aggregate increasing the number of focal adhesions in order to decrease the stresses induced on the cell membrane.

Several phenomena associated with reactive chemical species, charged species, or charging can possibly effect the cell migration [82]. In the following model, it will be demonstrated that cell membrane *charge change* can affect formation of the focal adhesions. The change in charge that caused by plasma at the cell might lead to conformational changes in the extracellular domain of the integrins ligands [83] and, therefore, integrins activation. Consider the direct charge transfer from the plasma to the cell. Recall that while cells are covered by media before the treatment plasma jet action lead to removal of the liquid above the treated cells during the treatment.

Charge that is changed at the cell can be calculated based on the floating potential argument [84]. In the case of a cell with radius of $\sim 10^{-5}$ m, the charge is $\sim 10^4$ electrons [84].

The charge will induce the force

$$\sim \frac{Q^2}{4\pi\varepsilon R^2}$$

where R is the cell radius and thus the elastic stress will be formed along the cell membrane. According to the thermodynamic argument, the aggregation (i.e., focal adhesion (FA) centers formation) occurs if the chemical potential in the aggregate is smaller than without focal adhesions. One can estimate the force due to the formed focal adhesions as $N_A f_A$, where N_A is the number of the focal adhesions and f_A is the resisting force of the anchor. Thus, the change in the thermodynamic potential can be expressed as

$$\Delta\mu(Q, N_A) = \Delta\mu_0 - \left(\frac{Q^2}{4\pi\varepsilon R^2} - f_A N_A \right) \tag{7.2}$$

where $\Delta\mu(Q, N_A)$ is the change in the chemical potential, $\Delta\mu_0$ is the change in the chemical potential in the absence of induced charge, $((Q^2/4\pi\varepsilon R^2) - f_A N_A)$ is the resulting tension between the electric field force and resistance of the FA appeared at the cell surface due to the interaction with plasma. Estimate the f_A for the unperturbed cell-surface charge Q_0 as $f_{A0} = (Q_0^2/4\pi\varepsilon R^2 N_{A0})$ (N_{A0} is the number of focal adhesions in the cell without plasma-induced charge), then change in the chemical potential can be rewritten as

$$\frac{\Delta\mu(Q, N_A)}{\Delta\mu_0} = 1 - \alpha\left(\frac{Q^2}{Q_0^2} - \frac{N_A}{N_{A0}}\right) \tag{7.3}$$

with $\alpha = (Q_0^2/4\pi\varepsilon R^2 \Delta\mu_0)$. It can be seen that increase in the number of focal adhesions leads to decrease in the chemical potential. Calculations based on this model are presented in Figure 7.18. One can see that according to this model, an induction of the extra charge at the cell surface will lead to the formation of the new focal adhesions (see Figure 7.18). This can explain decrease of the cell motility observed experimentally after plasma treatment.

7.3 Application of CAP in cancer therapy

In Section 7.2, it was demonstrated that CAP has the nonaggressive nature during the interaction with cells [85]. It was shown, albeit indirectly, that plasma can interact with organic materials without causing thermal/electric damage to the cell surface, and several biological applications were examined [7]. As evidence accumulates, it is becoming clear that low-temperature or cold plasmas have an increasing role to play in biomedical engineering.

There is still some controversy with respect to the mechanism of plasma–cell interaction. Some authors are of the opinion that ion species play the most important role in plasma–cell interactions by triggering intracellular biochemistry [17]. Alternatively, others have suggested that neutral species play the primary role in some plasma–cell interaction pathways [18]. Furthermore, the effects of various ion species may be highly selective; different species can have either "plasma killing" (such as O) or "plasma healing" (such as NO) effects [7,86]. The role of other species, such as O_3 and OH, are not yet clear.

One of the important and perhaps most promising application of CAP is the cancer treatment [87–91].

The complex nature of cancer makes it difficult to develop effective treatments. Targeted cancer therapies are defined as substances or methods that block tumor growth by interacting with specific molecules or that focus on treatments to cancer specific molecular and cellular changes. Several targeted cancer therapies exist today. Examples include chemotherapy, radiotherapy, and molecularly targeted drugs [92–94]. One way to target cancer cells is to interfere with the cell cycle. Cancer cells proliferate at a faster rate than normal cells [95]. Combination therapies are proved to be most effective; they impact cancer cell biology at multiple signal transduction pathways resulting in a synergetic effect. The goal is to trigger cells to execute what has been referred to as "the cell death pathway" or apoptosis. A major hurdle in treating cancers is preserving surrounding normal tissue while inducing death of the cancer cells within a tumor. While progress is being made at developing treatments that are selective for cancer cells, cancer treatment damage to normal cells within tissues remains a major problem in oncology.

In this section, we will describe the known (and published) application of CAP. In particular, the therapeutic potential of CAP with a focus on selective tumor cell eradication capabilities and signaling pathway deregulation will be described.

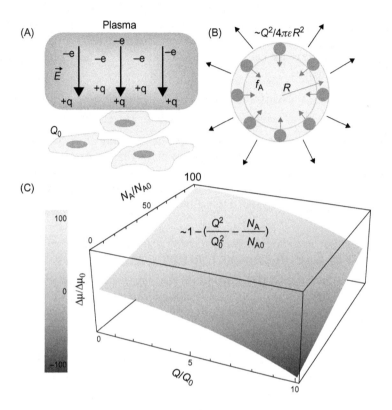

FIGURE 7.18

Thermodynamic model. (A) Schematic representation of the cold plasma interaction with a cell layer is shown. (B) Spherical approximation of the cell with induced charge Q and as a result resisting forces f_A in the points of the formation of the focal adhesions (red) are shown. (C) The qualitative plot of the chemical potential ($\Delta\mu/\Delta\mu_0$) as a function of the charge change at the cell (Q/Q_0) and the change of the number of focal adhesions (N_A/N_{A0}) is shown. One can see as charge increases, the chemical potential becomes negative. (For interpretation of the references to color in this figure legend, the reader is referred to the web version of this book.)

Source: *Reprinted with permission from [72]. Copyright (2012) by Institute of Physics.*

Demonstration of plasma antitumor activity has been demonstrated using various plasma delivery systems [89,90]. In particular, *in vivo* studies revealed that the cold plasma leads to decrease of tumor volume after plasma treatment.

7.3.1 Cold plasma selectivity

A strong selective effect was observed, the resulting 60−70% of SW900 (lung) cancer cells were detached from the plate in the zone treated with plasma, while no detachment was observed in the treated zone for the normal NHBE (lung) cells under same treatment conditions. Images of treated and untreated NHBE and SW900 cells are shown in Figure 7.19. Plasma treatment leads to a significant reduction in SW900 cell count, while NHBE cell count is practically unchanged.

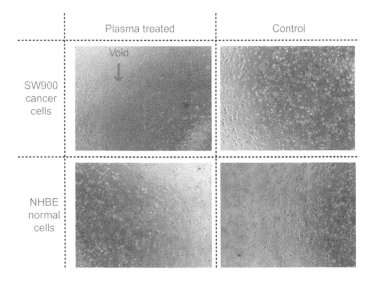

FIGURE 7.19

Selectivity effect of plasma treatment: SW900 cancer cells were detached from the plate in the zone treated with plasma, while no detachment was observed in the treated zone for the normal NHBE cells [22].

Both murine macrophages and B16 melanoma cells were treated with the cold plasma device for 0, 30, 60, and 120 s. Annexin *V* and 7-AAD staining was performed for flow cytometry analysis at 24 and 48 h after treatment. As seen in Figure 7.20, a clear dose—response to cold plasma treatment is seen in the murine melanoma cells at both 24 and 48 h ($P < 0.0001$), while the treated murine macrophages do not differ from control at either 24 or 48 h ($P = 0.1350$ and 0.1630, respectively). These findings suggest that the cold plasma jet has a more selective effect on murine melanoma cells.

In order to determine the cold plasma effect *in vivo*, we applied the cold plasma jet to nude mice bearing subcutaneous bladder cancer tumors (SCaBER). We examined the mouse skin after cold plasma treatment and did not see any damage to the skin after 2—5 min of treatment. Tumor models treated by cold plasma are shown in Figure 7.21. The plasma jet is shown in Figure 7.21A. One can see that a single plasma treatment leads to tumor ablation with neighboring tumors unaffected (see Figure 7.21B). These experiments were performed on 10 mice with the same outcome. We found that tumors of ~5 mm in diameter are ablated after 2 min of single time plasma treatment (see Figure 7.21B and E), while larger tumors decreased in size. Interestingly, ablated tumors did not grow back while partially affected tumors started growing back a week after treatment, although they did not reach the original size even 3 weeks after treatment.

Next step is to evaluate the cold plasma device for *in vivo* efficacy in a murine melanoma model. While tumors eventually recurred, a single transcutaneous cold plasma treatment induced ablation of the tumor through the overlying skin. As demonstrated in Figure 7.22, tumor growth rates were markedly decreased after cold plasma treatment. Notably, this resulted in a markedly improved survival in the treatment group ($P = 0.0067$), with a median survival of 33.5 vs. 24.5 days as shown in Figure 7.23.

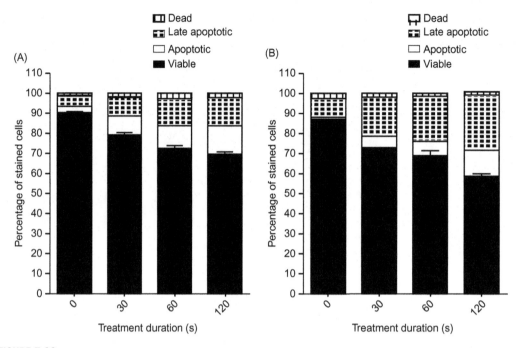

FIGURE 7.20

Selectivity effect of plasma treatment: B16 melanoma cells treated with the cold plasma device for 0, 30, 60, and 120 s: (A) 24 and (B) 48 h. Annexin V and 7-AAD staining was performed for flow cytometry analysis at 24 and 48 h after treatment. Four-quadrant analysis of the results characterizes the cells as viable (unstained), apoptotic (Annexin V positive), late-apoptotic (double positive), and dead (7-AAD positive) [22].

The skin temperature during plasma treatment was measured using an infrared thermometer to assess whether the cold plasma effect on cancer tissue is associated with thermal damage. Cold plasma treatment produced an increase in skin temperature of ~2°C above room temperature, which is below the temperature needed for thermal damage.

7.3.2 Gene expression analysis

The beta values of all probes on the Illumina BeadChip arrays were subjected to log 10 transformation and then normalized to the average in order to generate a heatmap of selected genes based on unsupervised hierarchical clustering with the Spotfire® software (Somerville, MA). The clustering was based on the unweighted average method using correlation as the similarity measure and ordering by average values. The red color was selected to represent upregulated genes and the green color to represent downregulated genes. Genes were selected for clustering if they were four times upregulated or downregulated after treatment with cold plasma. Figure 7.24 depicts the most upregulated genes (left panel) and the most downregulated genes (right panel) after cold plasma treatments. The list on upregulated and downregulated genes is shown in Table 1 of Ref. [22].

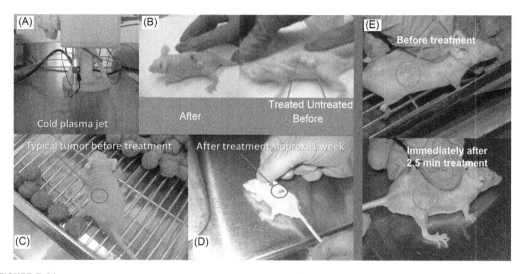

FIGURE 7.21

(A) Cold plasma device; (B) typical image of mice with three tumors before and after treatment (shown after 24 h); (C and D) typical image of mice with a single tumor before and ~1 week after treatment; (E) tumor before and immediately after 2.5 min treatment with cold plasma jet [22].

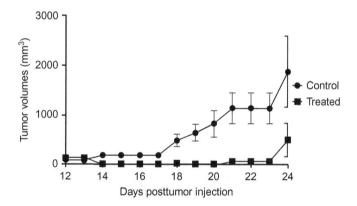

FIGURE 7.22

Cold plasma treatment effect on the growth of established tumor in a murine melanoma model [22].

Differences between genes that were up- or downregulated in treated or untreated cells were analyzed for biological significance using Geneontology (Spotfire®) and Ingenuity Pathway Analysis (IPA®). Differences in gene expression were found to be associated with pathways intimately related to cell adhesion, cell proliferation, growth regulation, and cell death ($P < 0.05$).

The experiments described above demonstrate potent effects of cold plasma treatment on cancerous tissue both *in vitro* and *in vivo*. Studies have offered several potential mechanisms for

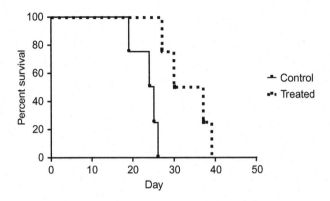

FIGURE 7.23

Cold plasma treatment effect on the mice survival in a murine melanoma model [22].

cold plasma's effect, including development of ROS, RNS, charged particles, heat, pressure gradients, and electrostatic and electromagnetic fields [7,18,22,26,86–90,96]. Notably, plasma has minimal impact on ambient cellular conditions. For instance, media pH-levels remain unchanged after treatment, and some study confirms that thermal effects associated with cold plasma are negligible as it was mentioned above. Beyond the direct external influence of the jet, cold plasma may induce living cells to produce their own ROS/RNS. Thus, described results suggest that multiple pathways involved in cancer processes, including cell adhesion, cell proliferation, growth regulation, and cell death, are selectively deregulated by cold plasma treatment in cancer cells. Some of these pathways may likely be responsible for tumor ablation. Perhaps consequently, it was further demonstrated that an induction of cellular apoptosis in treated cells, as manifested by both expression of cell-surface markers and gene expression, confirmed results of previous studies [7,18,22,26,86–90,97]. Most importantly, these findings are translated to *in vivo* models of cancer therapy, with marked reductions in tumor volumes and improved survival.

Given these findings, it can be suggested that cold plasma represents a promising new adjunct for cancer therapy, offering the ability to directly target and selectively kill neoplastic tissue. Notably, CAP device provides a method for practical administration of this cancer therapy. Plasma therapy could potentially target internal malignancies via an endoscopic delivery system, thus enabling this technology to serve as either a standalone treatment option or, more realistically, an adjuvant to existing therapies.

In summary, proof-of-principle studies shows new *in vitro* and *in vivo* response of cancer cells upon treatment with cold plasma jets. These very surprising preliminary results suggest that the cold plasma jet can selectively ablate some cancer cells (such as melanoma and bladder), while leaving their corresponding normal cells essentially unaffected. The two best known cold plasma effects, plasma-induced apoptosis and the decrease of cell migration velocity [22,96], can have important implications in cancer treatment by localizing the affected area of the tissue and by decreasing metastasis development. Moreover, the selective effect of cold plasma on different cell types suggests that it is possible to find the right conditions with plasma treatment affecting only

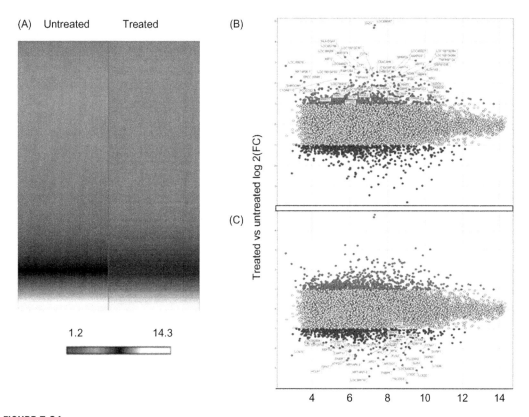

FIGURE 7.24

(A) A heatmap of the normalized log 2 signal intensity values in the illumina expression array for the untreated and treated sample. The yellow color was selected to represent high log 2 signal intensity values and the color blue to represent low log 2 signal intensity values. (B) Plot of upregulated genes (red) in a treated sample compared to an untreated sample. The values on the Y-axis represent the ratio of treated/untreated log 2 fold change. The values on the X-axis represent the average log 2 signal intensity. (C) Plot of downregulated genes (blue) in a treated sample compared to an untreated sample. The values on the Y-axis represent the ratio of treated/untreated log 2 fold change. The values on the X-axis represent the average log 2 signal intensity. (For interpretation of the references to color in this figure legend, the reader is referred to the web version of this book.) [22].

cancer cells, while leaving normal cells essentially unharmed. Finally, mid-sized tumors in nude mice were destroyed after a 2 min single time treatment by cold plasma without visible thermal damage of the skin. As such, we expect that the development of cold plasma treatment will cause a paradigm shift in cancer therapy. Such promising results warrant further efforts aim in understanding the mechanism of CAP action. Some progress has been achieved in this respect as will be described in the next section.

7.3.3 Targeting the cancer cell cycle by CAP

In this section, the correlation between the changes seen in cell motility observed previously in normal primary skin cells with viability and cell cycle progression under CAP treatment will be discussed. Furthermore, it will be hypothesized that cancer cells are more susceptible to the effects of CAP because a greater percentage of cells are in the S-phase. To test validity of this hypothesis, the normal cells derived from mouse skin as well as two mouse skin cancer cell lines will be treated by CAP.

It should be pointed out that the skin is the largest human organ and includes 15−20% of total body mass. The outer layer of the skin, the epidermis, is composed primarily of keratinizing stratified squamous epithelial cells. Two competing processes sustain the thickness of the epidermis: (i) proliferation and (ii) terminal differentiation of keratinocytes; both processes are driven by specific signaling pathways. Disturbances of cell−matrix, cell−cell adhesion, or terminal differentiation can lead to homeostatic imbalance in various diseases including skin cancers. Reduced terminal differentiation of skin cells allows the cells to continue proliferating and is a common feature of skin cancers [98,99]. Below the responses on CAP treatment of normal primary mouse keratinocytes with those of two well-studied mouse skin cancer cell lines that were derived from normal mouse keratinocytes will be discussed. While the two cell lines consist of transformed keratinocytes, they vary in their ability to induce tumors when injected into mice, a property referred to as tumorigenicity. One line (308) generates benign tumors called papillomas, whereas the other (PAM212) generates carcinomas [64,100,101].

7.3.3.1 The cell cycle

The cancer is defined as a disease in which one cell or the group of cells acquired the ability of uncontrolled indefinite proliferation and to invade distant site and organs. The current somatic mutation theory of carcinogenesis and metastases firmly defines the cancer as a cell-based disease [102]. However, in 1962, some argument existed that cancer should be studied as a problem of organismal disorganization [103]. The overwhelming evidence of the genetic origin of cancer is mutations, i.e., carcinogens (also mutagens) or some deficiencies related to DNA-repair process [104,105]. But still both cancer and normal cells have two fundamental behavioral properties, *proliferation* and *motility*. Nowadays it was identified the important role of the microenvironment causing the cancer [106]. The cancer is complicated; tumor development from an "initiating" cell is described by a conceptual model proposed by Nowell in 1986 [107]. It states the selectivity in generation and growth of variant (mutant cells) within the tumor toward more autonomous cells to become dominant subpopulation and, thus, leading the tumor progression. This model was developed from Foulds studies and later confirmed by Heppner et al. studies, where it was demonstrated that the growth and development of the cells inside the tumor is tightly regulated by interaction between the cells and extracellular environment [108]. Here, it is necessary to mention about the cancer immunoediting model, when immune system controls not only tumor quantity but also the tumor quality (immunogenicity), i.e., stresses the dual host-protective and tumor-promoting actions of immunity on developing tumors [109].

Cell proliferation is tightly regulated within tissues. And in normal self-renewal tissue, there is a homeostatic balance, i.e., cell proliferation rate is equal to the cell death rate. In early cancers and during the malignant progression, cell proliferation exceeds the cell death. One of the

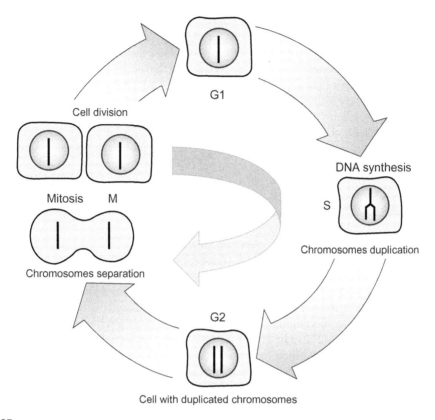

FIGURE 7.25

Cell cycle progression in the mammalian cell.

hallmarks of cancer is the deregulation of the molecular mechanisms controlling cell cycle [110]. The cell cycle is a "life of cell"; it describes the different steps that take place as a cell proliferates (Figure 7.25). The DNA synthesis occurs only in specific period of cell cycle, which is called S-phase, the morphological changes in cell refer to the mitotic phase (M-phase), a phase of the cell division. The gaps between M and S and between M and S phases are called G1 and G2 phases. Cell cycle is shown schematically in Figure 7.25. Nonproliferating cells are in the G1 (ground) phase of the cell cycle. When cells proliferate, they move progressively from G1 to S (synthesis of DNA), G2, and then M (mitosis) phase where they divide. Cells revert back to G1 after S. Cells in G1 may enter quiescence (G0), terminally differentiate, or, in response to specific signals, be induced to proliferate. Most cells in normal adult tissues are in quiescent G0-phase. Between S and G2 and between G2 and M phases, there are what have been called "checkpoints."

The proper progression through the cell cycle is monitored and controlled by checkpoints. These checkpoints verify whether the processes at each phase of the cell cycle have been accurately completed before progression into the next phase.

Activation of the checkpoints induces the cell cycle arrest by activation of the checkpoints through modulation of a cyclin-dependent kinases (CDKs) activity, a special group of protein kinases with evolutionary conserved function of the cell cycle control. Cell cycle arrest allows cells to repair possible defects through the complex system of DNA-repair enzymes. Only whose cells those DNA has been replicated correctly enter M-phase and divide. If errors are detected, a cell death pathway will be executed [111]. Whereas cells within most tissues of the body proliferate infrequently and are in either G0 or G1, the cancer cells in a tumor or in the circulation are more likely to be in the S-phase of the cell cycle. One way to specifically target cancer cells is to target the S-phase of the cell cycle. This is because cancer cells have mutations that allow them to evade the checkpoints and enter into the S-phase.

7.3.3.2 CAP effect on the cell migration and velocity distribution among the chosen cell population

Let us start with description of the CAP effect on migration of normal and transformed epithelial cells. The right panel of Figure 7.26A−C shows the dependence of cell migration rates on the CAP treatment time. For each time point, ∼50−60 cells were analyzed. The cell velocities/migration rates and cell numbers were normalized. Statistical analysis of the data showed that wild-type keratinocytes (WTK) (Figure 7.26A) and PAM212 cells (skin cancer line) (Figure 7.26C) have a statistically significant change in the migration rates beginning 40 s of CAP treatment. The migration rates for WTK cells and PAM212 cells after 60 s of CAP is ∼30% and ∼20% less than untreated cells; these changes are well identified by velocity distribution function of the cells and become even more statistically significant. The left panel of Figure 7.26A−C shows the probability distribution functions (PDF) of the cell migration rates (distribution of cell velocities) for untreated cells (control, blue) and cells treated with CAP for 60 s (plasma treated, red). The experimental data (∼250 cells of each cell type were analyzed for each condition) are fitted with inverse Gaussian distribution, which describes the first passage in Brownian motion with positive drift [112].

For normal and transformed (PAM212) cells, longer CAP treatments increased the reduction of cell migration. CAP treatment for ∼120 s gave the upper threshold in the cell migration reduction for the chosen time treatment range (see insets in Figure 7.26A and C). Unlike normal and PAM212 cells, 308 cells did not show any significant change in their migration rate after plasma treatment. It is important to note that 308 cells have very slow motility; their average velocity is ∼10-fold less than WTK and PAM212 cells.

The cell velocity distribution of WTK cells showed the most apparent negative shift from control under the mild CAP treatment, unlike PAM212 cells. The cells were also treated with helium only for 60 s. No significant differences in migration rates were found ($P > 0.05$, data is shown in Figure 7.29).

The experiments presented in Figure 7.26 indicate that cell migration rates of normal and PAM212 cells are sensitive to CAP. By plotting the data using PDF, data show that after CAP, migration rates of normal and PAM212 cells cluster more tightly indicating reduced variability among cells within the population. To determine whether there were differences in cell viability among adhered normal WTK, 308, and PAM212 cells, an MTT assay was performed.

The MTT assay results are shown in Figure 7.27A−C for WTK, 308, and PAM212 cells, respectively. The assay was performed 4, 24, and 48 h after two threshold regimes of CAP treatment: 60 s (low threshold) and 120 s (upper threshold) were used to identify changes in cell

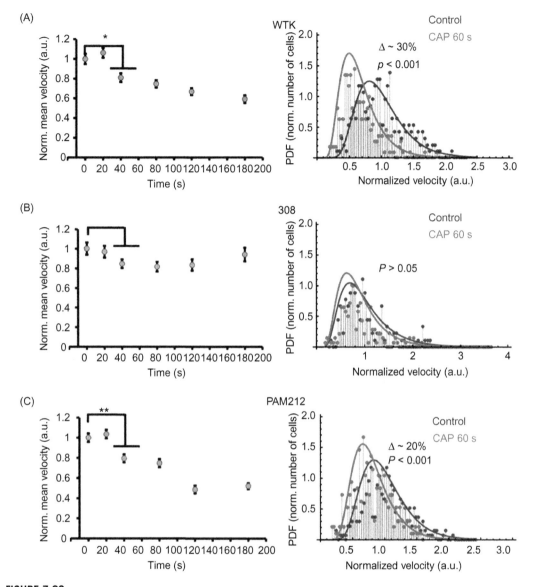

FIGURE 7.26

CAP effect on the cell migration and velocity distribution among the chosen cell population. The dependence of the cells migration rates on the CAP treatment time from 0 to 180 s are shown for WTK, 308, and PAM212 cells (A–C, left panel). SEM (standard error of mean) is shown with an error bars. For each time point ∼50–60 cells were analyzed. The difference between groups were considered statistically not significant for $P > 0.05$ (no mark) and for $P < 0.05$ (single star), $P < 0.01$ (double star) and $P < 0.001$ (triple star) are considered significant. The PDF of the cell migration rates (distribution of the cell velocities) for untreated cells (control, blue) and cells treated with CAP for 60 s (plasma treated, red) are shown for WTK, 308, and PAM212 cells (A–C, right panel). The experimental data (∼250 cells of each cell type were analyzed for each condition) is fitted with inverse Gaussian distributions.

Source: *Reprinted with permission from [91]. Copyright (2012) by Macmillan Publishers Ltd.*

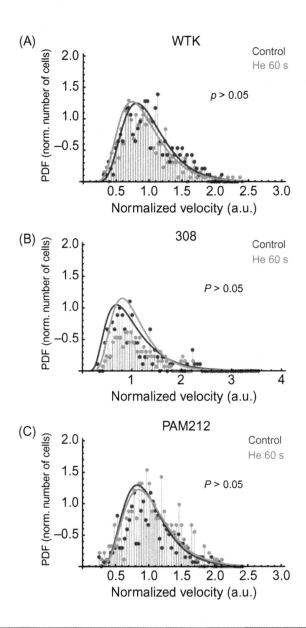

FIGURE 7.27

Helium does not affect the cell migration and velocity distribution among the chosen cell population. (A–C)
The PDF of the cell migration rates (distribution of the cell velocities) for untreated cells (control, blue) and cells
treated helium only for 60 s (helium treated, orange) are shown for WTK, 308, and PAM212 cells. The
experimental data (~250 cells of each cell type were analyzed for each condition) is fitted with inverse Gaussian
distributions. The difference between groups were considered statistically not significant for $P > 0.05$. (For
interpretation of the references to color in this figure legend, the reader is referred to the web version of this book.)

Source: *Reprinted with permission from [91]. Copyright (2012) by Macmillan Publishers Ltd.*

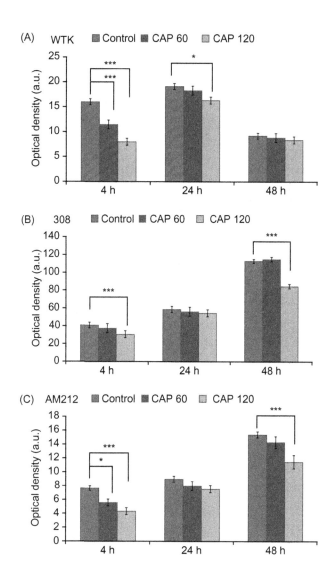

FIGURE 7.28

Cell viability studies in two threshold regimes. MTT assay was performed on: (A) WTK; (B) 308 cells (papilloma); and (C) PAM212 cells (carcinoma). Control (not treated) is shown in blue, CAP treated for 60 s—in red and CAP treated for 120 s—in gray cells at 0, 4, and 24 h time points. SEM is shown with an error bar. The experiment was repeated three times for each cell type and condition. Three stars denote extremely significant statistical difference ($P < 0.001$), one star—statistically significant difference ($P < 0.05$), data was considered statistically not different for $P > 0.05$. (For interpretation of the references to color in this figure legend, the reader is referred to the web version of this book.)

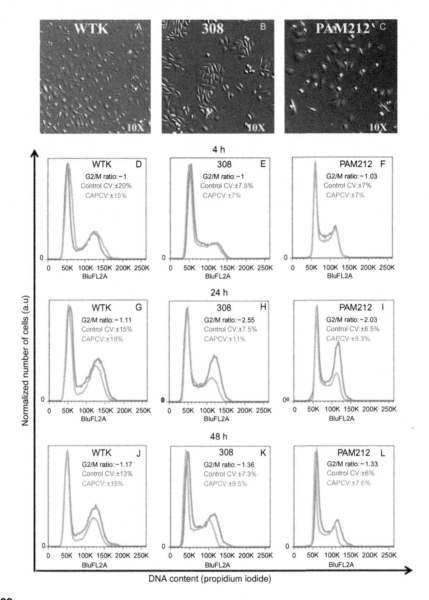

FIGURE 7.29

Identification of the cell cycle change in G2/M-phase. (A–C) Bright-field images of WTK, epidermal papilloma (308 cells), and epidermal carcinoma (PAM212 cells) cells are shown with a magnification 10 ×. (D–L) show cell cycle studies: Propidium iodide content (horizontal axis) and normalized number of cells (vertical axis) are shown. The controls are shown in blue and cells CAP treated for 60 s are in red. The ratio of the number of cells (treated to untreated) in G2/M-phase with coefficients of variation (CV, in percents) is shown in the right top corner of each figure. (D–F) shows cell cycle measurements in ~4 h; (G–I) in ~24 h; and (J–L) in ~48 h after the CAP treatment for WTK, 308, and PAM212 cells respectively. The data is shown for ~25,000–50,000 cells for each experimental condition. The measurements were repeated two to three times. (For interpretation of the references to color in this figure legend, the reader is referred to the web version of this book.)

Source: Reprinted with permission from [91]. Copyright (2012) by Macmillan Publishers Ltd.

viability. WTK cells viability decreased by 30% and 50% after 60 and 120 s of CAP treatment in ~4 h after treatment. By 24 h, cell viability after CAP recovers after both treatment lengths.

The viability of 308 cells was not changed 60 s of CAP treatment but was significantly reduced after 120 s of CAP treatment in ~4 and 48 h (Figure 7.28B). PAM212 cells responded to CAP treatment with a 30% and 40% decrease in viability after 60 and 120 s, respectively 4 h after treatment. While 308 cell viability recovered 24 h after treatment, similar to 308 cells, PAM212 cells showed reduced viability 48 h after 120 s of treatment (Figure 7.28B and C). No viability differences were observed after 60 s of CAP treatment during later ~24 and 48 h observation for all types of cells. Based on these data, CAP treatment for 60 s was chosen. This treatment regime induces a change in cell migration in normal and PAM212 cells and causes no significant effect on cell viability during 24 and 48 h after 60 s treatment, thus allowing determination of the initiation of the CAP effect on the cells. So the following studies were made for 60 s of CAP treatment.

7.3.3.3 Identification of the cell cycle changes in G2/M-phase

In addition, CAP effect on the cell cycle of normal, 308, and PAM212 cells was studied. Figure 7.29A−C are bright-field images with $10\times$ magnification of WTK, 308, and PAM212 cells morphology, respectively. Figure 7.29D−L shows DNA content measurement of control (untreated) cells in blue and cells treated with CAP for 60 s in red. The cells were analyzed in ~4 h (Figure 7.29D−F), ~24 h (Figure 7.29G−I), and ~48 h (Figure 7.29J−L) after 60 s of CAP treatment. No significant shifts in the G1−G2 peaks positions were observed for all of the cell types. The data were characterized by the fraction of the cells in the G2/M-phase as well as the coefficient of variation (CV) for control (untreated) and CAP treated cells. CAP-induced robust G2/M-cell cycle increases in both carcinoma (Figure 7.29I) and papilloma (Figure 7.29H) cells (two- to three-fold increase in ~24 h after CAP treatment), whereas normal keratinocytes (Figure 7.29C) showed a more modest cell cycle increase. This change diminished in ~48 h. The CV (a normalized measure of dispersion of a probability distribution) was used to characterize the data.

For all three time points, wild-type cells showed ~15−20% CV values for the cells in G2/M-phase, while transformed cells had CV value of ~6−10%.

7.3.3.4 Studies of the cell population's distribution during the cell cycle under CAP treatment

In order to determine if increased proliferation of the cells might result in the above observed G2/M-increase 24 h after CAP treatment, an additional cell cycle analysis was carried out. Cells were treated with the nucleotide analog EdU that is incorporated into DNA as it is being replicated. After staining EdU treated cells with a dye that binds to DNA (propidium iodide), flow cytometry was used to assess the numbers of cells in the three distinct phases of the cell cycle (G0/G1, S, and G2/M) simultaneously for all three cell types; data are shown in Figure 7.30A−F.

Results are presented for the time point 24 h after CAP treated cells for 60 s for the control (untreated) cells. As expected, fewer normal cells were in S-phase (~10%) compared to the two cancer cell lines (transformed cells are highly proliferative): ~50% for 308 cells and ~45% for PAM212 cells. No increase in the fraction of cells in the S-phase after CAP treatment was observed for the three cell types: their number either remained the same or decreased. However, we did observe an increase in the ranging of standard deviation value of CAP treated cells in S-phase of around ~20% for all three cell types suggesting that not all cells within the population of cells

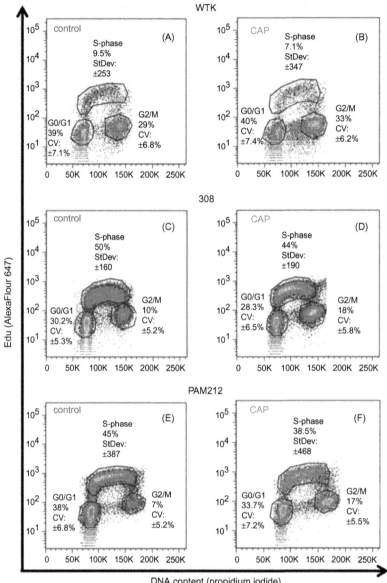

FIGURE 7.30

The cell distribution during the cell cycle after CAP treatment. The detailed studies of the cell cycles for control and CAP treated for 60 s cells are shown. The correlation between DNA content (propidium iodide) and DNA-replicating cells (EdU-component) is shown for: (A and B)—normal cells (WTK); (C and D)—papilloma cells (308 cells); and (E and F)—carcinoma cells (PAM212 cells) in ~24 h after CAP treatment. CV was used to characterize the propidium iodide data (linear scale) and standard deviation (StDev) was used for EdU-data (log scale) in each phase. Fractions of the number of cells in each cell-cycle phase are shown in percents. The data is shown for ~25,000 cells for each experimental condition. The experiments were repeated two to three times.

Source: *Reprinted with permission from [91]. Copyright (2012) by Macmillan Publishers Ltd.*

responded the same to CAP treatment. While there is no significant difference in the numbers of cells in the S-phase of the cell cycle, one can see that the number of cells in the G2/M fraction increased by ~25% for normal cells and two- to threefold for transformed cells.

The increase in the fraction of cells at the G2/M-phase of the cell cycle is accompanied a decrease in the number of cells in the G0/G1 fraction. This study also shows the general trend in the distribution of cells within the cell cycle and indicates that timing of the cell cycle is different for chosen cells.

7.3.3.5 Cell synchronization effect on the G2/M-peak increase

The next experiment was designed to determine whether cells synchronized at the G2/M-phase of the cell cycle and released would also demonstrate an increase in accumulation at the G2/M checkpoint. 308 cells were chosen for these studies because they showed a robust G2/M-peak increase after CAP treatment. On the other hand, 308 cells did not express significant changes in the migration after CAP was applied. This makes 308 cells most appropriate example to distinguish between the cell migration and cell cycle effects of CAP. Cells were synchronized with nocodazole, a drug that interferes with microtubule polymerization and causes mitotic spindle arrest, thereby preventing cells from entering mitosis; the effect is reversed upon its removal. The drug was removed and cells were treated by CAP in ~2 h after removal of nocodazole. Figure 7.31A shows the time dynamics of the cell cycle immediately after nocodazole removal (0 h point), at the ~4 h time point (it is around the time of CAP treatment) and in ~24 h after CAP treatment when G2/M-increase was observed. Figure 7.31B shows the detailed distribution of cells among phases of the cell cycle 24 h time point of the cells pretreated with nocodazole including CAP treatment. Cell synchronization did not impact the G2/M-pause induced by CAP.

The CV holds ~around 5−7%, while the fraction of the cells in G2/M-phase doubled in ~24 h after CAP treatment of nocodazole pretreated cells. The cell cycle of cells pretreated with nocodazole only restored to control (not treated) cells profile in ~24 h. The variation of the data spread for cells in S-phase for CAP treated cells was again ~20% more than cells pretreated with nocodazole only. This data also suggested that dynamics of the cell cycle in 308 cells is rapid. Nocodazole caused ~fourfold increase in the number of cells in the G2/M-phase of the cell cycle. When nocodazole is removed, by ~4 h cells begin progressing in synchrony from G2/M to G0/G1 and the ratio of the cells in G2/M-phase falls to ~twofold increase with the accumulated DNA-replicating cells in S-phase. Thus, from this result is not obvious whether or not CAP impacts cells within S-phase or G2/M-phase. However, it can be stated that G2/M-pause after CAP occurs whether or not cells have been synchronized.

7.3.3.6 CAP targets the cell cycle

To directly test whether CAP impacts some cell-cycle phase specifically, experiments were designed to assess the expression of γH2A.X (pSer139) at different phases of the cell cycle. Based on the earlier experimental results shown above, it was surmised that the G2/M-peak increase could be due to effects exerted on cells in the S and/or G2/M phases. It was previously reported in Ref. [22] that CAP based on helium flow may induce ROS response in the cells. Thus, we choose γH2A.X (pSer139) as an oxidative stress reporter of S-phase damage. This histone comprises around ~2−25% of the evolutionarily conserved variant of histone H2A; it goes through a posttranslational modifications in response to ionizing radiation, UV-light, or ROS [113,114]. It can be used in combination with

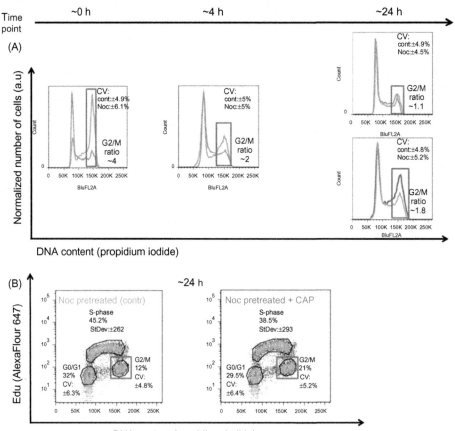

FIGURE 7.31

Cell synchronization for G2/M-increase studies. 308 cells were synchronized with nocodazole and analysis of the cell cycle was carried out for various time points. (A) shows the time evolution of the 308 cells: the control cells (not treated cells) are shown in blue, the cells pretreated with nocodazole are in orange, and nocodazole pretreated cells which were CAP treated for 60 s—in red. The cell cycle evolution after nocodazole removal is shown at ~0, 4 and 24 h points. The cells were treated with CAP at time point ~2 h after nocodazole removal (the state of the cell cycle for this time point is not shown). The final state of the systems control (not treated cells), cells pretreated with nocodazole only and cells pretreated with nocodazole and treated with CAP is shown for time point ~24 h. The change in ratio between G2/M fractions of the cells at each experimental condition is marked with a red square. (B) shows detailed cell cycle studies of 308 cells for chosen time point ~24 h. The DNA content (propidium iodide, linear scale) and DNA-replicating cells (EdU-component, log scale) characterized by CV (in percents) and StDev (a.u.) values, respectively. The data is shown for ~25,000 cells. The measurements were repeated two to three times. (For interpretation of the references to color in this figure legend, the reader is referred to the web version of this book.)

Source: *Reprinted with permission from [91]. Copyright (2012) by Macmillan Publishers Ltd.*

EdU-labeling to determine if the oxidative stress is induced in S-phase [115]. Figure 7.32 shows the results of time-sensitive 308 cell response studies to CAP treatment of 60 s: DNA content in cells, DNA-replicating cells, and γH2A.X expression were measured simultaneously. Figure 7.32A shows the cell cycle dynamics (EdU/DNA content) of 308 cells immediately following and at 4 and 24 h after CAP treatment. Note the very significant change in the S-phase signal (EdU) at $\sim$0 and $\sim$4 h: almost twofold decrease in the median value with decrease of variation range value, while still no significant changes were observed in G0/G1 and G2/M; at the $\sim$24 h time point, the S-phase recovered and the number of cells in G2/M 24 h after CAP treatment was increased. Figure 7.32B−D represent the correlation between γH2A.X signal and DNA content in G0/G1-phase (B); with EdU incorporation to highlight S-phase cells (C), and with DNA content to highlight cells in G2/M-phase (D). There were no significant differences in γH2A.X (pSer139) expression in cells at G0/G1 at any time point (Figure 7.32B). There was an increase in γH2A.X (pSer139) expression (maximum difference $D_{max}\sim$30% at time point 4 h) correlated with $\sim$twofold decrease in the EdU signal in cells mostly in the S-phase immediately (0 h) and 4 h after CAP treatment (Figure 7.32C). Differences in γH2A.X (pSer139) expression were less apparent 24 h after CAP treatment: G2/M-phase showed some increased variation of γH2A.X (increased ranging value of standard deviation with some increase in the median value) at the time point $\sim$24 h, which correlated with the increase of the number of cells (G2/M-increase, corroborating earlier observations, Figure 7.32D).

7.3.3.7 On the mechanism of CAP in cancer therapy

In the previous sections, we described how effects of cold plasma on cells depend on the time cells are treated with a specific power, the chemical composition (helium flow based jet), and the number of cells within the population treated that are in the S-phase. Normal skin cells are less likely to be in S-phase of the cell cycle compared to cancer cells. Normal skin cells do not induce tumors when injected into mice; two different cancer-causing cell lines do cause tumors [89]. More tumorigenic cell line (PAM212), like normal cells, also shows reduced migration rates and viability after mild CAP treatment, but the other cell line (less tumorigenic papilloma cells) shows neither reduced migration rates nor reduced cell viability. To determine the underlying causes of these differences, detailed cell cycle assessments were conducted on the CAP treated cells. These data show that CAP affects all stages of the cell cycle, however depending on the phase of the cell cycle in which a cell exists at the outcome of the CAP treatment or the cell response is different. We track only those cells that do not undergo mitosis over the 16 h and 40 min time period; the cells are subjected to time lapse were most likely in G0/G1- or S-phase. Experiments using dermal fibroblasts and epithelial cells suggest that differences in integrin expression and/or activation (conformational changes on the cell surface) cause the cell migration effects induced by CAP [72]. The migration rate of 308 cells is very slow ($\sim$10-fold less) compared to normal keratinocytes and PAM212 cells. Despite the fact that CAP had no effect on cell migration in 308 cells, our detailed analyses of the cell cycle and expression of γH2A.X (pSer139) in these cells show that they do respond to CAP treatment. Like normal cells and PAM212 cells, 308 cells undergo transient G2/M-pause after CAP.

The experiments presented show that the phosphorylation of histone γH2A.X was most evident in cells in S-phase (median rise up to $\sim$35%); however, a not-quite mathematically significant increase was also observed in G2/M. The increase of γH2A.X expression correlated with an abrupt decrease in the median value of EdU signal that started immediately after CAP treatment and continuing at the 4 h time point. Thus, CAP causes reduction in progression of cells through S-phase

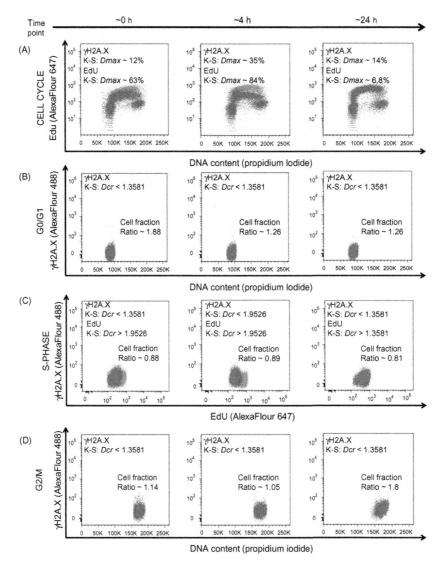

FIGURE 7.32

CAP targets the cell-cycle phase. The time-sensitive studies of the 308 cells cell cycles are shown for the time points ~0, ~4, and ~24 h points after 60 s of CAP treatment. The control (not treated cells) is shown in blue, CAP treated cells are in red. (A) shows the time evolution of the cell cycles (DNA content vs. DNA-replicating cells) of control and CAP treated cell. The changes in EdU and γH2A.X signals are characterized by K-S maximum difference D_{max} and D_{cr} values at the selected time point. (B) shows the correlation between DNA content and γH2A.X reporter for G0/G1 cell phase. (C) shows correlation between DNA-replicating cells (EdU) and γH2A.X for 308 cells in S-phase. (D) shows correlation between DNA content and γH2A.X for the 308 for G2/M-phase for ~0, ~4, and ~24 h time points. Approximately 25,000 cells are shown for each experimental condition. Changes in the fraction of cells between the control (not treated) and CAP treated cells are shown for each cell phase: cell number decreases if the ratio is <1. The statistical description of the signals shown in the (A) is a maximum differences between two distributions in the S-phase with confidence interval 99.9%; in the (B–D) D_{cr} value is shown: $D_{cr} < 1.3581$ ($P > 0.05$) is considered not statistically significant and $D_{cr} > 1.9526$ ($P < 0.001$) is extremely statistically different. (For interpretation of the references to color in this figure legend, the reader is referred to the web version of this book.)

Source: *Reprinted with permission from [91]. Copyright (2012) by Macmillan Publishers Ltd.*

(DNA-replication slows down), induces oxidative damage to DNA, indicated by phosphorylation of histone γH2A.X on serine residue 139, leading to the accumulation of cells at the G2/M checkpoint caused by the DNA damage.

The kinetics of γH2A.X phosphorylation suggests that CAP treatment most likely is not related to double-strand breaks [116] confirming earlier experimental findings. It has also been shown that phosphorylated H2A.X forms nuclear foci at the sites of stalled replication forks in response to exposure of S-phase cells to UV radiation, and chromatin changes or modifications play a role in the replication block pathway [117]. The spectral analysis of CAP jet showed that most of species were excited at the UV range (280−450 nm), providing a possible explanation for induction of S-phase damage; however, ROS stresses have to be taken into account as well. With the mild CAP treatment, only a ~10% decrease in the number of S-phase cells was detected, so this treatment (as MTT also showed) does not kill skin cells effectively. But as shown in Figure 7.32, increasing the treatment length to 120 s decreases the viability of the cells. The fact that most of the reduction appeared in ~48 h could be explained by the initiation of apoptotic pathway in those cells in S-phase which subsequently did not recover.

In ~24 h, restoration of the S-phase of the cell cycle was accompanied by an increase in G2/M-peak, indicating that the CAP treated cells progressed into the G2/M-phase. We did not observe significant differences in G2/M in normal keratinocytes where ~10−15% of cells within the population are in the S-phase; ~40−50% of cancer cells are in the S-phase (DNA synthesis phase).

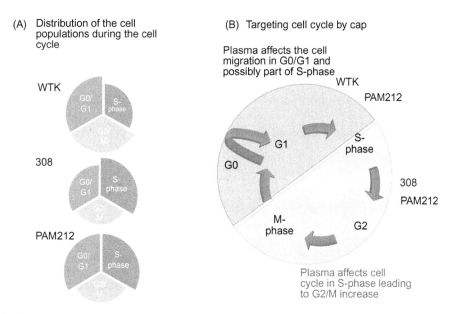

FIGURE 7.33

The schematic model of the CAP impact on the cell cycle. (A) is the schematic representation of the cells distribution during the cell-cycle phase for analyzed types of cells: WTK, 308, and PAM212. (B) is schematic of the CAP affect on the cell cycle with consequent cell response.

Figure 7.33 shows schematically how CAP affects various types of cells. The 308 cells are different in many ways from the PAM212 cells despite that fact that both cell lines arose from mouse keratinocytes in the same laboratory at the NCI at the NIH. It would be necessary to study in detail several additional sets of normal and transformed cells before we could definitively conclude that increased susceptibility to CAP correlates with tumorigenicity of the cells. The progression of these two cell types through the cell cycle is different and one should take into account that the strength of the treatment effect correlates with the time point chosen for analysis. However, the data we produce here goes a long way toward proving that mild CAP impacts DNA synthesis and causes a transient cell cycle pause at the G2/M-phase of the cell cycle; and that overall CAP impact on the migration rates, viability, and cell cycle increases on the cancer cells as they are more tumorigenic. It is important to note here that other cancer cell lines and *in vivo* models tested with the similar CAP treatment conditions showed different resistance and timing for the cell cycle effects, which is determined by their differences.

Homework problems
Section 1

1. **Give a definition of cold plasma.** What are the different approaches to form cold plasmas? Describe what diagnostic technique can be used for electron density measurements in cold plasmas.
2. **Explain concept of "plasma bullets"** applied for cold atmospheric plasma jet. What time resolution of high-speed camera is needed to measure "plasma bullets" speed.
3. **Explain concept of cell migration.** What is the possible mechanism? How plasma treatment can affect cell migration?
4. **What is the possible mechanism of selective effect of the cold plasma** toward the cancer cells?
5. **Explain the effect of cold plasma on the cell cycle.** What is the possible contribution of ROS?

References

[1] K.H. Becker, K.H. Shoenbach, J.G. Eden, Microplasma and applications, J. Phys. D 39 (2006) R55−R70.
[2] E. Stoffels, I.E. Kieft, R.E.J. Sladek, L.J.M. van den Bedem, E.P. van der Laan, M. Steinbuch, Plasma needle for *in vivo* medical treatment: recent developments and perspectives, Plasma Source. Sci. Technol. 15 (2006) S169−S180.
[3] G. Fridman, G. Friedman, A. Gutsol, A.B. Shekhter, V.N. Vasilets, A. Fridman, Applied plasma medicine, Plasma Process. Polym. 5 (2008) 503.
[4] Available from: http://usmedinnovations.com.
[5] J. Canady, K. Wiley, B. Ravo, Argon plasma coagulation and the future applications for dualmode endoscopic probes, Rev. Gastroenterol. Disord. 6 (2006) 1.

[6] E. Stoffels, A.J. Flikweert, W.W. Stoffels, G.M.W. Kroesen, Plasma needle: a non-destructive atmospheric plasma source for fine surface treatment of (bio)materials, Plasma Source. Sci. Technol. 11 (2002) 383.

[7] E. Stoffels, Y. Sakiyama, D. Graves, Cold atmospheric plasma: charged species and their interaction with cells and tissues, IEEE Trans. Plasma Sci. 36 (4) (2008) 1441−1457.

[8] K. Stalder, D. McMillen, J. Woloszko, Electrosurgical plasmas, J. Phys. D Appl. Phys. 38 (2005) 1728−1738.

[9] M. Laroussi, X. Lu, Room-temperature atmospheric pressure plasma plume for biomedical applications, Appl. Phys. Lett. 87 (2005) 113902.

[10] M. Laroussi, W. Hynes, T. Akan, X. Lu, C. Tendero, The plasma pencil: a source of hypersonic cold plasma bullets for biomedical applications, IEEE Trans. Plasma Sci. (June 2008).

[11] X. Lu, Z. Jiang, Q. Xiong, Z. Tang, Y. Pan, A single electrode room-temperature plasma jet device for biomedical applications, Appl. Phys. Lett. 151504 (2008).

[12] X. Lu, Z. Jiang, Q. Xiong, Z. Tang, X Hu, Y. Pan, An 11 cm long atmospheric pressure cold plasma plume for applications of plasma medicine, Appl. Phys. Lett. 081502 (2008).

[13] D. Mariotti, Non-equilibrium and effect of gas mixtures in an atmospheric microplasma, Appl. Phys. Lett. 92 (2008) 151505.

[14] J.F. Kolb, A.A.H. Mohamed, R.O. Price, R.J. Swanson, A. Bowman, R.L. Chiavarini, et al., Cold atmospheric pressure air plasma jet for medical applications, Appl. Phys. Lett. 92 (2008) 241501.

[15] G. Fridman, A. Shereshevsky, M. Peddinghaus, A. Gutsol, V. Vasilets, A. Brooks, et al., Drexel University, Philadelphia, PA, Bio-medical applications of non-thermal atmospheric pressure plasma, AIAA-2006-2902, in: 37th AIAA Plasma Dynamics and Lasers Conference, San Francisco, CA, 2006.

[16] E.N. Johnson, et al., Infectious complications of open type III tibial fractures among combat casualties, Clin. Infect Dis. 45 (4) (2007) 409−415.

[17] G. Fridman, D. Dobrynin, S. Kalghatgi, A. Brooks, G. Friedman, A. Fridman, Physical and biological mechanisms of plasma interaction with living tissue, in: 36th International Conference on Plasma Science, San Diego, CA, 2009.

[18] S. Kalghatgi, A. Fridman, G. Friedman, A. Morss-Clyne, Non-thermal plasma enhances endothelial cell proliferation through fibroblast growth factor-2 release, in: 36th International Conference on Plasma Science, San Diego, CA, 2009.

[19] M.G. Kong, G. Kroesen, G. Morfill, T. Nosenko, T. Shimizu, H. van Dijk, et al., Plasma medicine: an introductory review, New J. Phys. 11 (2009) 115012.

[20] G. Fridman, G. Friedman, A. Gutsol, A.B. Shekhter, V.N. Vasilets, A. Fridman, Applied plasma medicine, Plasma Process. Polym. 5 (2008) 503−533.

[21] E. Stoffels, Y. Sakiyama, D.B. Graves, Cold atmospheric plasma: charged species and their interactions with cells and tissues, IEEE Trans. Plasma Sci. 36 (4) (2008) 1441−145735pp.

[22] M. Keidar, R. Walk, A. Shashurin, P. Srinivasan, A. Sandler, S. Dasgupta, et al., Cold plasma selectivity and the possibility of a paradigm shift in cancer therapy, British J. Cancer 105 (2011) 1295−1301.

[23] M. Vandamme, E. Robert, S. Lerondel, V. Sarron, D. Ries, S. Dozias, et al., ROS implication in a new antitumor strategy based on non-thermal plasma, Int. J. Cancer 130 (2012) 2185−2194.

[24] A. Shashurin, M. Keidar, S. Bronnikov, R.A. Jurjus, M.A. Stepp, Living tissue under treatment of cold plasma atmospheric jet, Appl. Phys. Lett. 92 (2008) 181501.

[25] N. Georgescu, A.R. Lupu, Tumoral and normal cells treatment with high-voltage pulsed cold atmospheric plasma jets, IEEE Trans. Plasma Sci. 38 (2010) 1.

[26] J.L. Zirnheld, S.N. Zucker, T.M. DiSanto, R. Berezney, K. Etemadi, Nonthermal plasma needle: development and targeting of melanoma cells, IEEE Trans. Plasma Sci. 38 (2010) 948.

[27] M. Vandamme, E. Robert, S. Pesnel, E. Barbosa, S. Dozias, J. Sobilo, et al., Antitumor effect of plasma treatment on U87 glioma xenografts: preliminary results, Plasma Process. Polym. 7 (2010) 264−273.

[28] G. Isbary, G. Morfill, H.-U. Schmidt, M. Georgi, K Ramrath, J. Heinlin, et al., A first prospective randomized controlled trial to decrease bacterial load using cold atmospheric argon plasma on chronic wounds in patients, British J. Dermatol. 163 (2010) 78−82.

[29] K. Kim, J.D. Choi, Y.C. Hong, G. Kim, F.J. Noh, J.-S. Lee, et al., Atmospheric-pressure plasma-jet from micronozzle array and its biological effects on living cells for cancer therapy, Appl. Phys. Lett. 98 (2011) 0737013 pp.

[30] G.E. Morfill1, T. Shimizu, B. Steffes, H.-U. Schmidt, Nosocomial infections—a new approach towards preventive medicine using plasmas, New J. Phys. 11 (2009) 11501910 pp.

[31] A. Shashurin, M.N. Shneider, A. Dogariu, R.B. Miles, M. Keidar, Temporary-resolved measurement of electron density in small atmospheric plasmas, Appl. Phys. Lett. 96 (17) (2010) 171502.

[32] J. Park, I. Henins, H.W. Herrmann, G.S. Selwyn, R.F. Hicks, Discharge phenomena of an atmospheric pressure radio-frequency capacitive plasma source, J. Appl. Phys. 89 (2001) 20−28.

[33] X. Lu, M. Laroussi, Dynamics of an atmospheric pressure plasma plume generated by submicrosecond voltage pulses, J. Appl. Phys. 100 (2006) 0633026 pp.

[34] K. Ostrikov, Colloquium: Reactive plasmas as a versatile nanofabrication tool, Rev. Mod. Phys. 77 (2005) 489.

[35] M. Laroussi, The biomedical applications of plasma: a brief history of the development of a new field of research, IEEE Trans. Plasma Sci. 36 (4) (2008) 1612−1614.

[36] M.N. Shneider, R.B. Miles, Microwave diagnostics of small plasma objects, J. Appl. Phys. 98 (2005) 033301.

[37] Z. Zhang, M.N. Shneider, R.B. Miles, Microwave diagnostics of laser-induced avalanche ionization in air, J. Appl. Phys. 100 (2006) 074912.

[38] Z. Zhang, M.N. Shneider, R.B. Miles, Coherent microwave rayleigh scattering from resonance-enhanced multiphoton ionization in argon, Phys. Rev. Lett. 98 (2007) 265005.

[39] W.J.M. Brok, M.D. Bowden, J. van Dijk, J.J.A.M. van der Mullen, G.M.W. Kroesen, Numerical description of discharge characteristics of the plasma needle, J. Appl. Phys. 98 (1) (2005) 013302.

[40] Y. Sakiyama, D.B. Graves, Finite element analysis of an atmospheric pressure RF-excited plasma needle, J. Phys. D Appl. Phys. 39 (16) (2006) 3451−3456.

[41] Y. Sakiyama, D.B. Graves, Corona-glow transition in the atmospheric pressure RF-excited plasma needle, J. Phys. D Appl. Phys. 39 (16) (2006) 3644−3652.

[42] Y. Sakiyama, D.B. Graves, Nonthermal atmospheric rf plasma in one-dimensional spherical coordinates: asymmetric sheath structure and the discharge mechanism, J. Appl. Phys. 101 (7) (2007) 073306.

[43] F. Shi, D. Wang, C. Ren, Simulations of atmospheric pressure discharge in a high-voltage nanosecond pulse using the particle-in-cell Monte Carlo collision model in noble gases, Phys. Plasmas (2008).

[44] Y.J. Hong, S.M. Lee, G.C Kim, J.K. Lee, Modeling high-pressure microplasmas: comparison of fluid modeling and particle-in-cell Monte Carlo Collision modeling, Plasma Proc. Polym. 5 (2008) 583−592.

[45] F. Iza, J. Walsh, M.G. Kong, From submicrosecond to nanosecond pulsed atmospheric-pressure plasmas, IEEE Trans. Plasma Sci. (2009) in press.

[46] Z. Xiong, E. Robert, V. Sarron, J.-M. Pouvesle, M.J. Kushner, Dynamics of ionization wave splitting and merging of atmospheric pressure plasmas in branched dielectric tubes and channels, J. Phys. D 45 (2012) 275201.

[47] Z. Xiong, M.J. Kushner, Atmospheric pressure ionization waves propagating through a flexible capillary channel and impinging upon a target, Plasma Source. Sci. Technol. 21 (2012) 034001.

[48] A. Shashurin, M.N. Shneider, A. Dogariu, R.B. Miles, M. Keidar, Appl. Phys. Lett. 94 (2009) 231504.

[49] P.A. Rizzi, Microwave Engineering: Passive Circuits, Prentice Hall, Englewood Cliffs, NJ, 1988.

[50] S.Y. Liao, Microwave Devices and Circuits, Prentice Hall, Englewood Cliffs, NJ, 1980.

[51] W.H. Sutton, Microwave Processing of Ceramic Materials, Ceram. Bull. 68 (1989) 376.

[52] A. Shashurin, M.N. Shneider, M. Keidar, Measurements of streamer head potential and conductivity of streamer column in the cold nonequilibrium atmospheric plasmas, Plasma Source. Sci. Technol. 21 (2012) 034006.

[53] Y.P. Raizer, G.M Milikh, M.N. Shneider, S.V. Novakovski, Long streamers in the upper atmosphere above thundercloud, J. Phys. D Appl. Phys. 31 (1998) 3255.

[54] G. Gurtner, S. Werner, Y. Barrandon, M.T. Longaker, Wound repair and regeneration, Nature 453 (2008) 314.

[55] R.O. Hynes, Integrins: a family of cell surface receptors, Cell **48** (4) (1987) 549−554.

[56] A. Fridman, Plasma Chemistry, Cambridge University Press, New York, NY, 2008, pp. 157−257.

[57] X. Lu, Y. Cao, P. Yang, Q. Xiong, Z. Xiong, Y. Xian, et al., An RC plasma device for sterilization of root canal of teeth, IEEE Trans. Plasma Sci. 37 (5) (2009) 668−673.

[58] E. Stoffels, I.E. Kieft, R.E.J. Sladek, L.J.M. Van den Bedem, E.P. Van der Laan, M. Steinbuch, Plasma needle for *in vivo* medical treatment: recent developments and perspectives, Plasma Sources Sci. Technol. 15 (2006) S169−S180.

[59] X. Zhang, M. Li, R. Zhou, K. Feng, S. Yang, Ablation of liver cancer cells *in vitro* by a plasma needle, Appl. Phys. Lett. 92 (2008) 0215023 pp.

[60] A. Shashurin, M.A. Stepp, T.S. Hawley, S. Pal-Ghosh, L. Brieda, S. Bronnikov, et al., Influence of cold plasma atmospheric jet on surface integrin expression of living cells, Plasma Process. Polym. 7 (2010) 294−300.

[61] A. Shashurin, M. Keidar, S. Bronnikov, R.A. Jurius, M.A. Stepp, Living tissue under treatment of cold plasma atmospheric jet, Appl. Phys. Lett. 93 (2008) 181501.

[62] C. Bokel, N.H. Brown, Integrins in development: moving on, responding to, and sticking to the extracellular matrix, Dev. Cell 3 (3) (2002) 311−321.

[63] J.D. Humphries, A. Byron, M.J. Humphries, Integrin ligands at a glance, J. Cell Sci. 119 (19) (2006) 3901−3903.

[64] S.H. Yuspa, P. Hawley-Nelson, B. Koehler, J.R. Stanley, A survey of transformation markers in differentiating epidermal cell lines in culture, Cancer Res. 40 (1980) 4694−4703.

[65] R.A. Jurjus, Y. Liu, S. Pal-Ghosh, G. Tadvalkar, M.A. Stepp, Primary dermal fibroblasts derived from sdc-1 deficient mice migrate faster and have altered integrin function, Wound Repair Regenerat. (2008) in press.

[66] Available from: http://www.sas.com/.

[67] R.O. Hynes, Integrins: bidirectional, allosteric signaling machines, Cell 110 (6) (2002) 673−687.

[68] S.E. LaFlamme, B. Nieves, D. Colello, C.G. Reverte, Integrins as regulators of the mitotic machinery, Current Opin. Cell Biol. 20 (5) (2008) 576−582.

[69] D.S. Harburger, D.A. Calderwood, Integrin signalling at a glance, J. Cell Sci. 122 (Pt 2) (2009) 159−163.

[70] O. Volotskova, A. Shashurin, M.A. Stepp, S. Pal-Ghosh, M. Keidar, Plasma-controlled cell migration: localization of cold plasma−cell interaction region, Plasma Med. 1 (1) (2010) 83−93.

[71] R.A.F. Clark, The Molecular and Cellular Biology of Wound Repair, second ed., Plenum Press, New York and London, 1996, pp. 311−338

[72] O. Volotskova, M.A. Stepp, M. Keidar, Integrin activation by a cold atmospheric plasma jet, New J. Phys. 14 (2012) 053019.

[73] I.K. Gipson, S. Spurr-Michaud, P. Argueso, A. Tisdale, Ng.T. Fong, C. Leigh Russo, Mucin gene expression in immortalized human corneal−limbal and conjunctival epithelial cell lines, Invest. Ophthalmol. Visual Sci. 44 (6) (2003) 2496−2506.

[74] A.P. Mould, J.A. Askari, S. Barton, A.D. Kline, P.A. McEwan, S.E. Craig, et al., Integrin activation involves a conformational change in the alpha 1 helix of the beta subunit A-domain, J. Biol. Chem. 277 (2002) 19800−19805.

[75] F. De Toni-Costes, M. Despeaux, J. Bertrand, E. Bourogaa, L. Ysebaert, B. Payrastre, et al., A new alpha5beta1 integrin-dependent survival pathway through GSK3beta activation in leukemic cells, PLoS One **23** 5 (3) (2010) e9807.

[76] J. Humphries, P. Wang, C. Streuli, B. Geiger, M.J. Humphries, C. Ballestrem, Vinculin controls focal adhesion formation by direct interactions with talin and actin, J. Cell Biol. 179 (5) (2007) 1043−1057.

[77] M.A. Stepp, Corneal integrins and their functions, Exp. Eye Res. 83 (2006) 3−15.

[78] G. Bazzoni, L. Ma, M.L. Blue, M.E. Hemler, Divalent cations and ligands induce conformational changes that are highly divergent among b1 integrins, J. Biol. Chem. 273 (1998) 6670−6678.

[79] E.H.J. Danen, J. Van Rheenen, W. Franken, S. Huveneers, P. Sonneveld, K. Jalink, et al., Integrins control motile strategy through a Rho−cofilin pathway, J. Cell Biol. 169 (2005) 515−526.

[80] A. Byron, M.R. Morgan, M.J. Humphries, Adhesion signalling complexes, Current Biol. 20 (2010) R1063−1067.

[81] T. Shemesh, B. Geiger, A.D. Bershadsky, M.M. Kozlov, Focal adhesion as mechanosensors: a physical mechanism, PNAS 102 (2005) 12383−12388.

[82] M.G. Kong, M. Keidar, K. Ostrikov, Plasmas meet nanoparticles—where synergies can advance the frontier of medicine, J. Phys. D Appl. Phys. 44 (2011) 17401814pp.

[83] J.W. Weisel, C. Nagaswami, G. Vilaire, J.S. Bennett, Examination of the platelet membrane glycoprotein IIb−IIIa complex and its interaction with fibrinogen and other ligands by electron microscopy, J. Biol. Chem. 267 (1992) 16637−16643.

[84] M. Keidar, I.I. Beilis, R.L. Boxman, S. Goldsmith, Nonstationary macroparticle charging in an arc plasma jet, IEEE Trans. Plasma Sci. 23 (1995) 902−908.

[85] E. Stoffels, A.J. Flikweert, W.W. Stoffels, G.M.W. Kroesen, Plasma needle: a non-destructive atmospheric plasma source for fine surface treatment of (bio)materials, Plasma Source Sci. Technol. 11 (2002) 383.

[86] M. Kong, Plasma medicine—opportunities and challenges for science and healthcare, plenary talk, in: 36th International Conference on Plasma Science, San Diego, CA, 2009.

[87] N. Georgescu, A.R. Lupu, Tumoral and normal cells treatment with high-voltage pulsed cold atmospheric plasma jets, IEEE Trans. Plasma Sci. 38 (2010) 1949−1956.

[88] J.Y. Kim, S.O. Kim, Y. Wei, J. Li, Flexible cold microplasma jet using biocompatible dielectric tubes for cancer therapy, Appl. Phys. Lett. 96 (2010) 203701.

[89] M. Vandamme, E. Robert, S. Pesnel, E. Barbosa, S. Dozias, J. Sobilo, et al., Antitumor effect of plasma treatment on U87 glioma xenografts: preliminary results, Plasma Proc. Polym. 7 (2010) 264−273.

[90] M. Vandamme, E. Robert, S. Lerondel, V. Sarron, D. Ries, S. Dozias, et al., ROS implication in a new antitumor strategy based on non-thermal plasma, 130 (2012) 2185−2194.

[91] O. Volotskova, T.S. Hawley, M.A. Stepp, M. Keidar, Targeting the cancer cell cycle by cold atmospheric plasma, Sci. Rep. 2 (2012) 636.

[92] C.J. Lord, A. Ashworth, The DNA damage response and cancer therapy, Nature 481 (2012) 287−294.

[93] P.P. Connell, S. Hellman, Advances in radiotherapy and implications for the next century: a historical perspective, Cancer Res. 69 (2009) 383−392.

[94] T. Tsuruo, M. Naito, A. Tomida, N. Fujita, T. Mashima, H. Sakamoto, et al., Molecular targeting therapy of cancer: drug resistance, apoptosis and survival signal, Cancer Sci. 94 (2003) 15−21.

[95] G.K. Schwartz, M.A. Shah, Targeting the cell cycle: a new approach to cancer therapy, J. Clin. Oncol. 23 (2005) 9408−9421.

[96] R.F. Furchgott, Endothelium-derived relaxing factor: discovery, early studies, and identification as nitric oxide (nobel lecture), Angew. Chem. Int. Ed. 38 (1999) 1870.

[97] I.E. Kieft, M. Kurdi, E. Stoffels, Reattachment and apoptosis after plasma-needle treatment of cultured cells, IEEE Trans. Plasma Sci. 34 (2006) 1331−1336.

[98] S. Lippens, G. Denecker, P. Ovaere, P. Vandenabeele, W. Declercq, Death penalty for keratinocytes: apoptosis versus cornification, Cell Death Differ. 12 (2005) 1497–1508.

[99] M.M. Suter, K. Schulze, W. Bergman, M. Welle, P. Roosje, E.J. Müller, The keratinocyte in epidermal renewal and defence, Vet. Dermatol. 20 (2009) 515–532.

[100] R.A. Jurjus, Y. Liu, S. Pal-Ghosh, G. Tadvalkar, M.A. Stepp, Primary dermal fibroblasts derived from sdc-1 deficient mice migrate faster and have altered αv integrin function, Wound Rep. Reg. 16 (2008) 649–660.

[101] H. Hennings, D. Michael, S. Yuspa, Response of carcinogen-altered mouse epidermal cells to phorbol ester tumor promoters and calcium, J. Invest. Dermatol. 88 (1987) 60–65.

[102] C. Sonnenschein, A.M. Soto, The death of the cancer cell, Cancer Res. 71 (2011) 4334–4337.

[103] D.W. Smithers, Cancer: an attack of cytologism, Lancet 1 (1962) 493–499.

[104] I.F. Tannok, R.P. Hill, R.G. Bristow, L. Harrington, The Basic Science of Oncology, fourth ed., McGraw-Hill, New York, NY, 1998, pp. 77–100, 167–194, 205–231.

[105] B.H. Ames, R. Catchcart, E. Schwiers, P. Hochtein, Uric acid provides an antioxidant defense in humans against oxidant and radical-caused aging and cancer: a hypothesis, PNAS 78 (1981) 6858–6862.

[106] D. Hanahan, R.A. Weinberg, Hallmarks of cancer: the next generation, Cell 144 (2011) 646–674.

[107] P. Nowell, Mechanisms of tumor progression, Cancer Res. 46 (1986) 2203–2207.

[108] G.H. Heppner, F.R. Miller, The cellular basis of tumor progression, Int. Rev. Cytology 177 (1998) 1–56.

[109] R.D. Schreiber, L.J. Old, M.J. Smyth, Cancer immunoediting: integrating immunity's roles in cancer suppression and promotion, Science 331 (2011) 1565–1570.

[110] D.L. Longo, D. Longo, Harrison's Hematology and Oncology, McGraw-Hill Medical, New York, NY, 2010, pp. 294–318.

[111] M. Malumbres, M. Barbacid, Cell cycle, CDKs and cancer: a changing paradigm, Nat. Rev. Cancer 9 (2009) 153–166.

[112] J.L. Folks, R.S. Chhikar, The inverse Gaussian distribution and its statistical application—a review, J. Royal Stat. Soc. Ser. B 40 (3) (1978) 263–289.

[113] W.M. Bonner, C.E. Redon, J.S. Dickey, A.J. Nakamura, O.A. Sedelnikova, S. Solier, et al., GammaH2AX and cancer, Nat. Rev. Cancer 8 (2008) 957–967.

[114] T.M. Marti, E. Hefner, L. Feeney, V. Natale, J.E. Cleaver, H2AX phosphorylation within the G1 phase after UV irradiation depends on nucleotide excision repair and not DNA double-strand breaks, PNAS 103 (2006) 9891–9896.

[115] H. Zhao, J. Dobrucki, P. Rybak, F. Traganos, H.D. Halicka, Z. Darzynkiewicz, Induction of DNA damage signaling by oxidative stress in relation to DNA replication as detected using "Click Chemistry", Cytometry Part A 79A (2011) 897–902.

[116] M.P. Svetlova, L.V. Solovjeva, N.V. Tomilin, Mechanism of elimination of phosphorylated histone H2AX from chromatin after repair of DNA double-strand breaks, Mutation Res. 685 (2010) 54–60.

[117] I.M. Ward, J. Chen, Histone H2AX is phosphorylated in an ATR-dependent manner in response to replicational stress, J. Biol. Chem. 276 (2001) 47759–47762.

[118] A. Shashurin, M.N. Shneider, A. Dogariu, R.B. Miles, M. Keidar, Temporary-resolved measurement of electron density in small atmospheric plasmas, Applied Physics Letters 96 (17) (2010) 171502.

[119] A. Shashurin, M.N. Schneider, A. Dogariu, R.B. Miles, M. Keidar, Temporal behavior of cold atmospheric plasma jet, Applied Physics Letters 94 (235104) (2009).

Appendix: Physical Constants in SI

	Symbol	Value	Units
Elementary charge	e	1.602×10^{-19}	C
Electron mass	m_e	9.109×10^{-31}	kg
Proton mass	m_p	1.673×10^{-27}	kg
Planck constant	h	6.626×10^{-34}	J s
Speed of light	c	2.998×10^{8}	m/s
Permittivity of free space	ε_0	8.854×10^{-12}	F/m
Permeability of free space	μ_0	$4\pi \times 10^{-7}$	H/m
Temperature associated with 1 eV	e/k_B	11,604	K
Boltzmann contact	k_B	1.381×10^{-23}	J/K
Stefan–Boltzmann constant	σ	5.671×10^{-8}	$W/(m^2K^4)$

A.1 Ionization potentials

Xenon	12.13 eV
Helium	24.59 eV
Argon	15.76 eV
Nitrogen	14.53 eV

A.2 Work function

Material	Work Function (eV)
Al	4.25
C	4.7
Cu	4.4
Mo	4.3
Ni	4.5
W	4.54
LaB_6	2.3–2.4

A.3 Ion-induced secondary emission coefficients [1]

	Ion	Energy (eV)	SEE
Si	He	100	0.168
Al	H	300	0.0015
Al	H	1000	0.08
Cu	H	1000	0.11
Au	H	1000	0.12

A.3.1 Glow discharge
A.3.1.1 Normal cathode fall in various gases

Cathode Material	Air	O_2	H_2	Ar	He	N_2	Ne
Al	229	311	170	100	140	180	120
Cu	370		214	130	177	208	220
Fe	269	290	250	165	150	215	150
Ni	226		211	131	158	197	140
Zn	277	354	184	119	143	216	

A.3.1.2 A, B coefficients in equation for breakdown voltage

$$\left(\alpha/p = A \exp\left(\frac{-Bp}{E} \right) \right)$$

	A	B
Air	14.6	365
Ar	13.6	235
H_2	5	130
H_2O	12.9	289
He	2.8	34

A.4 Vacuum arcs

Table A.1 Ion velocity, Average Ion Charge and Erosion Rate

Cathode Material	Ion Velocity, $\times 10^4$ m/s	Average Ion Charge State [2]	Erosion Rate, $\times 10^{-4}$ g/C
Cu	1.28 [3]	2	1.15 [4]
	1.25 [5]		0.35 [6]
Mo	1.74 [2]	3.1	0.47 [5]
	1.55 [3]		0.36 [6]
W	1.05 [2]	3.1	0.62 [5]
			0.55 [6]
Ti	2.22 [2]	2.1	0.52 [5]
	1.3 [3]		0.3 [6]
Mg	3.06 [2]	1.5	0.31 [6]

A.5 Arc burning voltage

Cathode Material	Arc Burning Voltage, V
Ti	21.3
Cu	23.4
Mo	29.3
Ni	20.5

A.6 Sputtering yield (xenon ions) [7]

Ion Energy (eV)	Incidence Angle (deg)	BN Sputtering Coefficient (mm³/C)	Kapton Sputtering Coefficient (mm³/C)	Quartz Sputtering Coefficient (mm³/C)
100	0	0.0123		0.0098
250	0	0.0283	0.0084	0.0367
350	0	0.0252	0.0143	0.0398
500	0	0.0423	0.0297	0.0706
100	15	0.0122		0.0061
250	15	0.0234	0.0094	0.0333
350	15	0.02	0.0295	0.0461
500	15	0.0473	0.0372	0.0628
100	30	0.0130		0.0126
250	30	0.0287	0.0112	0.0547
350	30	0.0243	0.0201	0.0869
500	30	0.0485	0.0372	0.116
100	45	0.0153		0.0154
250	45	0.0258	0.0136	0.0875
350	45	0.0308	0.0278	0.131
500	45	0.0545	0.0445	0.151

A.7 Cross sections for helium and nitrogen [8,9]

Reaction	Rates
$He + e \rightarrow He^* + e$	$3.88 \times 10^{-16} \exp(-1.40 \times 10^6/E)$
$He + e \rightarrow He^+ + 2e$	$4.75 \times 10^{-16} \exp(-2.31 \times 10^6/E)$
$He^* + e \rightarrow He^+ + 2e$	$2.02 \times 10^{-13} \exp(-3.10 \times 10^5/E)$
$He^* + 2He \rightarrow He_2^* + He$	2.0×10^{-46} [m⁶/s]
$He^+ + 2He \rightarrow He_2^+ + He$	1.1×10^{-43} [m⁶/s]
$He_2^* + M \rightarrow 2He + M$	1.0×10^4 [s⁻¹]
$2He^* \rightarrow He_2^+ + e$	1.5×10^{-15}
$2He_2^* \rightarrow He_2^+ + 2He + e$	1.5×10^{-15}
$He_2^+ + e \rightarrow He^* + He$	$8.9 \times 10^{-15}(T_e/T_g)^{-1.5}$
$He^* + N_2 \rightarrow N_2^+ + He + e$	5.0×10^{-17}
$He_2^* + N_2 \rightarrow N_2^+ + 2He + e$	3.0×10^{-17}
$He_2^+ + N_2 \rightarrow N_2^+ + He_2^*$	1.4×10^{-15}

$$N_2^+ + e \rightarrow N_2 \qquad\qquad 4.8 \times 10^{-13}(T_e/T_g)^{-0.5}$$
$$O_2 + e \rightarrow O_2^* + e \qquad\qquad 1.7 \times 10^{-15} \exp(-3.1/T_e)$$
$$O_2^* + e \rightarrow O_2 + e \qquad\qquad 5.6 \times 10^{-15} \exp(-2.2/T_e)$$
$$O_2 + e + He \rightarrow O_2^- + He \qquad 10^{-37}$$
$$O^- + O_2^+ \rightarrow 3O \qquad\qquad 10^{-13}$$
$$O^- + O^+ \rightarrow 2O \qquad\qquad 2 \times 10^{-13}(300/T)^{0.5}$$
$$O_2^+ + e \rightarrow 2O \qquad\qquad 2.2 \times 10^{-14}T_e^{0.5}$$
$$O_2^* + O^- \rightarrow O_3 + e \qquad\qquad 3 \times 10^{-16}$$

Rate coefficients are in $[m^3/s]$ unless noted otherwise. T_e is electron temperature in [eV], T is the neutral temperature in [K].

Full summary of reaction rates for helium/oxygen plasma can be found in Ref. [10].

A compilation of mathematical and scientific formulas, physical parameters pertinent to a variety of plasma regimes, ranging from laboratory devices to astrophysical objects, can be found in NRL Plasma Formulary [11]. This booklet has been the mini-Bible of plasma physicists for the past 25 years.

References

[1] K. Ohya, T. Ishitani, Target material dependence of secondary electron images induced by focused ion beams, Surf. Coatings Technol. 158−159 (2002) 8−13.

[2] I.G. Brown, Vacuum arc ion sources, Rev. Sci. Instrum. 65 (1994) 3061.

[3] G. Yushkov, A. Anders, E. Oks, I.G. Brown, Ion velocities in vacuum arc plasmas, J. Appl. Phys. 88 (2000) 5618.

[4] J. Kutzner, H.C. Miller, Integrated ion flux emitted from the cathode spot region of a diffuse vacuum arc, J. Phys. D Appl. Phys. 25 (1992) 686−693.

[5] C.W. Kimblin, J. Appl. Phys. 44 (1973) 3074.

[6] I.G. Brown, H. Shiraishi, Cathode erosion rates in vacuum-arc discharges, IEEE Trans. Plasma Sci. 18 (1990) 170.

[7] A.P. Yalin, B. Rubin, S.R. Domingue, Z. Glueckert, J.D. Williams, Differential sputter yields of boron nitride, quartz, and Kapton due to low energy Xe$^+$ bombardment, in: 43rd AIAA Joint Propulsion Conference, Cincinnati, OH, AIAA paper 2007−5314, 2007.

[8] G.J.M. Hagelaar, L.C. Pitchford, Solving the Boltzmann equation to obtain electron transport coefficients and rate coefficients for fluid models, Plasma Sources Sci. Technol. 14 (2005) 722.

[9] Y.B. Golubovskii, V.A. Maiorov, J. Behnke, J.F. Behnke, Modelling of the homogeneous barrier discharge in helium at atmospheric pressure, J. Phys. D 36 (2003) 39.

[10] G. Park, H. Lee, G. Kim, J.K. Lee, Global model of He/O_2 and Ar/O_2 atmospheric pressure glow discharges, Plasma Processes Polym. 5 (2008) 569−576.

[11] NRL Plasma Formulary. Naval Research Laboratory, Washington, DC. <http://wwwppd.nrl.navy .mil/nrlformulary/>, 2011.

Index

Printed in the United States
By Bookmasters